# Electrostatic Discharge Control

Owen J. McAteer
*Westinghouse
Baltimore, Maryland*

**McGraw-Hill Publishing Company**
New York  St. Louis  San Francisco  Auckland  Bogotá
Caracas  Hamburg  Lisbon  London  Madrid  Mexico
Milan  Montreal  New Delhi  Oklahoma City
Paris  San Juan  São Paulo  Singapore
Sydney  Tokyo  Toronto

**Library of Congress Cataloging-in-Publication Data**

McAteer, Owen J.
  Electrostatic discharge control / Owen J. McAteer.
    p. cm.
  ISBN 0-07-044838-8
  1. Electronic apparatus and appliances—Protection. 2. Electric discharges. 3. Electrostatics. I. Title.
  TK7870.M397  1989
  621.381—dc20                                          89-37275

Copyright © 1990 by Owen J. McAteer. All rights reserved. Printed in the United States of America. Except as permitted under the United States Copyright Act of 1976, no part of this publication may be reproduced or distributed in any form or by any means, or stored in a data base or retrieval system, without the prior written permission of the publisher.

1234567890  DOC/DOC  8954321098

ISBN 0-07-044838-8

*The sponsoring editor for this book was Daniel A. Gonneau, the editing supervisor was Valerie A. Rothlein, the designer was Naomi Auerbach, and the production supervisor was Suzanne W. Babeuf. This book was set in Century Schoolbook. It was composed by the McGraw-Hill Publishing Company Professional & Reference Division composition unit.*

*Printed and bound by R. R. Donnelley & Sons Company.*

> Information contained in this work has been obtained by McGraw-Hill, Inc., from sources believed to be reliable. However, neither McGraw-Hill nor its authors guarantees the accuracy or completeness of any information published herein and neither McGraw-Hill nor its authors shall be responsible for any errors, omissions, or damages arising out of use of this information. This work is published with the understanding that McGraw-Hill and its authors are supplying information but are not attempting to render engineering or other professional services. If such services are required, the assistance of an appropriate professional should be sought.

*For more information about other McGraw-Hill materials, call 1-800-2-MCGRAW in the United States. In other countries, call your nearest McGraw-Hill office.*

*This book is dedicated to Professor Emeritus A. D. Moore of the University of Michigan, who has inspired many in the field of electrostatics. He authored books entitled "Electrostatics" (Doubleday, 1968) and "Invention, Discovery and Creativity" (Doubleday, 1969) and Professor Moore was editor of "Electrostatics and its Applications" (Wiley, 1973). He founded the Electrostatics Society of America, appropriately known as the "Friendly Society" in 1970. Professor Moore was devoted to educating others, particularly the young, in his favorite topic of electrostatics. He voluntarily maintained his University of Michigan laboratory following his retirement in 1963 at the age of 68. He continued his laboratory efforts and his famous electrostatics road show until 1984.*

PROFESSOR A. D. MOORE and OWEN MCATEER *(at the 1981 EOS/ESD Symposium)*

# Contents

Foreword    xi
Preface    xiii

**Chapter 1. Introduction and Extent of the ESD Problem**    1

    1.1   The Electrostatic Discharge Problem    2
    1.2   Electrostatic Discharge (ESD) Definition    3
    1.3   Synopsis of ESD in the Electronics Industry    4
    1.4   ESD Control Success Stories    9
    1.5   Sources of Information    11
          References    13

**Chapter 2. ESD Awareness**    15

    2.1   Frequent Existence of Doubt    15
    2.2   Reasons for Disbelief    16
    2.3   Who Needs to be Convinced    20
    2.4   Content and Method of Presentation    22
    2.5   Ongoing ESD Training and Certification    27
    2.6   Reliability Effects and Customer Image    27
          Reference    27

**Chapter 3. History of Electrostatics**    29

    3.1   Destructive Force of Lightning    29
    3.2   Attractive Force of Static Electricity    30
    3.3   Static Generating Apparatus    32
    3.4   Conductors and Nonconductors    34
    3.5   Different Types of Electricity    36
    3.6   The First Capacitors    37
    3.7   The First Batteries    38
    3.8   Polarities of Static Electricity    39
    3.9   The Inverse-Square Relationship    41
    3.10   Induction    42
    3.11   Electroscope    42

| | | |
|---|---|---|
| 3.12 | The Faraday Cage | 42 |
| 3.13 | Maxwell's Equations | 43 |
| 3.14 | Other Developments of the Nineteenth and early Twenthieth Centuries | 43 |
| 3.15 | Static Problems in Other Industries | 44 |
| 3.16 | History of Static Problems in the Electronics Industry | 45 |
| | References | 46 |

## Chapter 4. Basic Principles of Electrostatics 47

| | | |
|---|---|---|
| 4.1 | Charge | 47 |
| 4.2 | Contact Electrification | 48 |
| 4.3 | Triboelectric Generation | 48 |
| 4.4 | Conductivity | 50 |
| 4.5 | Resistivity | 51 |
| 4.6 | Faraday Cage | 55 |
| 4.7 | Voltage | 57 |
| 4.8 | Work, Power, and Energy | 57 |
| 4.9 | Electric Field | 57 |
| 4.10 | Induction | 58 |
| 4.11 | Capacitance | 58 |
| 4.12 | Electrophorous Generator | 63 |
| 4.13 | Van de Graff Generator | 64 |
| 4.14 | Ionization | 65 |
| 4.15 | Breakdown of Air and Other Gasses | 66 |
| | References | 66 |

## Chapter 5. Theoretical and Mathematical Fundamentals of Electrostatics 67

| | | |
|---|---|---|
| 5.1 | Electrostatic Force | 67 |
| 5.2 | Electric Field | 68 |
| 5.3 | Work to Move Charges | 71 |
| 5.4 | Conservative Nature of Electric Field | 71 |
| 5.5 | Potential Difference as a Function of Electric Field | 73 |
| 5.6 | Capacitance | 74 |
| 5.7 | Energy | 74 |
| 5.8 | Conductors in a Field | 75 |
| 5.9 | Displacement Field | 81 |
| 5.10 | Fields Produced by Selected Charge Configurations | 83 |
| 5.11 | Combinations of Capacitors | 93 |
| 5.12 | Green's Theorem of Reciprocity | 93 |
| 5.13 | Boundary Conditions of Insulators | 99 |
| 5.14 | Maxwell's Equations | 101 |
| | References | 102 |

## Chapter 6. Physical Concepts of Static Electrification — 103

- 6.1 Static Electrification — 103
- 6.2 Molecular Interfaces — 103
- 6.3 Band Structure of Solids — 104
- 6.4 Fermi Levels — 105
- 6.5 Work Function — 106
- 6.6 Contact Potential Difference or Volta Potential — 107
- 6.7 Contact Electrification — 108
- 6.8 Other Electrification Processes — 111
- 6.9 Polarization — 114
- 6.10 Relaxation Time and Decay Time — 118
- 6.11 Discharging Effects of Points — 119
- 6.12 Sparks — 120
- 6.13 Breakdown in Fluids — 122
- 6.14 Lightning — 123
- References — 123

## Chapter 7. Nature of Static Damage — 125

- 7.1 Characteristic Traits of ESD Failures — 125
- 7.2 Mathematical Models of Failure Mechanisms — 125
- 7.3 MOS Damage — 125
- 7.4 Junction Damage — 130
- 7.5 Metallization Damage — 139
- 7.6 Confusion Traits between ESD and EOS — 146
- 7.7 Wire Damage — 147
- 7.8 Passive Device Damage — 149
- 7.9 Effects on Very Thin Oxides — 152
- 7.10 Reversible Damage — 153
- 7.11 Schottky Diode Damage — 161
- 7.12 ESD Damage to Gallium-Arsenide Devices — 162
- 7.13 Schottky and Advanced Schottky Device Susceptibility Levels — 163
- 7.14 Cumulative Damage — 164
- 7.15 Latent Effects — 165
- 7.16 Walking Wounded — 166
- References — 166

## Chapter 8. ESD Damage Models — 169

- 8.1 Introduction — 169
- 8.2 Human Body Model — 169
- 8.3 Charged-Device Model — 173
- 8.4 Field-Induced Model — 178
- 8.5 Machine Model — 185
- 8.6 Field-Enhanced Model or Body-Metallic Model — 187
- 8.7 Capacitive-Coupled Model — 187

viii    Contents

|  |  |  |
|---|---|---|
| 8.8 | Other Models | 188 |
| 8.9 | Conclusions | 190 |
| 8.10 | Susceptibility Classification Testing | 190 |
|  | References | 201 |

## Chapter 9. Analysis of ESD Failures — 203

| 9.1 | Introduction | 203 |
|---|---|---|
| 9.2 | Approach to Analysis of Static Failures | 203 |
| 9.3 | Failure-Analysis Elements | 204 |
| 9.4 | Notification | 205 |
| 9.5 | Fact Gathering | 205 |
| 9.6 | Part Analysis | 206 |
| 9.7 | Failure Cause Identification | 221 |
| 9.8 | Corrective Action | 224 |
| 9.9 | Key Elements | 224 |
|  | References | 225 |

## Chapter 10. Failure Analysis Case Histories — 227

| 10.1 | Introduction | 227 |
|---|---|---|
| 10.2 | Selection of Case Histories | 227 |
| 10.3 | CMOS Devices Contained in a Metal Chassis Assembly | 228 |
| 10.4 | Catastrophic and Subtle Damage to Bipolar Operational Amplifier Inputs | 234 |
| 10.5 | Emitter-Coupled Logic (ECL) Failures | 238 |
| 10.6 | Shorted Hybrid Substrates | 241 |
| 10.7 | Hybrid Breakdown/Vaporized Metal | 247 |
| 10.8 | Charge-Coupled Device (CCD) Failures in Hybrid Assemblies | 251 |
| 10.9 | Wafer Level Processing | 252 |
| 10.10 | Avoidance of Possible Latent Failures on IRAS Spacecraft | 253 |
| 10.11 | Importance of Failure Analysis to an ESD Control Program | 255 |
|  | References | 255 |

## Chapter 11. Latent ESD Failures — 257

| 11.1 | Introduction | 257 |
|---|---|---|
| 11.2 | Definition | 257 |
| 11.3 | Background of the Latency Question | 258 |
| 11.4 | Research Strategy and Rationale for NAVSEA and NASA Studies | 261 |
| 11.5 | NAVSEA Study Results | 267 |
| 11.6 | NASA Study | 276 |
| 11.7 | Analysis of Data from NAVSEA and NASA Studies | 284 |
| 11.8 | Results of Other Latency Studies | 296 |
| 11.9 | An Assessment of the Latent ESD Failure Controversy | 303 |
|  | References | 307 |

## Chapter 12. Design Techniques — 309

| 12.1 | The Design Problem | 309 |
|---|---|---|
| 12.2 | Definition of Design-for-ESD-Immunity Techniques | 309 |

| | | |
|---|---|---|
| 12.3 | On-Circuit Protection Devices | 310 |
| 12.4 | Application of Protection Devices | 319 |
| 12.5 | Geometric Considerations | 325 |
| 12.6 | Processing Variations and Susceptibility | 328 |
| 12.7 | Special Input Protection Devices | 329 |
| 12.8 | Merged Input Protection Circuits | 335 |
| 12.9 | Protection Against the Charged Device Model | 336 |
| 12.10 | Adverse Effects of Protective Circuits | 337 |
| 12.11 | Assembly Protection | 337 |
| 12.12 | Electrical Design Protection at the Assembly Level | 339 |
| | References | 340 |

## Chapter 13. Factory Workplace Considerations and Problem Examples — 343

| | | |
|---|---|---|
| 13.1 | Insulative Materials in General | 343 |
| 13.2 | Contributing Factors | 344 |
| 13.3 | Possible Static Problems | 344 |
| 13.4 | Static Hazardous Factory Processes | 347 |
| 13.5 | Frequently Observed Conditions | 351 |
| 13.6 | Inspection | 354 |
| 13.7 | Test and Troubleshooting | 354 |
| 13.8 | Shipping | 355 |
| 13.9 | Engineering Labs | 355 |
| 13.10 | Environmental Considerations | 355 |
| 13.11 | Unusual Observances | 355 |
| 13.12 | Representative Problem Examples | 356 |
| | References | 362 |

## Chapter 14. ESD Control Management — 363

| | | |
|---|---|---|
| 14.1 | Introduction | 363 |
| 14.2 | The Control Problem | 363 |
| 14.3 | ESD Control Funding | 364 |
| 14.4 | Relating to ESD Experiences of Others | 364 |
| 14.5 | Determination of ESD Control Needs | 365 |
| 14.6 | ESD Control Management | 369 |
| 14.7 | Convincing Upper Management | 371 |
| 14.8 | Establishing the ESD Director Position | 371 |
| 14.9 | Additional Personnel Needed | 375 |
| 14.10 | Application of Knowledge to ESD Control Decisions | 375 |
| 14.11 | Return on Investment from Assigned ESD Personnel | 377 |
| 14.12 | Establishing an ESD Control Program | 382 |
| 14.13 | Periodic ESD Committee Meetings | 384 |

## Chapter 15. ESD Control — 385

| | | |
|---|---|---|
| 15.1 | Introduction | 385 |
| 15.2 | Tailoring Factory Controls Based on Susceptibility | 385 |

| | | |
|---|---|---:|
| 15.3 | Primary Controls | 387 |
| 15.4 | Secondary Controls | 388 |
| 15.5 | Tertiary Controls | 388 |
| 15.6 | Importance of Controls from All Three Categories | 388 |
| 15.7 | Primary Control Considerations | 389 |
| 15.8 | Secondary Control Considerations | 400 |
| 15.9 | Tertiary ESD Considerations | 409 |
| 15.10 | ESD Control in Automated Facilities | 415 |
| 15.11 | Clean Room Considerations | 417 |
| 15.12 | Safety Concerns | 417 |
| 15.14 | Customer Image Enhancement | 419 |
| | References | 419 |

## Chapter 16. Control Product Considerations and Evaluations 421

| | | |
|---|---|---:|
| 16.1 | Factors Affecting ESD Control Product Tests | 422 |
| 16.2 | Resistance and Resistivity Measurements | 429 |
| 16.3 | Triboelectric Generation Measurements | 429 |
| 16.4 | Static Decay Testing | 429 |
| 16.5 | Dissipative Materials | 430 |
| 16.6 | Antistatic Materials | 430 |
| 16.7 | Voltage-Monitored Activity | 431 |
| 16.8 | Evaluation of Flooring and Floor Finishes | 432 |
| 16.9 | Protective Package Considerations | 436 |
| 16.10 | Chairs and Carts | 442 |
| 16.11 | Ionization Evaluations | 443 |
| 16.12 | Antistatic Treatments | 444 |
| 16.13 | Miscellaneous | 444 |
| | References | 444 |

## Chapter 17. Standards 445

| | | |
|---|---|---:|
| 17.1 | Military Standards | 446 |
| 17.2 | Existing Government and Industry Standards Applied to ESD Control Products | 460 |
| 17.3 | Electrical Overstress/Electrostatic Discharge Standards | 462 |
| 17.4 | Conclusion and Future Needs | 465 |

Index   467

# Foreword

One of the most insidious and dangerous problems in modern electronic systems is that of electrostatic discharge (ESD). With the rapid increase in complexity and functional density of electronic systems and devices, the problem can only become more severe. Add to this the fact that ESD is often unrecognized or misunderstood, and it is easy to realize that education and guidelines regarding ESD problems and their management and control is critical to the reliability of present and future electronic systems.

The author of this book brings a wealth of personal experience to the fore in dealing with ESD-related problems. He is best known for his work in failure analysis, ESD awareness training, and latent ESD studies. In regard to the often raised question of how to identify ESD failures, he has numerous pertinent papers to his credit. As far back as 1970, he introduced the very concept of failure cause identification by failure signature recognition. In "Characteristic Traits of Semiconductor Failures," presented at the Annual Symposium on Reliability, he described rather precisely the expected physical damage to a bipolar junction from a 1-$\mu$s transient, which happens to be the characteristic of the typical human body ESD. He received awards for "An Effective ESD Awareness Training Program" and "Latent ESD Failures" presented at the 1979 and 1982 EOS/ESD Symposia, respectively. The latter paper described results of research directed by Mr. McAteer which ended the controversy over the existence of time-dependent latent ESD failures.

This book by an industry leader and pioneer in the field will fulfill a great need in industry at a critical period of growth. The book is well-organized and clearly presented in a manner which will enable readers to readily become acquainted with ESD awareness training, fundamentals of electrostatic phenomena, case histories, as well as design and manufacturing guidelines. In line with the author's extensive experience in these areas are comprehensive chapters on analysis of ESD failures, time-dependent latent ESD degradation, and ESD control program management. Three chapters addressing different aspects of electrostatic fundamentals are preceded by a chapter discussing the historical development of precepts of static electricity so important to modern industry. The topics of ESD control programs and evaluation of ESD control materials are fortified by illustrative case histories and cost effectiveness is stressed. Present-day government and industry standards and standard development needs are addressed with guidance as to proper application.

As the editor of numerous publications in the electronics field, I can assure all of those concerned with the design and manufacture of high-performance electronic systems that this book will become one of their most valuable reference sources and guidelines.

CHARLES A. HARPER
*President*
*Technology Seminars, Inc.*
*Lutherville, Md.*

# Preface

The acronym ESD (electrostatic discharge) is widely recognized throughout the world. Just a few years ago, the situation was quite different. The hazard of static electricity to electronic products was commonly disregarded, if not completely unknown. The many reported case histories of yield losses due to ESD in the literature have changed the overall perspective of the problem. Most of these reports included data measuring the successes of ESD control programs initiated to correct the problems. As a result, industry, for the most part, has come to realize that product failure and degradation from ESD warrants attention and corrective measures. Progress in achieving such a degree of ESD awareness may have seemed painfully slow to those closely involved with the process. ESD crusaders have often described their efforts in dealing with this insidious problem as most difficult and sometimes thankless. Nevertheless the improvement in overall awareness is gratifying.

As one might have expected, this increased awareness has brought about tremendous changes in virtually every manner of electronics processing intended to better cope with ESD. Such changes have included extensive purchases of ESD control equipment, modification of handling procedures, and periodic control-program monitoring. Proper selection from the available options in such costly measures requires a high degree of knowledge of the subject. Unfortunately, detailed knowledge in areas related to ESD has not always kept pace with the overall awareness of the problem's importance. A likely consequence is that in spite of increased expenditures, overall control-program effectiveness may not necessarily improve.

ESD control will be a way of life in the future. The economics of failure prevention in a high-tech industry leaves no other choice. Control measures will be instituted to meet customer demands if for no other reason. A goal of this book is to convey the need for a broad knowledge base on the part of decision makers. One cannot digest a significant portion of this book without delving into many diverse areas. This may help dispel the popular conviction that ESD control is a "narrow" field. The author predicts that out of shear necessity, industry will soon recognize the role of the "ESD control specialist" to provide the needed direction of ESD control activities. Leading companies have already begun the trend by instituting ESD control groups to test part sensitivities, evaluate control products, provide design assistance, solve problems, maintain up-to-date procedures, and perform ESD research and development.

A suggested starting point for a newcomer to ESD control is to read Chaps. 1 through 4, followed by Chaps. 13 through 17 before pursuing other areas of interest. The first three chapters contain the introduction and extent of the problem, static awareness, and the history of electrostatics. The fundamentals most needed to cope with static electricity in the factory and to understand subsequent chapters are contained in Chap. 4. Chapters 5 and 6 are included for those inclined to delve into the more technical aspects and electrostatics. The controversial subject of latent ESD failures discussed in Chap. 11 is expected to be of general interest. Chapters 7, 8, and 9, on damage, models, and failure analysis are specialized in nature, but will provide a broad insight into the ESD problem, and are thus recommended for all. The design techniques in Chap. 12 are intended primarily for those in design, reliability, and parts engineering. For problem examples refer to Chap. 1 (extent of the ESD problem), 10 (failure analysis case histories), and 13 (factory workplace examples). Chapters 14 and 15 cover ESD control management and the details of ESD control. Chapter 16 addresses the problems encountered in evaluating ESD control products. Chapter 17 discusses ESD control standards and future ESD control needs.

## Acknowledgments

The authors extends grateful acknowledgments to the many associates and peers involved with electrostatic phenomenon and ESD control. In particular, members of such organizations as the Electrical Overstress/Electrostatic Discharge Association, the Electrostatics Society of America, and the International Institute of Physics are to be commended for their many efforts in increasing knowledge on the subject. Special acknowledgment and gratitude is extended to the following individuals:

Thomas Speakman of AT&T for his review and encouraging comments on the first draft of this book.

Dr. Henry Charles of the University of Maryland and Dr. William Greason of the University of Western Ontario for their reviews and comments on drafts of selected chapters.

J. W. Anderson of Naval Weapons Support Center, Crane, Indiana for providing Fig. 16.4.

Ben Baumgartner of Lockheed; John Kolyer of Rockwell International Corporation; Denis Renaud of IBM; Art Trigonis of JPL; and Burton Unger of Bell Comunications Research for permissions to use figures from previously published works, and supplying prints as needed.

Richard Brooks and Ron Twist of Westinghouse Electric Corporation for their excellent performance in supporting ESD laboratory analysis and research efforts.

Rick Slayton and Pete Sullivan for their help in preparing some of the figures—particularly 2.1, 5.9, 5.14, 6.1 and 4.1, 4.7, 4.8, 5.13, 6.10, respectively.

OWEN J. MCATEER

## ABOUT THE AUTHOR

Owen J. McAteer is an advisory engineer in design operations with the Westinghouse Electric Corporation's Electronic Systems Group. His past experience includes engineering and management positions in electrical design, reliability engineering, failure analysis, and ESD consulting. One of the industry's most respected authorities on electrostatic discharge, Mr. McAteer was the first President of the EOS/ESD Association, past General Chairman of the EOS/ESD Symposium, and Standards Committee Chairman. His current activities with this organization include participating on the Administrative Committee and the Symposium Board of Directors, and chairing the Professional Development Committee. He has authored over 35 papers on ESD, failure analysis, and related topics. He has received two best-paper awards and has given several symposia keynote addresses. He earned both his B.S.E.E. and M.S.E.E. from the Johns Hopkins University, is a senior member of the IEEE and a member of the Electrostatics Society of America, and is listed in *Who's Who in Finance and Industry*.

# Chapter 1

# Introduction and Extent of the ESD Problem

Electrostatic discharge (ESD) damage to parts and assemblies has been a major problem of the electronics industry in recent years. This book reviews the history of the problem and presents guidance for establishing proper ESD control measures. In the broad sense the ESD problem includes questions concerning design, procedures, equipment, and monitoring apparatus as well as the failure mechanisms and static phenomena involved. Some issues are controversial and the industry is faced with a number of challenges to overcome.

In order to compete on the international market, the electronics industry has placed a growing emphasis on reliability and quality assurance. Improved statistical data accumulation and analyses are used to identify problem trends. Enhanced methods of physical analysis of defective parts have been introduced to better understand the failure mechanisms involved. In spite of these problem identification measures, ESD failures often elude the net and their incidence and impact go underestimated.

Even when duly acknowledged as important, ESD is often regarded as a simple problem with an easy fix. Sometimes there is a tendency to consider only the charged human body source as problematic. Static controls can appear straightforwardly simple, requiring limited knowledge in a narrow discipline. On the other hand, as familiarity with the subject grows, the spectrum broadens. One can then become overwhelmed by the many facets of ESD control, especially in light of the fact that static electricity can be generated on so many items in the industrial environment. Even more perplexing are such destructive factors as:

Charged items other than the human body

Damage to ungrounded items

Damage to items with no physical contact

Destruction due to the static charge on the damaged item itself

Destruction due to specially purchased static control measures

Such means of damage make the prospect of latent ESD failures occurring at some indeterminate time following exposure to ESD even more frightening. The existence of such failures has been a controversial issue because of the small amount of well-documented evidence in this regard. Theories of the hypothetical physical nature of latent ESD failures are included in Chap. 11 as well as details of experimental studies showing conclusive evidence of their existence.

The trends of microminiaturization toward submicron geometries and higher speeds indicate that ESD is likely to be a continuing threat to the industry with increasing severity. An effective long-term strategy is needed to deal with static electricity in our work and usage environments. The electrostatic discharge problem has two possible cures:

1. Extensive ESD control measures usage
2. Design of equipment with ESD immunity

The best alternative is a combination of the two remedies since practical issues preclude either being a panacea.

In order to provide guidance and insight to assist in placing ESD in a more appropriate perspective this chapter will

1. Define ESD
2. Highlight the evolvement of the problem
3. Discuss selected problems to indicate the extent
4. Address yield losses and latent ESD failures
5. Highlight examples of ESD control success stories
6. Discuss some major sources of additional information

## 1.1 The Electrostatic Discharge Problem

Electrostatic discharge has been a significant cause of failures and consternation within the electronics industry. The classic example is the release of charge accumulated on a human body through direct contact with a susceptible part connected to ground. High failure incidence is to a large extent rooted in the extensive use of integrated circuits. The small geometries of individual junction areas and/or thin insulative oxides that make up these circuits are so delicate that commonly encountered charged items can cause physical damage to the devices. The problem is aggravated by the fact that in addition to integrated circuits, certain other parts are vulnerable to ESD. Assemblies containing susceptible parts can also be at risk.

Although the human body has been perhaps the most common ESD source, charges on items such as clothing, furniture, and plastics can be damaging as well. Static generation on such items is much more prevalent in modern manufacturing facilities than in older plants and factories. This is due mainly to the synthetic materials that are used extensively to improve cleanliness, cost, longevity, weight reduction, and appearance. Other contributing factors are high-speed operations and reduced relative humidity from air conditioning and heating. Synthetic clothing is common today both for street wear and for many special factory garments. In particular, the increasing use of rubberlike shoe soles tends to raise average personnel static potentials.

Automation trends are reducing the threat from charged human body contact with susceptible equipment. However, it is common practice to use insulative composite materials and ungrounded metal pieces in robotics and similar handling equipments. Without proper corrective attention to such detail, ESD failures will result.

## 1.2 Electrostatic Discharge (ESD) Definition

Electrostatic discharge is defined as the transfer of electrostatic charge between bodies at different electrostatic potentials caused by direct contact or induced by an electrostatic field.

### 1.2.1 ESD event

An ESD event is the unplanned occurrence of electrostatic discharge. In the context of this book the inference is generally to an ESD event occurring in the handling or processing of electronic parts and/or assemblies.

### 1.2.2 Capacitive discharge

The usual ESD event consists of the release of stored charge from a capacitive item, either directly or indirectly. The capacitive item in the most fundamental form is the human body. Other typical charge-storing bodies include carts, chassis, trucks, chairs, printed circuit tracks, and the part itself. The degree of damage caused by ESD is related to the size of the charged body as well as the voltage levels. These relationships are discussed in Chap. 4 and in greater detail in Chap. 8.

### 1.2.3 Field-strength breakdowns

Certain part structures are susceptible to failure when exposed to high electric fields. A classic example is the metal-oxide-semiconductor (MOS) gate oxide. Failures of such devices can occur whenever the field strength exceeds the dielectric breakdown limit. Details of dielectric breakdown failure mechanisms are discussed in Chap. 7 and in conjunction with ESD protection circuitry in Chap. 12.

## 1.3 Synopsis of ESD in the Electronics Industry

One might have expected solutions to the ESD problem in the electronics industry to have been quite easy considering the amount of information available from other industries. Additionally, such a simple problem might have been expected to present little challenge to the many electrical-electronic engineers and technicians working in the field. Nevertheless, in many instances those most familiar and best trained in electronics appeared to have most difficulty in believing that there was an ESD problem with anything other than unprotected MOS devices. Little attention is given to electrostatics in many modern electrical engineering curricula. Perhaps this explains why, as we approach the twenty-first century, so many engineers seem unfamiliar with basic electrostatic principles so well documented by William Gilbert, Stephen Gray, Benjamin Franklin, and Michael Faraday during the sixteenth, seventeenth, eighteenth, and nineteenth centuries, respectively.

### 1.3.1 Early ESD problems

Although ESD came to widespread attention in the 1970s and 1980s, the problem has been experienced by the electronics industry for much longer. As far back as the 1940s, point-contact diodes were found to be extremely sensitive to static electricity. Users of necessity learned to take extreme precautions including grounding and protective packaging of these devices. Figure 1.1 shows an early detector diode used in World War II communications and other applications. The conductive lead case in the photo is one of the earliest static shielding packages used in the electronics industry. There were some static failure problems encountered by users of discrete bipolar junction diodes and tran-

**Figure 1.1** Static susceptible point-contact diode used in the 1940s (with protective case).

sistors in the late 1950s. These problems were rare and received little attention.

### 1.3.2 Thick-film resistors

Early thick-film resistors were noted as being sensitive to static electricity. Static discharge or similar high-voltage transients were even used to trim resistor values or to recover out-of-tolerance resistors. Recent progress has greatly improved thick-film materials and processes, and ESD immunity with these parts has been improved significantly. The static vulnerability of thick-film resistors did not receive much attention, however, or result in any significant efforts to control ESD.

### 1.3.3 Thin-film resistors

Thin-film resistors are even more susceptible to electrostatic discharge than thick-film resistors. In particular selected thin-film resistance with very tight (0.01 to 0.1 percent) tolerances, are vulnerable to the resultant resistance changes from ESD. Tests show that the resistance change can be positive or negative depending upon the exposure level. Users sometimes discount the 1 percent or so resistive change from ESD as a minor degradation. The point is that if the circuit can tolerate a 1 percent change then a 0.01 to 0.1 percent tolerance need not have been used.

### 1.3.4  Metal-oxide-semiconductor failures

The static electricity vulnerability of MOS devices became evident from the initial production runs. Even during fabrication, shorted gates were occurring at final test after having functioned normally at wafer level tests. Users were detecting failures upon receipt and after assembly processing and handling. The structure of these devices was so vulnerable that failures occurred at very low energy levels as long as the gate oxide breakdown potential, typically about 100 V, was exceeded.

As a result a certain degree of ESD control became common to those handling MOS products. Some even went to the extent of using personnel-grounding wrist straps and grounded work surfaces in addition to protective packaging. Such controls, where existent, were usually applied at the part-handling level only. When unprotected MOS circuitry was installed in assemblies, there was often enough buffering by bipolar and resistive capacitive circuitry to raise the assembly level threshold considerably for most devices. Users could handle the assemblies with more abandon than the devices with a much smaller amount of resultant failures. This helped to set the foundation for a myth that susceptible items were somehow immune to ESD after installation on assemblies such as a printed-circuit board.

**1.3.4.1  Protected MOS circuitry.**  With such early recognition of ESD susceptibility with MOS devices, the industry of necessity soon began to incorporate protective circuits to shunt the static transient away from the gate. Up until about 1970 these primitive protection circuits were capable of raising the threshold from 100 V up to about 400 V. This was a significant improvement and yields increased dramatically in spite of the still rather high susceptibility. The protected MOS was received by users with open arms, and a complacent attitude about ESD resulted from a false sense of security.

### 1.3.5  ESD from 1972 to 1974

The period from 1972 to 1974 was extremely difficult for the few engineers investigating ESD problems with electronics parts other than MOS. Most of the discoveries came from failure analysis engineers working for large equipment manufacturers. Device manufacturers were scaling down geometries to meet microminiaturization demands of such manufacturers. In the case of bipolar devices, no forethought was given to possible damage from electrostatic potentials on the human body or other charged objects. The few recorded cases of bipolar operational amplifier devices failing in the range of 400 to 800 V from the human body were not widely publicized. Much static susceptibility testing was done using the actual human body. Human body simulation circuits varied considerably, and early susceptibility data results range accordingly. Since there was no great notoriety on the topic of ESD, especially with regard to bipolar devices, investigators were faced with overwhelming doubt on the part of management and the broad engineering community.

During this period my personal work duties involved analysis of microcircuit failures at the Westinghouse Advanced Technology Laboratories (ATL) in Baltimore, Maryland. Custom integrated circuits fabricated at ATL included complementary metal-oxide semiconductor (CMOS) and bipolar devices including operational amplifier and comparator circuits. During the winter of 1972 detailed physical analysis disclosed that ESD accounted for about 75 percent of our assembly and test failures of the bipolar devices. Although the CMOS line was slightly more sensitive, overall acceptance of CMOS susceptibility to static electricity resulted in sufficient ESD controls to reduce failures. Failure analysis case histories from both the bipolar and CMOS lines are discussed in Chap. 10.

The 1974 International Reliability Physics Symposium (IRPS) included two significant technical papers on the sudden emergence of ESD as a threat to bipolar devices. Speakman's paper[1] presented a model for the damage threshold of bipolar devices from ESD that has now become a classic. His work was the result of actual failures experienced at Western Electric. A paper by E. R. Freeman and J. R. Beall[2] discussed static failures of junction field-effect transistors, MOS capacitor compensated operational amplifiers, and a low-power TTL hex-inverter. The hex-inverter example is discussed in Chap. 11, as it became relevant to later studies on time-dependent degradation due to ESD. Both papers contained good discussions on static generation and ESD control.

### 1.3.6 ESD from 1974 to 1979

**1.3.6.1 Phantom emitter.** After the two informative papers presented at the 1974 IRPS, many felt that all needed ESD topics had been exhausted, and the subject was rarely selected for discussion at future meetings. From 1975 through 1978 ESD got little attention with the exception of two papers on CMOS and one on bipolar devices at the 1977 IRPS Symposium. The bipolar paper[3] in 1977 by R. L. Minear and G. A. Dodson discussed the "phantom emitter" static protection structure. This innovation is discussed in greater detail in Chap. 12 on design protection.

**1.3.6.2 The first Electrical Overstress/Electrostatic Discharge (EOS/ESD) Symposium.** The lack of sufficient ESD publicity led to the first Electrical Overstress/Electrostatic Discharge (EOS/ESD) Symposium. In 1978 a group of engineers dedicated to reducing electrical overstress (EOS) and/or electrostatic discharge (ESD) failures met in Denver, Colorado, to discuss the need and to formulate plans for a symposium in September of 1979. There was some apprehension that the meeting was not likely to draw many attendees and that if a second symposium were to be held, it would likely be in 2 years and be primarily devoted to EOS.

This symposium provided a forum of exchange for interested participants. It also provided vendor exhibits of static control products, which opened communications between vendors and users and was a catalyst to product development. The added publicity began the trend toward increased "static awareness." Interest in ESD snowballed and attendance at the annual symposium grew from 384 attendees in 1979 to over 1200 since 1986.

In 1979 the lack of sufficient ESD awareness was considered the foremost challenge. In the author's "An Effective ESD Awareness Training Program,"[4] the difficulties of achieving ESD awareness were discussed and a suggested approach was given. A highlight of this paper was the recommendation for demonstration of actual failures with the use of a curve tracer as discussed in Secs. 2.4.2 and 2.4.3.

**1.3.7 More representative examples**

Verification that the cause of electronic malfunction is ESD requires good failure analysis capability and facilities. Thus insufficient visibility of the ESD problem might be expected for small companies without such resources to identify subtle failure causes. Case histories have been selected to illustrate the potential value of such investigations in reducing life-cycle costs and as indications of the extent of the ESD problem. These case histories of ESD failures are discussed in detail in later chapters. The examples include a variety of problem sources and include failures from:

1. Subtle bipolar operational amplifier damage
2. Shorts of interlayer metallization on hybrid assembly substrate
3. Capacitive coupling to a CMOS device inside a metal chassis
4. Touching lid of hybrid assembly
5. Touching shielding of coaxial cable tied to emitter-coupled logic (ECL) output

6. Wafer processing operation
7. Masking tape application and removal
8. Possible time dependent latent failures on a spacecraft program

In addition to the aforementioned, countless additional failure examples are discussed and/or used for illustration throughout this book.

## 1.4 ESD Control Success Stories

One side of the ESD story is the indisputable fact that static electricity has been a most serious problem in the electronics industry since about 1973. The other side is the question of return on investment or "payback" from expenditures to control static electricity in the industry. Readers are cautioned to measure the extent of their own ESD problems and improvements brought about by control measures. In recognition of the difficulty in accurately measuring either the problem or resultant improvements, reviewing the experience of others can be helpful. The following success stories are summarized to give a qualitative indication of the validity and possible payback from ESD controls.

### 1.4.1 Automotive industry example

In 1987 Edward C. Y. Lai and Jeffrey S. Plaster [5] reported on experiences at the Delco Electronics Corporation facility in Kokomo, Indiana. ESD failure data were compared on two distinct product lines during 1983 and 1984. The older product A contained significantly less static sensitive parts than the more modern product B. ESD controls for product B were thorough and consistent with the latest philosophies and practices, whereas processes for product A reflected minimal (or no) controls typical of earlier industry practices.

Cost effectiveness of thorough ESD controls was reflected in much fewer static failures in the more sensitive newer product. In 1983, static was attributed as the cause of 9.40 percent of failures of product A, but only 3.89 percent of failures of product B. Furthermore, in 1984 the rate for product A increased to 17.61 percent of total failures. This may be attributed to design changes and increased sensitivity of the product without concurrent improvements in ESD controls. At the same time the rate for product B improved slightly to 3.07 percent of total failures, perhaps reflecting growth and improvement in the ESD controls.

Additional data indicated an estimated $22 million in EOS/ESD damages over a 2-year period. Review of part usage showed a 500 percent increase in static sensitive parts in product A between 1980 and

1986. Even in product B, the initially high percentage of sensitive parts increased by 200 percent.

### 1.4.2 Western Electric's Denver Works

W. Y. McFarland[6] reported a 2300 percent return on investment from ESD control instituted at the Western Electric Denver Works. Controls included the usual grounded work surfaces, wrist straps, heel straps with conductive floor mats, and strategically placed ionized air blowers. In addition they used conductive stool covers, caution labels, and field service kits. Antistatic bags were chosen over conductive bags based on a cost analysis. The yield improvements attributed to these measures included over 1 percent of integrated circuits at incoming test, nearly 11 percent of printed-circuit test and repair, 0.81 and 0.75 at printed-circuit board and system burn-in, respectively, and over 4 percent in field repair.

### 1.4.3 Wafer-level controls

The question of whether or not ESD is a problem at the wafer level arises frequently. There have been a number of reported success stories in overcoming such problems by better ESD control. R. Euker[7] reported failures due to 8 kV acquired during high-speed wafer spin drying on teflon racks. Failures were due to subsequent tweezer handling and use of wafer break papers also charged to as much as 2 kV. Elimination of these two hazardous operations resulted in a yield improvement of 10 percent. The referenced paper states further that 15 percent of all parts diagnosed for cause of failure by the failure analysis lab at Hewlett Packard's Loveland, Colorado, facility were due to ESD. The actual percentage was believed to be higher because of possible masking by some of the 40 percent attributed to electrical overstress. Assuming the 15 percent level, an 800 percent return on investment for ESD controls was cited. Pre-ESD control showed a 6.5 percent instrument assembly reject rate during low humidity winter months with a 3.5 reject rate during summer months. After initiation of a comprehensive ESD control plan the reject rate was about 3.2 percent year-round. Similarly the integrated circuit (IC) assembly rejects in winter months dropped from 40 percent to 25 percent with ESD control.

### 1.4.4 Hybrid, printed-circuit board, and field improvements

Several experiments and case histories of ESD control payback were reported by G. T. Dangelmayer.[8] In one experiment, two samples consisting of 216 each printed-circuit boards were processed at relative humidity of 30 percent with and without the single control consisting of wrist straps. A 2.7 to 1.0 improvement in defect rates was experienced in the controlled to uncontrolled product line which contained bipolar integrated circuits and discrete devices.

A similar experiment was conducted on two samples of 1275 hybrid integrated circuits. One sample was processed with no ESD controls and the second sample was processed with just grounded wrist straps and antistatic tote trays. Results were factors of 1.9 and 2.4 times the number of defects in the uncontrolled sample at inprocess and final tests, respectively. The survivors from this hybrid-level experiment were then traced through to printed-circuit board processing. The printed-circuit board line had no ESD controls at the time. The hybrids that had previously had ESD controls were transported in conductive foam whereas the uncontrolled hybrids were transported in either polystyrene or styrofoam shipping trays. Results were 5.5 times the rejects of the uncontrolled sample over those controlled. These experiments were followed by comparisons of factory 3-month average yields in the entire hybrid processing lines before and after introduction in October, 1980, of wrist straps and antistatic trays. Results were factors of 1.8 and 3.3 at the inprocess and final tests, respectively, equal to a 950 percent return on investment. Factory data further showed that the printed-circuit processing showed a 2.1 to 1.0 reduction in rejects in 1981 after incorporation of controls in the hybrid shop. Controls initiated in board processing in 1982 included wrist straps, heel straps, and dissipative floor mats where applicable, and dissipative table mats. A further reduction of board rejects of 1.6 to 1.0 was realized in 1982 as a result.

## 1.5 Sources of Information

The amount of information available in the area of static control and ESD is growing at an exponential rate. This is due largely to the influences of governmental agencies, the Electrical Overstress/Electrostatic Discharge (EOS/ESD) Association, and growing interest in the subject. Information on certain aspects is controversial due to the rate of change in the state of the art.

### 1.5.1 General information

Broad and timely information on the subject is offered at the annual EOS/ESD Symposium jointly sponsored by the EOS/ESD Association and ITTRI/RAC. Military Standard 1686A can be used as a reference in establishing an ESD control program. The associated DOD Handbook 263 provides comprehensive information on ESD control. Other good sources are articles in periodicals and journals of professional societies. Numerous tutorials held throughout the world provide an opportunity for novices to ESD to learn from experienced individuals. An excellent source of specific information not to be overlooked is networking through personal contacts within the industry.

### 1.5.2 Susceptibility data

The quest for accurate susceptibility data is difficult. Available data can be invaluable, but the user should be aware of the limitations on accuracy and utility. Providers of such data will make efforts to improve accuracy and precision as new methods and information develop.

**1.5.2.1 RAC VZAP data base.** The best source of general data of part susceptibility is the VZAP data base[9] available from Reliability Analysis Center (RAC). The RAC has accumulated the available data from cooperative donors within industry and governmental agencies. Efforts are continuously underway to broaden and improve this data base.

As available, data presented includes part number, Mil Std and slash number, part type, manufacturer, Mil Std 1686A classification, and the method of classification. Backup information on donors' test results includes part date code, test procedure including model, number of exposures, sample size, pin combinations, test results, and data source. The current revision (VZAP-2) contains about 4300 test results relating to approximately 1600 ICs and 900 discrete device types.

The variety of sources of the data results in problems of accurate comparative quantification. A review of Chap. 8 will verify that many inconsistent results exist even when the same human body simulation model is used. The donors of the VZAP data used several different models including various Human body model (HBM) circuits as well as electromagnetic pulse (EMP) and charged-device model (CDM) circuits. A normalization technique (discussed in Chap. 8) was used by RAC to relate all data to the standard 100 pF, 1.5 k$\Omega$ human body model. Data provided was often limited to a single pin combination, perhaps in one polarity, tested in failure duplication efforts. Thus, the parts might be more susceptible at other pin combinations or at the opposite polarity.

### 1.5.3 Information on ESD control problems

The cooperative Government-Industry Data Exchange Program (GIDEP) includes ESD related information in its data base. Governmental and industrial participants in the program are provided access to engineering, failure rate, metrology, and failure-experience data banks. The ESD file has been included in the engineering data base. Efforts are underway to update this file with susceptibility data to avoid duplication of efforts. The GIDEP data base also includes applicable problems and solutions for ESD control materials.

## References

1. T. S. Speakman, "A Model for Failure of Bipolar Silicon Integrated Circuits Subjected to Electrostatic Discharge," *International Reliability Physics Symposium Proceedings,* 1974.
2. E. R. Freeman and J. R. Beall, "Control of Electrostatic Discharge Damage to Semiconductors," *International Reliability Physics Symposium Proceedings,* 1974.
3. R. L. Minear and G. A. Dodson, "Effects of Electrostatic Discharge on Linear Bipolar Circuits," *International Reliability Physics Symposium Proceedings,* 1977.
4. O. J. McAteer "An Effective ESD Awareness Program," *Electrical Overstress/Electrostatic Discharge Symposium Proceedings,* 1979.
5. Edward C. Y. Lai and Jeffrey S. Plaster, "ESD Control in the Automotive Electronics Industry," *Electrical Overstress/Electrostatic Discharge Symposium Proceedings,* 1987.
6. W. Y. McFarland, "The Economic Benefits of an Effective Electrostatic Discharge Awareness and Control Program—An Empirical Analysis," *Electrical Overstress/Electrostatic Discharge Symposium Proceedings,* 1981.
7. Rick Euker, "ESD in I.C. Assembly (A Base Line Solution)," *Electrical Overstress/Electrostatic Discharge Symposium Proceedings,* 1981.
8. G. T. Dangelmayer, "ESD—How Often Does It Happen?" *Electrical Overstress/Electrostatic Discharge Symposium Proceedings,* 1983.
9. "Electrostatic Discharge Susceptibility Data," VZAP-2, Vols. I and II, Available for purchase from: Reliability Analysis Center, P.O. Box 4700, Rome, NY 13440-8200. (Attention of William H. Crowell, phone 315-330-4151).

# Chapter 2

# ESD Awareness

The term "static awareness" has taken on a significant connotation in the ESD control field. A common saying among ESD specialists is that awareness is at least 85 percent of the battle toward conquering the ESD threat. The primary reason for difficulty in achieving effective ESD control in a timely fashion has been due largely to the lack of sufficient "ESD awareness." Many involved with ESD control find the awareness problem to persist even with the vastly increased amount of information available on the subject. This chapter will

1. Discuss the awareness problem
2. List reasons for disbelief in ESD
3. Identify those requiring training
4. Provide justification for awareness training
5. Suggest methods of training
6. Indicate types of training needed for different groups
7. Emphasize attainment of management commitment to ESD control

## 2.1 Frequent Existence of Doubt

The examples cited in Chap. 1 and other extensive documented reports are evidence of the widespread detrimental effects of electrostatic discharge. Yet why do so many behave like the bird shown in Fig. 2.1 when confronted with the ESD issue? This attitude is still quite common but not nearly to the degree of my first encounters with bipolar device static failures in 1972. At that time and for some years to follow most static investigators were faced with a great deal of skepticism on virtually all aspects related to ESD. Even today many doubt the severity of static electric damage to electronic equipment.

**Figure 2.1** Sandy-billed headhider.

In the early seventies I prepared numerous detailed failure analysis reports that included photomicrographs of damage sites as well as parametric test data before and after ESD exposure. Yet this evidence was often disregarded as doubts about static vulnerability of bipolar devices was deeply rooted. For several years this reaction perplexed me and communications with my peers in industry disclosed that they felt the same frustration.

## 2.2 Reasons for Disbelief

It is not that people want to discount the ESD problem out of contempt or bias. Some inherent differences require a better approach if others are to be convinced of its significance. In order to become a better teacher these differences need to be identified and overcome. I attempted to perceive the ESD phenomenon from the eyes of the many personnel involved. With this approach I have come to realize that there are more apparent reasons to doubt the seriousness of the ESD problem than the contrary. In summary, the primary reasons for disbelief follow:

1. Most ESD failures occur below the human sense of feeling to static discharge which begins at about 4000 V.

2. Many part types are sensitive to ESD. A variety of failures could be experienced with no indication of a trend.
3. ESD failures are most likely to be detected at the first electrical test, which places them on a low priority for failure analysis compared to higher assembly levels or system failures.
4. Physical damage is often very subtle and unlikely to be recognized in cursory failure analysis.
5. Static paths to susceptible nodes can be subtle, which often tend to discourage consideration of ESD as the cause of failure.
6. Damaging potentials are not always present so disregard for proper ESD control does not always result in failure. Repeated experiences of "successes" of this type tend to fortify disbelief already present.
7. Other extremely obvious failure causes tend to be recognized as far more important "yield detractors" and absorb a disproportionate amount of resources.

To reflect on these reasons in greater detail may provide a realization of the understandable proliferation of doubt that exists.

### 2.2.1 Most damage below threshold of feeling

The sense of feeling differs between people, but starts around the level of 4000 V. Table 2.1 lists one representative sample of susceptibility data for a group of part types. The plus or minus indicates the polarity of the first pin with respect to the second pin of the combination tested. There is little question that the majority of integrated circuits and many discrete parts fail at thresholds below 4000 V. The sensation of feeling from static discharge is a rare occurrence except under extremely dry atmospheric conditions. Therefore the conclusion is that most ESD failures occur at levels well below the threshold of feeling. Thus, factory operators have little reason to suspect they are causing failures. The rather hardy discrete 2N2222 transistor has the only threshold likely to have been felt by most people. Although the level of 15,000 V may be regarded as insensitive by most standards, readers are cautioned to consider the end-use environment. For instance, this level is not unusual on personnel in dry climates of the arctic region or in some desert areas.

### 2.2.2 Variety affected without trend indicated

As indicated by Table 2.1 a wide variety of part types are susceptible to ESD. The list of static-sensitive items is expanded in Table 2.2. The

TABLE 2.1 Representative Sample of Port Susceptibility Data

| Device no. | Technology | Susceptibility level | Pin combination |
|---|---|---|---|
| VN98AK | VMOS | 110 V | (+) Gate to source |
| 3N170 | MOSFET | 150 V | (−) Gate to source |
| Custom IC | CMOS | 150 V | (−) Input to gnd |
| 2N4416 | JFET | 220 V | (−) Gate to source |
| Custom IC | Bipolar op-amp | 400 V | (−) Input to gnd |
| 1N5711 | Schottky diode | 500 V | (−) Anode to cathode |
| MC1660 | ECL | 500 V | (+) Output to $V_{CC}$ |
| CD4001A | CMOS | 800 V | (−) Input to $V_{DD}$ |
| 54S04 | Schottky TTL | 1000 V | (+) Input to $V_{DD}$ |
| RNC50 | Thin-film resistor | 1000 V | Lead to lead |
| 5404 | TTL | 1600 V | (+) Input to $V_{DD}$ |
| 54L04 | Low-power TTL | 3500 V | (−) Input to gnd |
| 2N2222 | Bipolar transistor | 15000 V | (+) Emitter to base |

TABLE 2.2 Generalized List of Static-Sensitive Items

- Integrated circuits, including:

  MOS(N,C,P,H,V, etc.)
  Bipolar: op-amps, TTL, ECL, ASTTL
  Schottky, AS, ALS
  GaAs

- Microwave devices

- MOSFETs and JFETs

- SCRs and tunnel diodes

- Thin/thick-film resistors

- Some discrete diodes, transistors, and capacitors

- Metal-oxide metal structures

- Piezoelectric crystals

- Hybrids, PC boards, and other assemblies containing the above

last item is in contradiction to the popularized myth that susceptible items are no longer vulnerable once mounted on an assembly. This topic is addressed in greater detail in Chap. 7. In a typical factory situation a diverse assortment of part types is further diluted by placement on a second level of various assemblies. Static might affect a hodgepodge of assemblies in such a manner as to damage a diversity of part types. If this were the case, the common causal thread of ESD is not likely to be suspected and a failure trend would go unnoticed. Such failures are likely to be tolerated rather that analyzed. So ESD goes undetected or even suspected.

### 2.2.3 ESD failures unlikely to be analyzed thoroughly

In most assembly operations the maximum opportunity for ESD seems to occur at the first or lowest assembly level. At this level the part itself is processed frequently and much handling occurs at assembly points leading directly to susceptible nodes of parts. Since failure analysis is generally allocated on a prioritized basis, detailed failure analysis is generally limited to higher assembly, or even system level occurrences. First-level failures are not likely to be analyzed, or at best receive a cursory analysis. In either case, the subtle nature of the resultant physical damage may escape detection or proper identification.

### 2.2.4 Subtle damage paths

Static discharge failures often result where there is no apparent direct path. This is frequently the result of capacitive coupling or induction as discussed in Chap. 4. Frequently the path is difficult to visualize so static is not suspected.

### 2.2.5 Experience the best teacher

Experience is said to be the best teacher. Certainly things learned by experience are deeply rooted and difficult to forget. Unfortunately the degree of permanency is not related to the correctness of the item learned. In other words, experience can teach things that are wrong. In the case of ESD, experienced engineers and technicians are often the strongest and most influential doubters. Perhaps they remember handling parts and printed-circuit board assemblies in their bare hands while ungrounded with no resultant failure. In fact, the majority of such improper actions are unlikely to result in failure. A steadfast opinion that ESD isn't a threat often develops. At best it may be seen as a minor problem affecting only very sensitive items such as unprotected MOS devices. Such experienced and well-meaning individuals will often cite their 15 or so years of handling parts and assemblies with no failure occurrences.

The following important factors were overlooked in the foregoing discussion:

1. There probably was an occasional failure that they might have attributed to a "bad part" from an "unreliable vendor."
2. The parts or assemblies that continued to work may not have had the same performance parameters that would have prevailed with proper handling.

3. Some of the mishandled assemblies may not have functioned as long as they would have with proper handling.

### 2.2.6 Other more visible problems

It is difficult to compete for funds for combating ESD with the more visible problems in a production facility. The emphasis on statistical quality control that has recently become part of the electronics industry has revealed many yield detractors that had been heretofore tolerated. Among these are such mundane things as wrong, broken, or missing parts; solderability; faulty incoming parts; and a variety of poor workmanship defects. Such problems deserve and require a fair amount of attention in order to achieve higher quality levels. However the statistical data analysis applied is generally not sophisticated enough to perceive the yield and reliability detraction caused by ESD.

## 2.3 Who Needs to Be Convinced

ESD awareness training of some form is needed for management, factory operators, and technical personnel. The types of training required varies between these groups, but is of great importance for each. It is extremely important that ESD training is not limited to factory operators alone.

### 2.3.1 Management training need

Managers must be trained in several aspects of ESD. Without knowledge of the subject, inappropriate ESD control decisions are assured. A manager must know the procedures for handling susceptible items lest

1. Bad example be given
2. Damage is caused
3. Customers witness poor practices and attitudes

**2.3.1.1 Management commitment.** Management awareness training is mandatory in order to initiate an ESD control program. Funds for control measures including training of other personnel will not be committed until management is convinced of the seriousness of the problem. The term "management commitment" has become a recognized challenge that must be met by individuals dedicated to minimization of ESD problems. This term means management's recognition of ESD as a serious problem worthy of a concerted control effort. True management commitment is evidenced by the following:

1. Dedicated funds and appropriately trained personnel to establish an ESD control program
2. Continued recognition of the difficulties associated with maintaining an effective control program
3. Dedicated personnel for long-term monitoring and maintaining a dynamic control program
4. An ongoing ESD awareness training program for line management, factory operators, and technical personnel

**2.3.1.2 Line management training need.** Line management seems to have a natural tendency to take a passive and distant attitude toward ESD control. The apparent feeling is that control measures are in place and that should be enough as they have other pressing duties. Frequently these managers have not had ESD awareness training, whether or not their personnel have been trained.

For effective ESD control, line managers must be trained and concerned about the proper application of measures instituted at their work areas. Properly trained managers can be invaluable in motivating, monitoring, and disciplining their personnel in ESD matters as well as in sensing the effectiveness of certain control measures. An additional benefit is the greatly improved customer image that will most likely result.

## 2.3.2 Factory operator training need

The term factory operator is meant to include any person involved with processing of susceptible equipment. Normally included would be those from receiving, kitting, transportation and storage, assembly, test, inspection, repair, and shipping. Such personnel generally are dedicated to doing a good job while trying to achieve maximum output with minimal rejects. Sometimes the need, effectiveness, or even the purpose of certain imposed ESD controls is not apparent. A conscientious factory operator might show lack of diligence toward procedures which seem meaningless. For these and other reasons, factory operators deserve to be trained in the purpose of imposed controls rather than simply have them forced upon them by edict.

## 2.3.3 Technical personnel training need

Personnel such as engineers and technicians with prior technical training are not always cognizant of all the subtleties of ESD problems and require a specialized program. It is particularly important to train engineers in ESD matters in order that an appreciation for the need for immunity in designs be developed. The engineering community holds the best hope for future elimination of the ESD threat to the electronics industry. Breakthroughs in ESD control measures are as likely to be made by engineers as are improved system designs. An additional benefit is that concerned engineers and technicians will have a positive influence on other factory personnel.

## 2.4 Content and Method of Presentation

In order to best reach the three groups to be trained, emphasis must be shifted to each group's primary motivational interests. Good management is attentive to return on investments and long-term increased profits. Factory operators have an aversion for rework and high reject rates. Technical personnel are typically interested in the scientific aspects and the challenge to reduce susceptibility to ESD. All three groups must know how to use the control measures in place. Thus the operator training program, or at least an abridged version, must be presented to all.

### 2.4.1 Common myths to overcome

ESD awareness has grown significantly, but a number of myths prevail at all levels. Several common ones are listed in Table 2.3. These myths affect ESD control choices in a deleterious manner. The principles and examples discussed in this book should dispel any misconceptions of the types listed. An ESD training program with proper coverage of these items reduces these misunderstandings.

### 2.4.2 Attain management commitment

Management will ultimately be interested in estimates of costs of losses through reduced factory yields, degraded reliability of delivered

**TABLE 2.3 Common Myths about Static Electricity and ESD**

1. Conductive items are never involved in ESD.
2. Material imprinted with the word "antistatic" are nongenerative.
3. Antistatic material is protective of static-susceptible items.
4. ESD is primarily an MOS problem.
5. Devices with input-protection circuitry are immune to ESD.
6. Parts are insensitive to ESD after installation on an assembly.
7. Devices or printed-circuit boards with the connector pins shorted together are immune to ESD.
8. Antistatic, static-dissipative, and conductive materials are equivalent.
9. Handling printed-circuit boards by the edges prevents ESD damage.
10. Controlling relative humidity to >40 percent eliminates the static problem.
11. Room ionization eliminates the static problem.
12. We're not having any ESD problems.

product, and reduced contracts due to poor customer image. There will be a lack of interest in these items, however, until their attention is gotten and ESD is shown to be a real problem. After all, managers are likely to have been exposed to influential technical advisors and associates with tendencies to discount the ESD threat.

**2.4.2.1 An experience in attaining initial commitment.** Several years ago I was asked to review the static electricity conditions at a large electronics manufacturing facility and to recommend corrective measures if necessary. Many static-susceptible parts were being used and very little ESD controls were in place except for extremely vulnerable parts with known severe failure histories. Obviously static awareness was lacking. Realizing the possible frustrations in crusading for ESD controls, I carefully planned an approach. The final strategy was to attack the reasons for disbelief discussed in Sec. 2.2. The challenge had more appeal once these predispositions to be overcome were put into proper perspective.

The thought occurred to me that most of these reasons would be erased if one were to have the personal experience of causing an ESD failure while being able to monitor degradation as it happened. In all honesty I may have been a serious doubter of ESD failures also had my work experiences been different. In failure analysis efforts I had frequently watched devices degrade from ESD as I touched them while monitoring their characteristics by curve tracer. As further reinforcement, I had seen the resulting physical damage under microscopic examination.

In order to "get management's attention," a meeting was held with the director of manufacturing and operations and all of top supporting management team. Included were managers of manufacturing, test, quality assurance, and purchasing and their technical advisors. I brought a curve tracer and several parts and assemblies with me to start the meeting with some demonstrations. The most sensitive item was an operational amplifier circuit which would be catastrophically damaged at levels above 600 V or so. The device was packaged in a metal flat pack and touching the lid with an input tied to ground would result in failure. The second item was a metal chassis containing a CMOS device. The CMOS device could be destroyed by the touch of a human body charged to a voltage level of about 1000 V. The third item was a CMOS device in a metal package with a floating (non-connected electrically) lid. The device could be destroyed by breaking down the air gap of about 0.6 mm between the lid and the sensitive input node. This required about 2000 V from the human body. Static generation on actual personnel is so dependent upon the humidity, their clothing, especially footwear, and other variables. For these

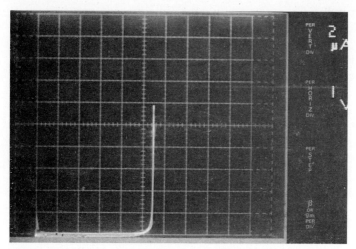

**Figure 2.2** Normal I-V characteristic of reverse-biased PN junction.

reasons a human body simulator was ready in the event that conditions precluded actual body demonstrations.

Initial attempts began with a good operational amplifier connected to the curve tracer so that all could see the normal I-V characteristic curve as in Fig. 2.2. The first volunteer was the manager of quality and reliability assurance. He walked about four steps from his seat to the device and touched the lid. The resultant characteristic is shown in Fig. 2.3. This one failure incident not only convinced a most influ-

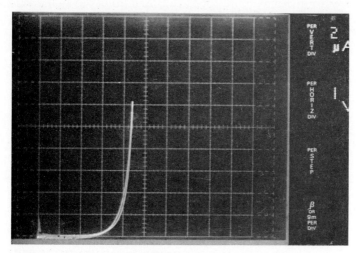

**Figure 2.3** Resultant "softened knee" of junction after ESD.

ential manager of nearly 1000 Q and RA personnel, but the observers as well. Demonstrations continued with additional 600-V operational amplifiers, then several 1000-V chassis, and finally three of the 2000-V CMOS devices. Conditions were apparently ideal and all susceptible item types failed through normal touches by management volunteers.

**2.4.2.2 Long-term management commitment.** Producing actual failure conditions in the presence of upper management is undoubtedly the best approach to achieving initial management commitment. Although normally too difficult and costly to repeat for all employees, the effort to select one or more susceptible items for a management demonstration is usually justified. Through this approach acute attention can be obtained in just a few minutes. Management is then ready to review the costs of static and control recommendations.

**2.4.2.2.1 Return on investment figures.** The approach to motivating management toward monetary expenditures for ESD control must be based upon return on investment analyses. If the facility has a good data-gathering system the task will be easier, especially when supported with an effective failure analysis laboratory. Data showing the amount and types of failures experienced can be used as evidence of a minimal level of cost of the problem. Normally, actual ESD failure amounts exceed those indicated by in-house data because of subtleties of the problem and the fact that first-level test failures are often not analyzed. Where data are incomplete, extrapolations and estimates of the extent of the problem must be exercised. Suggested sources of yield detraction data are as follows:

1. Failure data and failure analyses results as available
2. Actual and extrapolated data extending estimates to first-level and incoming tests
3. Conclusions based upon part sensitivity and static voltages present without the control measures in question
4. Increases in failure rates at times of low humidity (a good indicator of ESD problems)
5. Published payback data from other companies, especially if end product and/or part technologies are similar

In addition to yield detraction, the potential costs of time-dependent latent ESD failures must be considered. Chapter 11 discusses these proven reliability detractors in detail. Additionally, the somewhat intangible benefits from good ESD control of improved cleanliness, enhanced customer image and possibly improved morale must not be underestimated.

### 2.4.3 Factory operator training

All factory operators require sufficient training to:

1. Convince them that ESD problems are a real concern
2. Give them an understanding of the basic principles discussed in Chap. 4
3. Instruct them in the proper identification of sensitive items in their work area
4. Instruct them how to properly use the ESD control measures in their work area
5. Demonstrate that these measures are effective and necessary
6. Instruct them in any self-checks or maintenance required in their job.

Of course a demonstration with actual part failures as described in Sec. 2.4.2.1 would be an excellent approach for a small number of employees. For large numbers this is usually not practical. A reasonable approximation to actual device-failure demonstrations can be obtained by using some commercially available sensitive item simulators that react to "damaging" ESD thresholds by visual and/or audio alarms. Some discussion of in-house case histories is likely to be effective in convincing them that the problem is *ours* also.

Video programs for ESD awareness training are available from a number of sources and can be selectively used for targeted groups. A list of tapes is available from the Reliability Analysis Center (RAC) of the Illinois Institute of Technology Research Institute in Rome, New York. Some of these programs are excellent, but it is strongly recommended that a portion of the training program be carried out in the presence of a qualified instructor. This will minimally provide the needed dialogue of a question and answer period. A further recommendation is that demonstrations of static principles and proper use of selected control measures be included.

### 2.4.4 Technical personnel training

The approach to training engineers and technicians is to develop specialized programs with more technical depth of content. A suggested underlying theme is motivation toward improving future designs and practices. Such a program should include demonstrations of the basic electrostatic principles discussed in Chap. 4. Some of the basics of static phenomena that may have been forgotten will be reinforced. Explanations can appropriately include technical terminology and analysis. Case histories can be effective, in particular if related to design parameters.

## 2.5 Ongoing ESD Training and Certification

An ongoing ESD training program must be established and continuously updated and maintained. Periodic refresher training is required (at least once a year) as is training for new employees. Where films or video programs are utilized, it is recommended that a selection be available and periodically rotated to maintain the viewer's interest. Additional sessions may be required to cover absentees. Such ongoing training must include management and technical personnel as well.

Employee certification of susceptible item handling qualifications and maintenance of records for future reference are essential. Some facilities require personnel to wear badges or carry certificates when working in controlled areas.

## 2.6 Reliability Effects and Customer Image

The awareness issue involves more than just being convinced that some of the yield loss is due to ESD. The degree of ESD knowledge can affect reliability, if not through time-dependent latent degradation, then through the cumulative damage of excessive rework cycles. Good ESD controls with the associated improved yields and reliability serves to improve customer image which leads to improved future business forecasts.

## References

1. O. J. McAteer, "An Effective ESD Awareness Program," *Electrical Overstress/ Electrostatic Discharge Symposium Proceedings*, 1979.

# Chapter 3

# History of Electrostatics

The early history of electrostatics is based largely upon the results of observation, theorization, and experimentation with a certain degree of iteration. This approach has brought about many great discoveries and although it is valid today it is too rarely used. By such investigations, the basic principles of electrostatics were developed and widely understood long before the development of semiconductors and the emergence of the ESD problem. This chapter highlights selected discoveries of earlier times in particular those related to understanding modern-day ESD control.

## 3.1 Destructive Force of Lightning

Our early ancestors no doubt were awed by the powerful displays and force of lightning and its associated roar of thunder. Occasional felling of trees and deaths of animals or humans by lightning strikes were not then understood as a form of electrostatic discharge.

### 3.1.1 The first ESD control practice

In medieval days many believed that lightning was created by demons that rode the storm. To warn the populace of these oncoming evil fiends, bells were customarily sounded at many Christian churches. Unfortunately bell ringing became a hazardous occupation as many were killed because the spires of churches were often struck (see Fig. 3.1). The likely path to ground was to the bell and down the moist rope which the ringer may have been touching at the time.

The practice of bell ringing is, to my knowledge, the first recorded attempt at static control: a constant monitoring system. A good under-

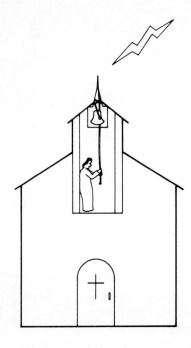

**Figure 3.1** The first known ESD control practice was instituted hastily.

standing of the basic principles of electrostatics had not yet been developed in these early times, so such a disastrous error on the part of the persons that established the warning system can be forgiven. However this example should serve as a reminder of the importance of understanding the fundamentals of electrostatics if one is involved with modern ESD control. Improperly designed control measures could result in costly product failures or could even compromise personnel safety.

## 3.2 Attractive Force of Static Electricity

### 3.2.1 Thales

Thales (640–546 B.C.), a Greek philosopher from Miletus, had studied the magnetic properties of lodestones from nearby fields. While he was involved in these studies, he became curious about the possibility of any similar nature of amber stones brought to the region by traders from northern Europe. Amber is a hard, yellow-to-brown fossil resin of prehistoric soft woods. The Greek word for amber is "elektron." He discovered that after rubbing it with flannel, the amber would attract bits of thread, straw, or feathers. Thales thus ran the first electrostatic experiments to be recorded in history.

Other than occasional sightings of St. Elmo's fire, little was recorded about electrostatics until the sixteenth century. St. Elmo's fire is a little understood ionization phenomenon evidenced by a glowing effect at or near ships' masts somewhat akin to ball lightning.

### 3.2.2 William Gilbert

William Gilbert (1540–1603) of Colchester, England, repeated the experiment of Thales. He then pondered what characteristic of amber, or elektron, would make it attract small items after being rubbed with certain other materials, which he expanded to include silk and fur as well as flannel. Amber is capable of taking on a fine polish and is often used for ornaments. Gilbert thought that this jewel-like property might be related to the attraction phenomenon. He therefore extended the experiments to certain jewels such as diamond, opal, and sapphire and found them to exhibit the attractive trait similar to amber (see Fig. 3.2). He termed these jewels "electrics," derived from the Greek word for amber.

Gilbert later added other items such as amethyst, glass, jet, rock crystal, sulfur, and hard sealing wax to his list of electrics. In order to better quantify the ability of such materials to attract, Gilbert invented the versorium. This first known electrical measuring instrument shown in Fig. 3.3 was simply a pivoted needle made of light material, usually wood or metal. He also reportedly noted that the attractive force as well as the crackling sound when discharges occurred was easier to produce on dry days.

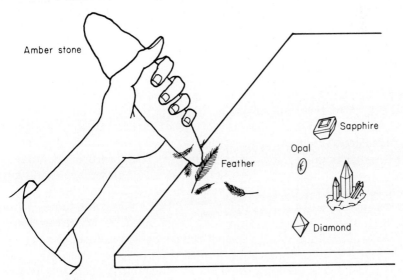

**Figure 3.2**  Gilbert's electrics.

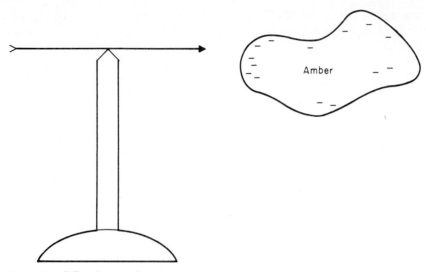

**Figure 3.3** Gilbert's versorium.

His experimentation with magnets culminated with a book entitled *De Magnete* (*Concerning Magnets*) in which he described the earth as a giant magnet. Gilbert became known as the "father of electricity" for his efforts in this area of science. It is interesting to note that this pioneer of electrostatics was also a man of many interests. As a matter of fact he was the attending physician of Queen Elizabeth I.

## 3.3 Static Generating Apparatus

### 3.3.1 Otto von Guericke

The first known machine for producing static electricity was made in 1678 by Otto von Guericke (1602–1686), a German scientist. He began by melting pulverized sulfur which he poured into a hollow glass sphere used as a mold. A wooden rod was then inserted through the center of the sulfur sphere to serve as an axle for turning the sphere. Once the sulfur solidified and returned to room temperature, the outer glass shell was broken away. His machine was completed by adding a crank to turn the axle through the sulfur sphere of approximately 6-in diameter. As the ball turned rapidly it rubbed against his hand or a cloth which caused static buildup. This machine depicted in Fig. 3.4 was capable of producing large sparks and probably produced the largest man-made static potentials up to that time.

Von Guericke was the burgermeister of Magdeburg, spending most of his time in politics, but was trained as an engineer. He also in-

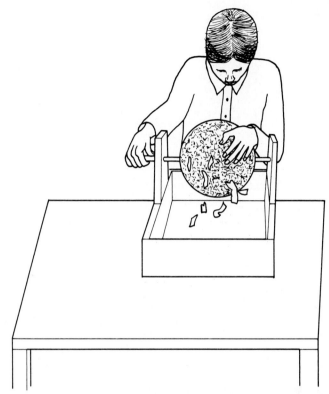

Figure 3.4  Von Guericke's sulfur-ball generator.

vented the first vacuum pump which he demonstrated by the inability of 16 horses to separate two evacuated hemispheres.

### 3.3.2 Francis Hauksbee

In 1709 Francis Hauksbee (1666–1712), an English scientist, fashioned a similar electric machine. His use of a glass sphere indicated that he may have realized that von Guericke need not have gone to such effort in producing his sulfur sphere. Hauksbee had recently invented an improved vacuum pump so he attempted to determine if there would be any electrical effect due to removing the air from the globe. The addition of a large wheel and belt system to turn the globe at a higher rate resulted in a significantly higher charge generation than that of prior friction machines. He calculated the speed of rotation to be 29 ft/s. When the globe was turned in this fashion and rubbed against wool, sufficient charge accumulated to cause the globe to glow. This electrostatic machine is illustrated in Fig. 3.5.

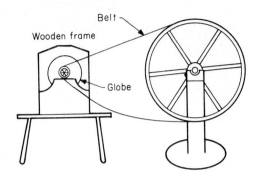

Figure 3.5 Hauksbee's glowing machine.

**3.3.2.1 Other discoveries.** Another observation was made by Hauksbee when he brought his face near the charged glass and felt a sensation like a breeze against his cheek. This is now known as an electric wind consisting of charged particulate in the air being repelled by the charged item. Hauksbee also observed that the ends of a bent straw would part from the force of repulsion that varied with the amount of charge present. This apparatus in Fig. 3.6 was the first electroscope.

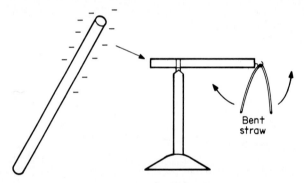

Figure 3.6 Hauksbee's straw electroscope.

**3.3.2.2 Inductive effect.** Hauksbee made another great discovery that was not fully appreciated until many years later. A globe that had been evacuated was rubbed by hand while rotating with a non-evacuated globe located within 1 in away. Both globes were observed to glow even though friction had been applied only to one. This phenomenon was not understood by Hauksbee, but was a dramatic demonstration of electrostatic induction.

## 3.4 Conductors and Nonconductors

### 3.4.1 Stephen Gray

The first published technical paper in 1720 of Stephen Gray (1666–1736), an Englishman, described experiments which added some new electrics, namely silk, wood, and hair (animal and human) to Gilbert's prior list. In this same paper he asserted that electrification experiments with a glass rod were more effective on dry days. He further described how stiff paper could even be electrified after heating almost to the flash point to dry it sufficiently prior to rubbing.

### 3.4.1.1 Transmission of attraction property of charged items.

The paper also described certain experimentation with a hollow glass tube about 1 m long. Gray found that after objects were contacted by the rubbed tube they apparently took on the same attractive property as the tube. Thus this mysterious property was transmitted. Mainly, this discovery came about accidentally after Gray inserted corks into the ends of the tube and observed that they also attracted light objects when the tube was rubbed. The original reason for inserting the corks was to maintain cleanliness in the hollow of the tube rather than to observe their electric properties.

Following the discoveries described in his first paper, Gray continued experimentation to better define the limits of the ability to transmit the attraction property. He first drilled a small hole through an ivory ball of about 1-in diameter and thrust a stick through the hole. With the other end of the stick attached to the cork in the glass tube he found that the ivory ball would also attract. He successfully extended this test using a wooden fishing pole 18 ft long.

Then he tried wire attached to the cork and again found that the wire was capable of transmitting the property. Next he tied the ivory ball to the cork with about 3 ft of hemp twine. The ivory ball still attracted, as illustrated in Fig. 3.7. To test the extent of this property further he went to the roof of his house and found that the ivory ball even attracted feathers from a height of about 34 ft. He showed also that wet string worked better than dry string.

Since there were no high cliffs or buildings in the area, Gray decided to try lengthening the string horizontally. He looped it back and forth across his room by hanging it above the beams. The attractive force was no longer transmitted. He concluded that the electric "virtue" had gone to the beam rather than the ivory ball.

A friend, Granvile Wheler, suggested that Gray try supports of silk thread. Gray agreed because he surmised that the smallness of the silk thread would impede the electric virtue loss. He gradually increased the length with success until it was about one tenth of a mile long, at which time the silk thread loops began to break. He then replaced the silk hangers with thin metal wires to increase the strength but found that the experiment no longer worked. Finally he used larger silk thread hangers as shown in Fig. 3.8, and the experiment again was repeated successfully. From all of this Gray determined

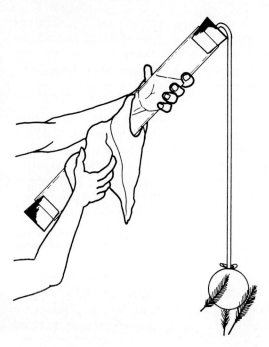

**Figure 3.7** Gray's experiment with glass tube, twine, and ivory ball.

that some materials, such as iron and copper, were conductors whereas other materials, such as silk, were nonconductors.

Gray also concluded that the human body was conductive, and this factor precluded certain materials from acquiring charges in his earlier experiments when he often held them in his hands. He finally determined that even metals could be "electrified" and that electrified bodies could be insulated by placing them on blocks of resin.

## 3.5  Different Types of Electricity

### 3.5.1  Charles Du Fay

Charles Du Fay, a Frenchman (1698–1739) was a contemporary of Gray. After repeating many experiments similar to Gray's, Du Fay concluded that any material could be electrified if insulated. Du Fay experimented with electrifying very thin pieces of gold leaf. He found that a piece of gold leaf electrified by contact with rubbed glass would repel a second leaf electrified in the same manner, but would be attracted to a second leaf that had been electrified by copal (a resin from tropical trees.) His conclusion was that there were two types of electrification: vitreous (coming from glass) and resinous. Du Fay did not

**Figure 3.8** Gray's hanging string.

arrive at the assignment of different polarities to these two types of charge.

## 3.6 The First Capacitors

### 3.6.1 Ewald von Kleist

In 1745 Ewald von Kleist (ca. 1700–1746) of Germany half filled a glass bottle with water and topped it with a cork. He then pushed a nail through the cork till it just touched the water. Next he connected a static generator allowing electricity to be conducted through the nail to the water. He observed that the bottle charged in this manner could store the charge till a later time when it could be used to attract items or to cause a spark during discharge. He later increased the size of discharge sparks by using other liquids, such as mercury or alcohol, instead of water. This bottle with a charge-storing ability was the first-known capacitor.

### 3.6.2 Pieter von Musschenbroek

A few years later, Pieter von Musschenbroek (1692–1761), a professor at the University of Leyden, Holland, repeated similar experiments.

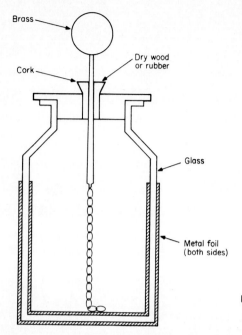

**Figure 3.9**  Leyden jar.

He eventually developed a jar with foil on the bottom half of both the inside and outside. A wire was inserted in a cork lid until it contacted the inner foil lining. The "Leyden jar" is shown in Fig. 3.9.

## 3.7  The First Batteries

### 3.7.1  Dr. Luigi Galvani

An anatomy professor at the University of Bologna, Dr. Luigi Galvani (1737–1798) discovered a type of electricity that was different from the previously known static potentials caused by frictional rubbing. Dr. Galvani observed that a frictional static machine would make frog legs twitch even though severed from the frog's body. A second discovery happened by accident one windy day in 1786 after he had hung several frogs' legs by copper hooks outside his window. The legs would twitch whenever the wind blew them so that they touched an iron support pole. Galvani recognized the reaction as being similar to the effect observed by attaching the legs to a static electric machine, but there was no apparent source of electricity present.

Experimenting in the laboratory, Galvani placed a frog's leg on an iron skillet, then touched a nerve with a copper wire, and the leg twitched. When the same nerve was touched with iron wire nothing happened. After trying other metal combinations, he concluded that when the extremes of the frog's leg were touched with two different metals the twitching occurred. The fact that two metals in electrical contact produces an electrical potential is now widely known. This phenomenon is known as the "Galvanic" or contact potential. The natural fluid in the frog's leg acted as a current-carrying electrolyte and produced what is now termed Galvanic electricity.

### 3.7.2 Alessandro Volta

**3.7.2.1 Voltaic cell.** An Italian professor of physics at the University of Padua named Alessandro Volta (1745–1827) later proved that a weak acid could serve as the current-carrying fluid. In about 1800 he produced the first voltaic cell by dipping plates of zinc and copper into a jar of weak acid. This was the first man-made battery.

**3.7.2.2 Voltaic pile.** Subsequent to his voltaic cell discovery, Volta produced a more powerful battery by stacking voltaic cells in series. This voltaic pile was produced by alternating discs of zinc and copper and placing salt dampened paper in between each pair. A spark was produced when a wire was brought into contact with each end of the pile. The volt, which is the unit of electromagnetic force, was named after him.

**3.7.2.3 Electrophorus generator.** Volta invented the electrophorous inductive-type electrostatic generator in 1775. With an electrophorous generator, a conductor can be inductively charged to a high voltage from a triboelectrically charged insulator. Volta did not have such a wide choice of materials to use for his electrophorous as is available today. The operation of a simple electrophorus generator built from modern materials is described in Chap. 4.

## 3.8. Polarities of Static Electricity

### 3.8.1 Benjamin Franklin

Benjamin Franklin (1706–1790) of Boston, performed experiments which provided great insight into the nature of electricity. He experimented with charges from different sources and concluded that there were two types of static electricity; positive and negative. One of his most famous experiments in this regard involved two people standing on insulating wax platforms and a third person standing on the floor. One insulated person was electrified from a glass rod, the second by the cloth that had rubbed the rod. A shock was felt if these individuals touched fingers. A lesser shock was felt if either individual touched a neutral person standing on the floor.

In 1747, Ben Franklin declared charge from rubbed glass as positive and charge from resin as negative. It is interesting to note that although this was a tremendous step forward, later science might rather have had Franklin's choices reversed. Had this been the case the charge of the electron could have been designated as positive rather than negative. Furthermore positive current would then have been consistent with electron movement rather than the opposite direction. In spite of the resultant inconsistency, Franklin's convention is still used today throughout the world. Franklin also concluded that static electric characteristics were not created by rubbing an electric but were due to a transfer of "electric fluid." Although his contention was not entirely correct the net effect was in agreement with the later developed concept of conservation of charge.

### 3.8.2 Verification of positive and negative polarities

The recent invention of the Leyden jar provided a new tool which became a vehicle leading to some of Franklin's greater discoveries. Franklin attached a wire to each of the inner and outer coatings of a Leyden jar and separated them at a certain distance above a table. A cork suspended by silk thread would swing to touch each wire alternately until the jar became deelectrified. This proved the existence of opposite polarities on the inner and outer coatings.

### 3.8.3 The nature of lightning

Franklin discovered that pointed objects were optimum for attracting electricity. He conducted extensive experiments with a pointed rod on a roof to determine the polarity and other characteristics of electricity in the clouds and air. He would use a Leyden jar to collect electricity gathered in this manner. This practice was later refined to increase storage capacity by collecting electricity on several Leyden jars connected in parallel. A somewhat secondary fallout of his experiments in studying the nature of lightning was the realization that pointed rods connected to ground might safely dissipate lightning strikes. He went into the business of making lightning rods and sold them throughout the colonies.

The famous kite experiment of Franklin is perhaps one of the least understood. Sometimes it seems that Franklin is portrayed as a foolhardy nitwit who flew a kite with a wire at the height of a lightning storm. Actually, the experiment was conducted as a storm was approaching one day in 1752. Franklin wanted conditions to be good for the likelihood of electricity to be transferred to a stiff wire attached to the top of his kite from the typically highly-charged air existing just prior to a thunderstorm. The light rain moistened his string which conducted electricity from the kite wire down to a key which hung close to a Leyden jar. As the storm approached sparks jumped continually from the key to the jar until the capacity of the jar was saturated with this "electric fluid."

Franklin recognized the possible hazard from experiments of this type and took reasonable safety precautions in his experiment. Mainly, he tied a silk ribbon to the string so that the ribbon could be used as a handle. The key was tied to the string above the ribbon. He also instructed that anyone attempting to repeat his kite experiment should stand in a doorway or window so that the silk ribbon was kept dry. In Russia, Georg Wilhelm Richmann (1711–1753), a German scientist, was killed attempting a similar lightning experiment. He approached an isolated rod during the height of a storm. His assistant was reportedly knocked unconscious.

### 3.8.4 The first electric motor

Franklin invented an electrostatic motor that was a precursor of motors used today. It consisted of an insulative wheel with two metal knobs at opposite ends of a diameter. Two oppositely charged knobs were placed in line with the knobs on the wheel so that they were almost touching. As one of the charged knobs repelled the uncharged metal knob on the wheel the other charged knob repelled the opposite knob of the wheel causing rotation. Continued alternate repelling causes the wheel to continue turning until the charges are dissipated.

## 3.9 The Inverse-Square Relationship

### 3.9.1 Joseph Priestley

An Englishman who later became an American citizen, Joseph Priestley (1733–1804) wrote a book entitled *History and Present State of Electricity* in 1767. In this book he repeated and confirmed an experiment that had already been conducted by Franklin. The experiment showed that there was no electrical charge on the inner surface nor any electrical force inside a closed metal box. He inferred this was due to charges of like polarity repelling each other with a force inversely proportional to the square of the distance between them.

### 3.9.2 John Robison and Charle de Coulomb

John Robison (1739–1805), a Scot, proved experimentally that the force of attraction or repulsion between two charged objects varied as the square of the distance between them. Since the results of Robison's work was not published until many years later, Charle Augustine de Coulomb (1736–1806) is credited with the inverse square law commonly known as Coulomb's law. A French engineer, Coulomb invented the torsion balance, the use of which helped him quantify the inverse-square relationship of distance between charged objects.

## 3.10 Induction

John Clanton (1718–1772) was inspired by Franklin's work in electrostatics. He found in 1753 that charged bodies would cause a separation of charges in a conductor brought near them. That is, the side nearest the charged object would be of opposite charge whereas the side farthest away would be of the same polarity as the charged item. We know this today as simple induction.

## 3.11 Electroscope

Abraham Bennet, an Englishman (ca. 1785), invented the gold leaf electroscope. An electroscope is constructed as shown in Fig. 3.10. A wire is attached to the metallic ball above the bottle's cork and thin conductive leaves of material, such as gold or aluminum foil, are attached to the bottom of the wire. Whenever a charge is placed upon the metal sphere the leaves will separate due to repulsion of like charges. The amount of separation is a crude means of measuring the amount of charge.

## 3.12 The Faraday Cage

Michael Faraday (1791–1867), an Englishman, was one of the greatest experimental scientists. His investigations into science crossed many fields especially chemistry, physics, and magnetism, as well as electrostatics. Michael Faraday introduced the concept of electrostatic lines of force around a charged object.

Of particular interest is his Faraday cage experiment depicted in Fig. 3.11, in which he sat in a metallic room and showed that an electrometer sensed no potential inside when large static discharges were applied to the outside walls. A similar experiment, the Faraday icepail experiment, showed that when a charged conductor was touched to the inside of a conductive pail, the charge was transferred to the outside of the pail. This transfer took place independent of any charge already present on the exterior of the pail.

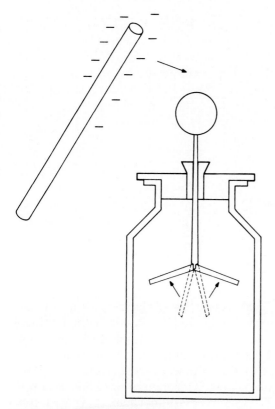

**Figure 3.10**  Metal-leaf electroscope.

## 3.13  Maxwell's Equations

James Clerk Maxwell (1831–1879), a Scot, was a professor of physics at Cambridge and is recognized among the greatest contributors to electrical science. Among his many accomplishments he verified the experimental work and theories of Faraday and refined them in mathematical form. He was able to describe the condition of any point under the influence of varying electric and magnetic fields through the use of only four equations. These equations are listed and discussed in Chap. 5.

## 3.14  Other Developments of the Nineteenth and Early Twentieth Centuries

During the late nineteenth to early twentieth centuries certain developments began to bring electrostatics out of the purely experimental and academic stage into more practical applications. Lord Kelvin in-

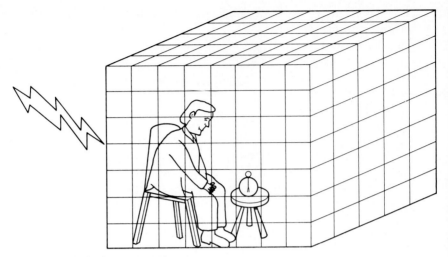

**Figure 3.11** Faraday's cage experiment.

vented the first electrostatic voltmeter. Better electrostatic generators were invented by C. F. Varley and James Wimhurst of England and Dr. Robert J. Van de Graff, an American. The Van de Graff generator has seen the most practical application, and is used widely by educators and experimenters. The first such machine was made in 1929 using tin cans and a belt of silk ribbon. This generator is based on several principles discussed in greater detail in Chap. 4: triboelectric generation, induction, and Faraday's ice-pail experiment (or Gauss' law).

### 3.15 Static Problems in Other Industries

The continuing industrial revolution of the twentieth century brought about a number of static problems in industries other than electronics. Most problems still exist to some extent. Many controls were instituted to circumvent these problems, which provide a reference base.

Conductive floors are used in hospital rooms to prevent explosive anesthetic vapors from igniting. Conductive shoes and special garments are worn. All equipment is grounded.

Items moved by conveyor belts have had a history of problems caused by excessive charging of the items transported. This has included dust and dirt accumulation as well as personnel shocks. These problems have been alleviated by newer belting materials less prone to static generation, brush accumulators, grounding, and ionization.

Trucks and automobiles have been known to accumulate static

charge and cause shocks to personnel. Some of this nuisance is due to passenger movement in a vehicle made up of many static generative materials. A large amount of vehicle charging, however, has been due to continual tire flexing. Dragging ground cables have been used with limited effectiveness to drain vehicle static charges to ground. Small wires can often be observed on the approach to toll booths to discharge the vehicle and prevent shocks between the driver and toll collector. Aircraft charging from precipitation has been known to generate sufficient noise to eradicate communications. Static bleedoff streamers can be observed on airplane wings as a means of reducing charge.

Vehicular charging, or even passenger static accumulation, have been known to cause sparks resulting in dire consequences when gunpowder, gasoline, or other explosive or flammable materials were transported. Static electricity induced explosions have been experienced at sugar mills, granaries, and sulfur mills. Similarly, static explosions have occurred during handling of powdered coal, gunpowder, and other such materials.

Static has also been a problem in processing highly insulative products such as plastics, textiles, and films. Even paper can generate tremendous voltages when processed at high speeds. The photographic industry has experienced problems with dendritic discharge patterns caused by static buildup as film passed over rollers and then sparked. The textile industry has had similar problems when materials were processed or transported by belts. One result was the attraction of dust and other particulate to the surface of the product. Such problems are now alleviated by increased relative humidity, strategically located ionizers, passive discharge apparatus, and other controls. Dust accumulation on phonograph records as well as video and computer disks can be a problem as well. Reportedly computer memory can be erased from floppy disks by high-static discharges.

## 3.16 History of Static Problems in the Electronics Industry

As discussed in Chap. 2 the electronics industry has had some isolated ESD problems as far back as the 1940s. In the late 1950s there were a few reported problems with discrete transistors and diodes. The 1960s brought the metal-oxide-semiconductor transistor which was notoriously sensitive to ESD. This MOS transistor demanded and got a certain degree of ESD control and even some limited design protection. Prolific use of integrated circuits with decreasing physical geometries during the 1970s brought a serious problem that caught the industry off guard and is the subject matter for the remainder of this book.

## References

1. Bernard Seemans, *The Story of Electricity and Magnetism,* Harvey House, New York, 1967.
2. George deLucenay Leon, *The Electricity Story,* Arco Publishing Company, Inc., New York, 1983.
3. Park Benjamin, Ph.D., L.L.B, *A History of Electricity,* John Wiley & Sons, Inc., New York, 1898.
4. Charles E. Jowett, *Electrostatics in the Electronics Environment,* John Wiley & Sons, Inc., New York, 1976.

# Chapter 4

# Basic Principles of Electrostatics

This chapter is intended to provide reference information on some of the basic principles involved in electrostatic phenomena. Electrostatic behavior in the factory environment is often puzzling and minimal understanding requires careful consideration of fundamentals and relationships discussed herein. The principles governing electrostatic effects often seem to be discordant with intuition. The inclination to favor one's intuition has resulted in widespread ignorance concerning some of the longest-known principles of physics. The tendency of all industries has been to invoke little curiosity and concentrate only on short-term immediate needs for problem solutions. After almost two decades of serious losses to ESD, the electronics industry is just beginning to vigorously investigate the details necessary to bring about effective ESD control.

## 4.1 Charge

The early discoveries of charged bodies as well as Benjamin Franklin's differentiation of polarities is discussed in Chap. 3. In understanding and controlling electrostatic phenomena, it is important to differentiate between positive and negative charges. All electrical particles have their own mass and charge potential. The smallest negative and positive charges are designated as elementary charge. The magnitude of elementary charge turns out to be the charge of one electron, although the mass varies depending upon the material.

A body is electrically neutral when elementary charges are in an equal or balanced position with respect to polarity. A charged object thus has an excess of charge, either positive or negative. Charles Augustin de Coulomb applied charges to small pith balls in order to determine the forces of point charges. He observed that the force is either attraction or repulsion depending on whether the charges are of different or the same polarity. The mathematical analysis of the forces from charges including Coulomb's law is covered in Chap. 5.

Charge is measured in a basic unit of coulombs. A coulomb is defined as the charge that passes through a conductor in 1 s in order to create 1 A of current. A negative charge of 1 C is equal to $6.28 \times 10^{18}$ electrons. Thus the charge of an electron, or one elementary charge, is equal to $(1/6.28) \times 10^{-18} = 1.6 \times 10^{-19}$ C.

### 4.1.1 Pith-ball electroscope

A pith ball suspended by silk thread as shown in Fig. 4.1 is a simple type of electroscope. Charged articles will attract the pith ball, then repel it as the pith ball takes on the same charge polarity as the original charged article such as a comb. Sometimes experimenters coat the pith ball with metal to increase its conductivity.

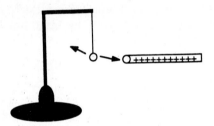

Figure 4.1  Pith-ball electroscope.

### 4.2  Contact Electrification

The simplest type of contact electrification is the sharing of electrons between two materials in contact when one of the items contacted is already charged. In this manner the uncharged item becomes charged to the same polarity as the initially charged item. Other types of contact electrification not dependent upon a prior charge on one of the materials is discussed in the section, "Triboelectric Generation," and in Chap. 6.

### 4.3  Triboelectric Generation

Electrostatic charges result from mechanical separation processes involving combinations of solids, or solids and liquids. Such solids include dust, and the liquids include mists. The surface of neutrally charged solids or liquids are electrically double-layered to varying degrees. Predominant in the outer part might be electrons balanced by a preponderance of positive charge internal to the material. For some materials the opposite may be true. When two such surfaces touch, electrons can be transferred, leaving one material negatively charged and the other positively charged. The material that acquires electrons becomes negative while the material that loses electrons becomes positive. The electron transfer takes place in the form of atomic or molecular ion formation and/or neutralization of previously charged ions.

This type of static generation is called triboelectric generation which is derived from the Greek word, "tribo," meaning to rub. Strictly speaking, the process is merely a charge transfer rather than generation. Furthermore, rubbing normally results in an increase over the amount transferred by simple contact and separation due to the greater number of contact sites involved.

### 4.3.1 Triboelectric series

Gilbert and many to follow have attempted to record the polar relationships of electrified materials resulting from contact and separation. It turns out that materials can generally be placed in an order called a triboelectric series from which the resultant polarities between any two materials on the list can be predicted. A representative list is given in Fig. 4.2. If a material A on the list contacts a material B further up the list, on separation A will be negatively charged and B will be positively charged. If, on the other hand, B had been further down the list, A would have become positive whereas B would have become negative.

In reality, a triboelectric series can be regarded only as a reasonably accurate predictor of charge polarity. Other variables can bring about results conflicting with the oversimplified results predicted by the list. Such variables include surface finish, preconditioning, contamination,

```
POSITIVE
    +           Acetate
                Glass
                Nylon
                Wool
                Silk
                Aluminum
                Polyester
                Paper
                Cotton
                Steel
                Nickel, copper, silver
                Zinc
                Rubber
                Acrylic
                Polyurethane foam      Figure 4.2  Triboelectric series.
                PVC (Vinyl)
    −           Teflon
NEGATIVE
```

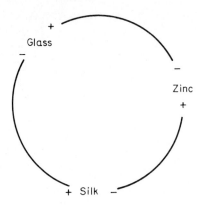

**Figure 4.3** Circular-charging relationship of glass, zinc, and silk.

and others. A common example not predicted by the series is the resultant charge transfer from rubbing two pieces of the same insulative material together. Even more perplexing are cases where three materials form an overlapping circular polar relationship rather than strictly following the pattern predicted by the triboelectric series. An example is shown in Fig. 4.3. The circle indicates that glass will charge positive with respect to zinc; zinc will charge positive with respect to silk; and silk will charge positive with respect to glass. Yet, glass is above silk, which is above zinc on the triboelectric series. This indicates that another charge transfer mechanism may be involved and illustrates that the triboelectric series is not to be taken as an absolute predictor of polarity.

## 4.4 Conductivity

The conductivity of the materials involved greatly affects the strength of the charge separation. For instance, two materials of high conductivity will permit immediate recombination of any charges that would have a tendency to become separated. Conductivity is a measurable parameter, generally as a volume characteristic, with units in mhos per centimeter (mhos/cm), the inverse of ohms-centimeter ($\Omega \cdot$ cm).

Even insulators can be characterized in terms of conductivity. Table 4.1 lists conductivities of some conductors and insulators. The data are typical of measurements taken under ambient conditions of 25°C and 50 percent relative humidity. For the purpose of ESD-related matters the primary interest is the manner in which conductors and insulators differ when charged and upon grounding. Figure 4.4a shows that, in the absence of external field influences, the charge on a conductor will be uniform in magnitude and polarity. As indicated in Fig. 4.4b, the charge on different locations of an insulator can vary in both magnitude and polarity. This difference in behavior is due to the high mobility of charges on conductors versus the virtual immobility of

TABLE 4.1 Conductivities of Selected Conductors and Insulators

| Material | Conductivity (mhos•/cm) | General category |
|---|---|---|
| Silver | $6.2 \times 10^5$ | Conductor |
| Copper | $5.8 \times 10^5$ | Conductor |
| Aluminum | $3.8 \times 10^5$ | Conductor |
| Nickel | $1.5 \times 10^5$ | Conductor |
| Nichrome | $1.0 \times 10^4$ | Conductor |
| Graphite | $3.0 \times 10^2$ | Conductor |
| Salt water | $4.0 \times 10^{-2}$ | Conductor |
| Distilled water | $1.0 \times 10^{-6}$ | Insulator |
| Bakelite | $1.0 \times 10^{-11}$ | Insulator |
| Mica | $1.0 \times 10^{-17}$ | Insulator |
| Quartz | $1.0 \times 10^{-19}$ | Insulator |

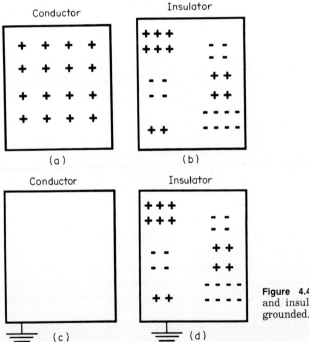

Figure 4.4 Charged conductor and insulator: ungrounded and grounded.

charges on insulators. This same difference in charge mobility accounts for why a conductor loses its charge when grounded and an insulator remains charged as shown in Fig. 4.4c and 4.4d.

## 4.5 Resistivity

The quantitative definitions of what constitute a conductor and nonconductor and the variety of intermediate categories have been changing over the years and are somewhat arbitrary. The reciprocal of conductivity (resistivity) is the most common parameter used in the ESD control field to differentiate between conductors and insulators.

### 4.5.1 Volume resistivity

For a solid material, resistance can be expressed as

$$R = \frac{\rho_v l}{A} \quad \Omega \tag{4.1}$$

where $\rho_v$ = volume resistivity
$l$ = length
$A$ = cross-sectional area

Since length, $A$, and resistance can be measured, the expression is more commonly written in the form

$$\rho_v = \frac{RA}{l} \quad \Omega \cdot \text{cm} \tag{4.2}$$

It turns out that in the MKS system of units $\rho_v$ is numerically equal to the resistance between the faces of a 1-cm cube of the material.

### 4.5.2 Surface resistivity

A common convention used in the industry is the concept of "surface resistivity." This parameter is commonly specified in characterization of the electrostatic properties. As the term implies, it is used to quantify the resistance of a well-defined geometry of the surface of a material.

Surface resistivity is at best an often misapplied parameter and to the purist the term is meaningless as it describes a nonexistent entity. The problem begins with the question "How thick is the surface?" Theoretically the surface is infinitesimal. In practice, the effective thickness during a surface resistivity measurement varies depending upon the sample. The parameter loses all meaning when the sample consists of nonhomogeneous layers. With this forewarning we shall now attempt to explain the term and indicate some of the associated problems. Further clarification will be brought out in Chaps. 16 and 17.

#### 4.5.2.1 Definition.
For a surface, the resistance decreases in proportion to the width, $W$, and increases proportionately with length.

$$R = \frac{kl}{W} \tag{4.3}$$

where $k$ is a proportionality constant.

In the case of a square, where $l = W$, the equation reduces to

$$R = k \tag{4.4}$$

This case is taken as the conventional way to measure the resistance of surfaces and $k$ for a square is called surface resistivity with units of ohms per square. As can be seen from the equation, if $l = W$, the size of $l$ and $W$ does not matter, so the size of the square is immaterial. Restated in the following way:

$$\rho_s = \text{surface resistivity} = k = \text{resistance when } l = W \tag{4.5}$$

Figure 4.5 is provided to illustrate the concept of surface resistivity independent of the size of the square. The resistance across a uniform thin layer of homogeneous material can be considered as a distributed resistance as shown. Suppose the resistance across an infinitesimally thin layer across a square of arbitrary $l = w = x$ is measured to be $y\ \Omega$ as shown in Fig. 4.5a. If the length is doubled to $2x$ as in Fig. 4.5b, the added series-distributed resistance would effectively result in $R_{AB} = 2y$. If the length is kept at $2x$ and the width is also doubled to $2x$, as in Fig. 4.5c, an effective resistance of 2y is added in parallel resulting in $R_{AB}$ of $y\ \Omega$. Since the same logic would apply to any multiples of $x$, the resistance is therefore independent of the size of the square.

**4.5.2.2 Comparative uniform sheet resistivity.** In light of the precautions against indiscriminate use of surface resistivity, I will suggest a more definitive term in hopes of allaying improper material characterizations. Surface resistivity measurements made to compare similar sheets of material, as on an assembly line, are a valid means of assessing that the proper sheet resistivity range is maintained. Therefore, I prefer to define the term "comparative uniform sheet resistivity" in lieu of the more commonly used "surface resistivity."

### 4.5.3 Categorization of materials

The terms surface resistivity and volume resistivity are commonly used to characterize materials. Jowett[1] defines the following in terms of volume resistivity:

$$\text{Conductive } k = 10^{-7} \leq 1\ \Omega \cdot \text{cm}$$

$$\text{Semiconductive } k = 1 \leq 10^6\ \Omega \cdot \text{cm}$$

$$\text{Insulators } k > 10^6\ \Omega \cdot \text{cm}$$

The reciprocal data for conductivity in Table 4.1 are consistent with this categorization. The terms for categorization of materials by surface resistivity are currently defined in the Department of Defense

54    Chapter Four

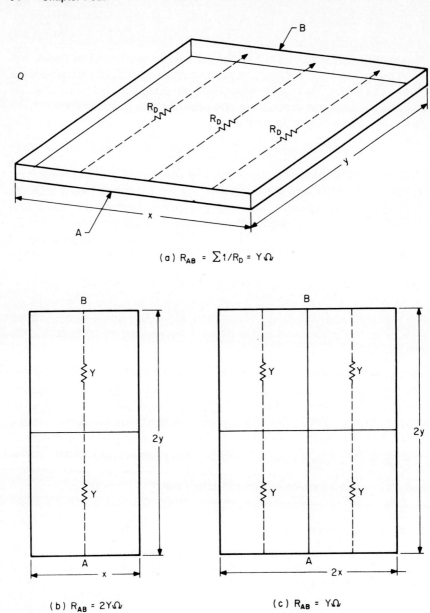

**Figure 4.5** Ohms per square.

DOD Handbook 263 (1980) as follows:

$$\text{Conductive} \leq 10^5 \ \Omega \text{ per square}$$
$$\text{Static dissipative} > 10^5 \text{ to } 10^9 \ \Omega \text{ per square}$$

Antistatic > $10^9$ to $10^{14}$ Ω per square

Insulative > $10^{14}$ Ω per square

Definitions of surface resistivity terms are presently undergoing some reexamination and are likely to change in the near future. In particular, most are finding that $10^{14}$ Ω per square is too high for acceptable static bleedoff or dissipative behavior. The antistatic on nongenerating property is not directly correlated to resistivity. These properties are discussed further in Chaps. 16 and 17.

### 4.5.4 Effects of humidity

Higher humidity will increase the moisture content of materials to varying degrees depending upon the material. This increased moisture content will reduce the resistivity. Since the humidity effects are very material dependent, a generalized mathematical relationship of static generation versus humidity is nonexistent. In the conglomerate however, static accumulation seems to vary exponentially with decreasing humidity.

## 4.6 Faraday Cage

Among the numerous experiments of Michael Faraday were several that showed that charges rest on the outer surface of conductors. This phenomenon is intrinsic to the Faraday shield effect. Electrostatic shielding is the basis for much of the protective packaging used within the electronics industry. Conductive containers totally enclosing susceptible items provide a blockage of external fields and any charge accumulation rests on the outside surface only. Such containers are ideal for storage and transportation of susceptible items when away from static controlled areas. The principles involved are related to Gauss' law which is addressed in Chap. 5.

### 4.6.1 Faraday pail

One of Faraday's most noteworthy experimental studies involved observations of the effects of charged objects on a metal ice pail that was insulated from ground. Faraday discovered that when a charged metal object was suspended inside the pail the outer surface of the pail acquired an induced charge equal in magnitude and polarity to the charge on the metal object. He further observed that if the charged metal object were removed from the pail the induced charge no longer existed. Perhaps more importantly, he discovered that when the charged object contacted the inside of the pail, all of its charge was transferred to the outside of the pail. This process could be repeated with additive increases in the charge on the pail's outer surface, limited only by leakage across the insulator between the pail and ground.

In Fig. 4.6a a conductive sphere with a small opening is used to illustrate the phenomenon. Such a sphere is preferable to a pail because of the optimal shape and negligible inconsistencies introduced by the small opening. Figure 4.6b shows a metallic sphere $A$ with a small

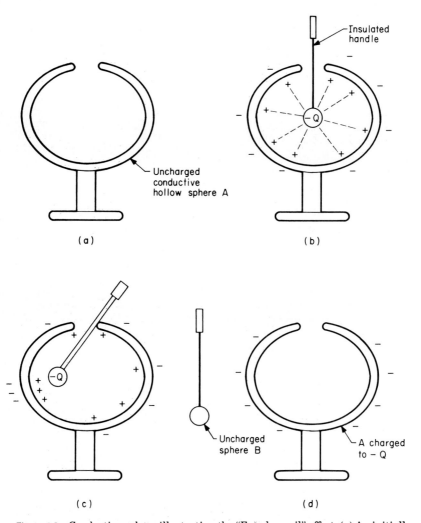

**Figure 4.6** Conductive sphere illustrating the "Faraday pail" effect. (a) An initially uncharged conductive sphere $A$; (b) insertion of a smaller negatively charged sphere $B$; (c) charged sphere $B$ inserted off-center; (d) condition following contact of $B$ with inner surface of $A$.

conductive sphere $B$ with charge $-Q$ at its center. The internal charge causes a charge separation on the outer sphere resulting in $-Q$ distributed around the outer surface and $+Q$ on its inner surface. If $B$ is removed, the induced charge neutralizes as the separated charges return to equilibrium. If $B$ is inserted off-center, as shown in Fig. 4.6c, the induced charges on the inner surface of $A$ are no longer symmetrical. If $B$ contacts the inner surface of $A$, a resulting transient current exists until the charged inner surface of $A$ and the oppositely charged $B$ neutralize to zero. Note in Fig. 4.6d that the added outer charge $-Q$ remains following contact and removal. The process could be repeated independent of the existing charge on the outer surface of $A$.

## 4.7 Voltage

Voltage is the unit used to designate the work required to move charges in an electric field. A volt is defined as the work to move a unit charge one centimeter in a field of 1 dyn/C. Conceptually, most people visualize volts in terms of potential difference more easily. The potential difference between two points $V_{ab}$ is defined as the work in joules that must be expended to move a positive point charge from point $b$ to point $a$ in an electric field. Conventionally, if $V_{ab}$ is positive then $a$ is at a higher potential than $b$. Thus the potential at a point, such as $a$, is dependent upon the point of reference, in this case $b$. It is common practice to reference potentials or voltages to ground, or the chassis of a system.

## 4.8 Work, Power, and Energy

The terms work, power, and energy are frequently encountered in electrostatics, and because of their close relationships are sometimes confused. Therefore, these relationships are stated here for clarification in future references to the terms.

Work = force × distance
Power = work per unit of time
Energy = the capacity to do work

## 4.9 Electric Field

Every static charge creates an electric field. All electric charges in a field are under influences of energies or field lines. These field lines are characterized by forces of attraction of opposite charges and repulsion of like charges. Thus mobile charges will be influenced and moved accordingly as an item is brought under the influence of a field.

In like manner, upon removal from the field there is a tendency toward charge recombination resulting in neutralization. It is the circumstances of such recombination that we are ultimately attempting to restrain in ESD control.

## 4.10 Induction

A conductor in the presence of an electric field will experience a charge separation. This phenomenon illustrated in Fig. 4.7 is normally called induction. For the sake of differentiation with other inductive effects we shall define this separation as simple induction. When the field influence is removed either by moving the conductor from the field or the field away from the conductor, the charges return to their neutral state. For the purposes of ESD concerns, the rapid movement of such charges can be a problem.

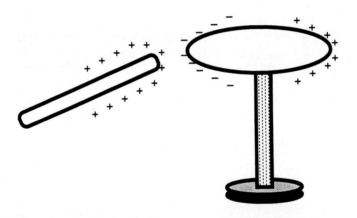

**Figure 4.7** Simple induction: separation of charges from field influence.

If one side of the conductor is grounded while under the influence of a field, and the field influence is subsequently removed, compound induction takes place. This leaves the conductor with a residual charge as shown in Fig. 4.8. Compound induction in the factory, as discussed in Chap. 13, often presents a double jeopardy to static-susceptible items.

## 4.11 Capacitance

Capacitance is sometimes referred to as capacity. The following discussion illustrates an example of the meaning of the term "capacity." If a charged plate (A) is kept insulated from ground while connected to

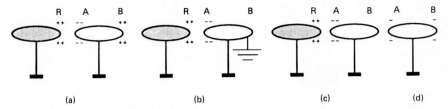

**Figure 4.8** Compound induction. (a) Simple induction; (b) grounding; (c) removal of ground; (d) removal of field.

a leaf electroscope, the leaves will separate. The charge $-q$ could be such that the leaf position is just short of *full* separation as in Fig. 4.9a. Now if a second grounded plate (B) is brought near the opposite side of the first plate A, the leaves will begin to go closer together as shown in Fig. 4.9b. The amount of charge on A has stayed the same, but the attractive force of the grounded plate has drawn much of the negative charge to the other side of the plate, thus changing the charge distribution. In order to cause the leaves to again become nearly *fully* extended much greater charge $-Q$ must be present on A as shown in Fig. 4.9c. Thus the *capacity* of A has been greatly increased. As a matter of fact the combination of two plates separated by an insulator (or dielectric) is known as a "capacitor." In this example, air is the dielectric. The capacity could be increased by enlarging the area of the plates, using a better dielectric as shown in Fig. 4.9d, or decreasing the distance between plates. Another way of explaining this example is to consider the charge on A as having an inductive effect on B and repelling like charges to ground leaving opposite charges on the side of B facing A.

### 4.11.1 Self-capacitance

The self-capacitance of an isolated conductor is a geometrically dependent constant defined as the ratio of charge to voltage on the conductor. That is, $C = Q/V$. This relationship, more frequently stated as $Q = CV$ is extremely important in dealing with ESD problems as will be apparent after discussions of problem examples in later chapters.

### 4.11.2 Energy and voltage relationships and ESD susceptibility

The energy of a charged capacitance is

$$E = \tfrac{1}{2}CV^2 \tag{4.6}$$

Thus the available electrostatic energy is proportional to the capaci-

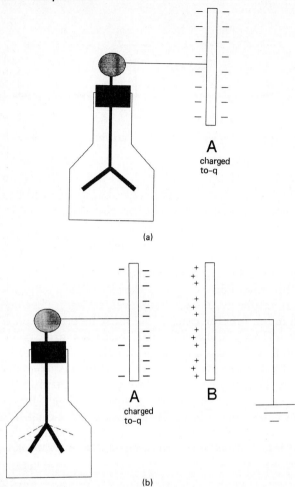

**Figure 4.9** The functioning of a capacitor. (a) A conductive plate A with sufficient negative charge to cause electroscope leaves to approach full separation; (b) a grounded plate B brought near the opposite side of A.

tance of the charged object. When the ESD-susceptible item is due to a contained element such as bipolar junction, metallization stripe, or thin-film resistor, heating is required to bring about failure. These types of parts are energy sensitive. Some prefer to use the term "current sensitivity" indicating the $I^2R$ heating factor which is equivalent to energy. Other parts that contain a dielectric that can break down at human body or similar static potentials are called voltage-sensitive devices. The unprotected MOS device is the classic example. A failure can occur once the breakdown potential is exceeded even if the charged capacitance is small. Damage to energy-sensitive parts,

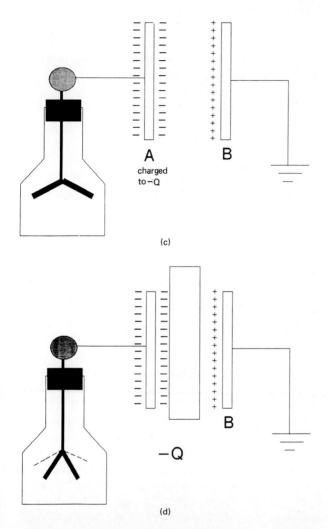

**Figure 4.9** (*Continued*) The functioning of a capacitor. (*c*) Charge on *A* must be increased in order to again approach full separation; (*d*) inserting a dielectric better than air between the plates results in decreased leaf separation, or increased charge required on *A* for full separation.

on the other hand, is much more dependent on the capacitance of the charged item. ESD damage and models for damage are discussed in greater detail in Chaps. 7 and 8.

### 4.11.3 Voltage suppression

Voltage suppression is a result of the relationship $Q = CV$. Suppose a charged conductive tote box is placed on a surface. The capacitance (or mutual capacitance between the box and table) is inversely proportional to the distance from the bottom of the box to the table. When at a significant distance, the voltage might be quite high for a given charge. Yet when placed on the surface, even if the surface is ungrounded or nonconductive, the voltage will be greatly reduced. This is because the mutual capacitance has been greatly increased and

$$V = \frac{Q}{C} \quad (4.7)$$

The voltage is said to have collapsed or been suppressed. If the tote box were raised without bleeding off the charge, a high voltage could again be measured on the box. This phenomenon, if not understood, can easily fool one in assessing effectiveness of work surfaces and other static-control items. The fact that various forms of voltage suppression frequently occur in normal factory operations accounts, to some extent, for the unpredictable nature of ESD events.

### 4.11.4  Circuit operation of a capacitor

Many find it perplexing to understand how a capacitor in a circuit can allow current since the dielectric medium between plates is an insulator. The discussion on induction in Sec. 4.10 shows that a charge on plate $A$ will tend to repel like charges to the opposite side of $B$ which is grounded in Fig. 4.9$b$. This leaves the side of $B$ nearest $A$ with an opposite charge. Thus far we have already admitted the existence of a current for the short time required for the charge movement to ground from the far side of plate $B$. If the potentials at $A$ and $B$ are not altered, a steady-state or dc condition exists and there is no current. If the potential differences at $A$ vary, the charge on $A$ must adjust in accordance to $Q = CV$ and the inductive effect on plate $B$ will alter $B$ accordingly resulting in a current during the period of voltage change. Since a changing voltage is required, capacitors are said to be conductive to ac, but nonconductive to dc. In other words a capacitor has infinite impedance to dc but the impedance approaches zero as the frequency or rate of voltage change increases. This is covered further in Chap. 5.

### 4.11.5  Capacitive coupling

Capacitance, to some extent, exists between any two conductors. The amount of capacitance is dependent upon the geometries and orientation of the conductors as well as the dielectric medium(s) and physical separation. The normally unplanned capacitance that occurs between conductors in a circuit is called stray or parasitic capacitance. For most conductors the separation and geometries are such that stray capacitance is insignificant. However, for the relatively fast ESD transients, some otherwise insignificant stray capacitances become low-impedance paths. These paths often become involved with ESD events as the stray capacitance provides a path to ground, or a path to a sensitive node that has an existing path to ground. This type of ESD event is called capacitive coupling. A sufficient capacitive path to ground can often be provided by such things as a printed-circuit track, a few inches of wiring in a test fixture, and with close proximity to ground or a large conductor, even a device lead itself. In this manner a sensitive item can fail from ESD exposure even when not *grounded*. Capacitive coupling into a sensitive node has been observed via closely spaced wires, coaxial cables, and even a metal chassis to an internal wire. Some examples of capacitive coupling are included in Chap. 10.

## 4.12 Electrophorous Generator

Allesandra Volta invented the electrophorous generator. Operation of this rather simple but sometimes awing apparatus is illustrated in Fig. 4.10a through 4.10d. In Fig. 4.10a the white insulative circle of Teflon is rubbed with a piece of nylon carpet. From the triboelectric series of Fig. 4.2, we expect the Teflon to take on a negative charge, which it does. Teflon at the bottom of our triboelectric series is not only a good insulator capable of generating and holding a large static charge, but is also little affected by humidity. So a large negative charge is easily acquired and held. If a slightly smaller circle of aluminum, manipulated by a Teflon insulated handle is brought into contact with the Teflon as shown in Fig. 4.10b, simple induction takes place. The high negative charge on the Teflon creates an electric field which causes charge separation in the aluminum as it approaches the Teflon. Like (in this case negative) charges are repelled to the top of the aluminum circle, while positive charges are attracted to the bottom side nearest the negatively charged Teflon. If one were to touch the aluminum at this time it is likely that a severe shock would be felt. Instead, in Fig. 4.10c, a grounded probe is touched to the top of the aluminum plate neutralizing the top surface and rendering it safe to touch at 0-V potential. At the same time, however, the bottom side has a high positive charge. If the aluminum plate is now lifted away from the Teflon so that the field influence is negated, the positive charge distributes equally over the surface of the aluminum. For demonstration purposes an insulated wire is usually connected from the top of the aluminum to a spherical electrode of about ½-in dia. As the top surface takes on the extremely high positive potential, an impres-

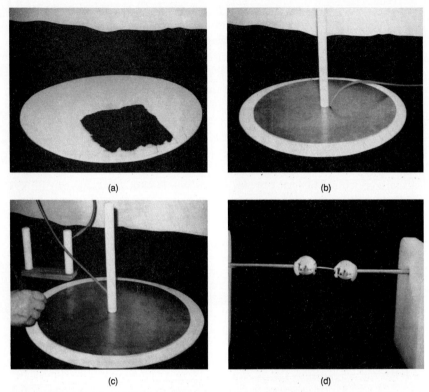

**Figure 4.10** The electrophorous generator. (a) The white insulative circle of Teflon is rubbed with a piece of nylon carpet; (b) a smaller circular aluminum plate is brought into contact with the negatively charged Teflon; (c) a grounded probe contacts the top surface of the aluminum plate; (d) resultant spark as the aluminum plate is raised away from the Teflon.

sive spark can be observed across a 1-in spacing to a similar grounded sphere as shown in Fig. 4.10d. An audible crack associated with such a violent spark breakdown adds to the effect.

## 4.13  Van de Graff Generator

The Van de Graff generator, mentioned in Chap. 3, is a popular inexpensive device used by many experimenters in electrostatics. A cross-sectional sketch is shown in Fig. 4.11 in order to illustrate the principles of operation. The insulative belt $A$ is rotated around pulleys $B$ and $C$. The metal brush $D$ is maintained at a potential high enough to ionize air around the brush points to spray ions onto the belt. A similar brush $E$ at the upper end is attached to the inside of a metal sphere $F$. The upper brush acts like a group of discharge corona points

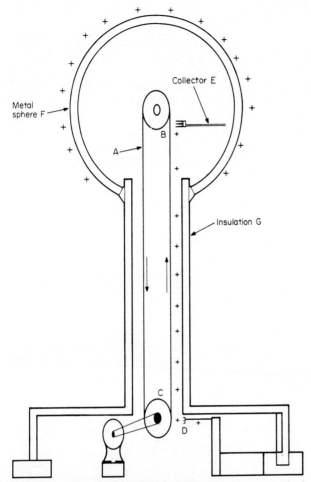

**Figure 4.11** The Van de Graff generator.

and collects most of the previously applied charge. As discussed in Sec. 4.6.1 the connection of the collector brush to the inside of the metal sphere transfers the charge to the outer surface of the sphere. Charge accumulation on the outer surface can continue until leakage across the insulative mount $G$ or ionization of the outside air occurs. With a properly chosen insulator, Van de Graff generators with a sphere of 15 cm in diameter can easily attain levels greater than 50 kV.

## 4.14 Ionization

Static charges on insulative materials can be effectively neutralized by oppositely charged airborne ions. This process which can be regarded as partially conducting air is called "air ionization." Ionization is discussed in detail in Chaps. 15, 16, and 17.

## 4.15 Breakdown of Air and Other Gasses

Many static problems involve breakdown across gaps containing air or other gasses. The breakdown of air at 760 mmHg and 20°C for parallel planes as a function of separation is given in Fig. 4.12. Table 4.2 lists selected air spark-gap voltages for needle points to spherical electrodes up to 2.5 cm in diameter. These figures are also good approximations for most gasses at standard pressure and temperature.

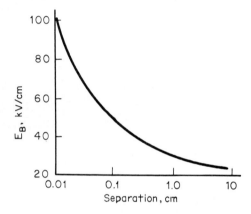

Figure 4.12 Breakdown of air across gap.

TABLE 4.2 Selected Spark-Gap Voltages for Air at 760 mmHg and 25°C

| Peak voltage (kV) | Diameter of spherical electrodes (cm) | | | | Needle points |
|---|---|---|---|---|---|
| | 2.5 | 5 | 10 | 25 | |
| 5  | 0.13 | 0.15 | 0.15 | 0.16 | 0.42 |
| 10 | 0.27 | 0.29 | 0.30 | 0.32 | 0.85 |
| 15 | 0.42 | 0.44 | 0.46 | 0.48 | 1.30 |
| 20 | 0.58 | 0.60 | 0.62 | 0.64 | 1.75 |
| 50 | 2.00 | 1.71 | 1.65 | 1.66 | 5.20 |
| 70 | 4.05 | 2.68 | 2.42 | 2.37 | 8.81 |

## References

1. Charles E. Jowett, *Electrostatics in the Electronics Environment*, John Wiley and Sons, New York, 1976.
2. Ralph Morrison, *Grounding and Shielding Techniques and Instrumentation*, 3d. ed., John Wiley and Sons, New York, 1986.
3. G. G. Skitek and S. V. Marshall, *Electromagnetic Concepts and Applications*, Prentice Hall, New York, 1982.

**Chapter**

# 5

# Theoretical and Mathematical Fundamentals of Electrostatics

This chapter addresses the fundamentals of electrostatics from a geometric modeling and mathematical approach. Examples discussed were chosen for applicability in the electronics industry environment.

## 5.1 Electrostatic Force

The electrostatic force (in newtons) between two charged bodies was verified in careful torsion-balance experiments conducted by Coulomb. Using small charged pith balls he found the magnitude of the force to be proportional to the product of their charges $Q_1$ and $Q_2$ and inversely proportional to the square of the distance between them.

$$f = K \times \frac{Q_1 Q_2}{r^2} \qquad (5.1)$$

where $Q$ = charge in Coulombs
 $r$ = distance in meters
 $K$ = a proportionality constant = $(4\pi\epsilon_0)^{-1}$
 $\epsilon_0$ = the permittivity of free space (vacuum)

The permittivity of free space can be determined experimentally by measuring the force between two small bodies charged at 1 C each separated by 1 m. This force has been determined to be $9 \times 10^9$ newtons.
Thus,

$$\epsilon_0 = (4\pi fr^2)^{-1} C^2 \tag{5.2}$$

$$\epsilon_0 = (36\pi \times 10^9 \times N \cdot m^2)^{-1} C^2 = 8.85 \times 10^{-12} \frac{C^2}{N \cdot m^2} \tag{5.3}$$

The permittivity of free space is most frequently written in farads per meter.

$$\epsilon_0 = 8.85 \times 10^{-12} \frac{F}{m} \tag{5.4}$$

where the farad $= \dfrac{C^2}{N \cdot m}$

Equation (5.1) is known as Coulomb's law. The force is repulsive for like signs and attractive for unlike signs. As it turns out, the proportionality constant is dependent upon the medium. Greatest forces are obtained in a vacuum. The greater the permittivity, the more the force magnitude is reduced. The dielectric constant is a characteristic defined as the relative permittivity of a medium.

$$k = \text{dielectric constant} = \text{relative permittivity} = \frac{\epsilon}{\epsilon_0} \tag{5.5}$$

Therefore the permittivity of a medium can be written

$$\epsilon = \epsilon_0 k \tag{5.6}$$

The relative permittivity of free space is, by definition,

$$k_0 = \frac{\epsilon_0}{\epsilon_0} = 1 \tag{5.7}$$

Coulomb's law in mks units is

$$f = \frac{Q_1 Q_2}{4\pi\epsilon_0 r^2} \quad \text{for a vacuum;} \tag{5.8}$$

or, in the more general case for any medium

$$f = \frac{Q_1 Q_2}{4\pi\epsilon r^2} = \frac{Q_1 Q_2}{4\pi k \epsilon_0 r^2} \tag{5.9}$$

## 5.2 Electric Field

Coulomb's law describes the force between two charged bodies. If one contemplates the nature of this force, a likely consideration is that each charged body somehow affects the space around it. This affected space is normally described as the electric field. The convention is that the direction of the field at any point in the affected space is congruent with the movement of a positive charge placed at that point. Some understanding of the nature and behavior of electric fields upon items of various geometries is important as it relates to analysis of real world static encounters.

### 5.2.1 Field intensity

If a point charge $+Q_1$ is placed in a vacuum, the force-field lines are all directed outward from $+Q_1$. The forces, or field intensity, at a given point within the field can be determined by the effect on a test charge $Q_t$ placed at the point in question. By definition:

$$E_t = \frac{f_t}{Q_t} \qquad (5.10)$$

From Coulomb's law,

$$f_t = \frac{Q_1 Q_t}{4\pi\epsilon_0 r^2}$$

Therefore,

$$|\mathbf{E}| = |\mathbf{E}_t| = \frac{Q_1}{4\pi\epsilon_0 r^2} = \frac{Q_1}{kr^2} \qquad (5.11)$$

where $\mathbf{E}$, a vector, is directed along $r$ out toward the point in question. The force $\mathbf{f}$ on a charge $Q_2$ in the field of $Q_1$ can be written as

$$\mathbf{f} = Q_2 \mathbf{E}_1 \qquad (5.12)$$

If $Q_2 = 1$, then $\mathbf{f} = \mathbf{E}_1$.

This can be interpreted as stating that the field is equivalent to the force on a unit charge.

### 5.2.2 Field intensity for a system of charges

Since the electric field at a point in space has both magnitude and direction, it is a vector force. In many of the discussions on fields the vector direction is geometrically obvious, such as when a single charge source is involved. With multiple charge sources, the obvious direction of each source can often be used to advantage to find the net field con-

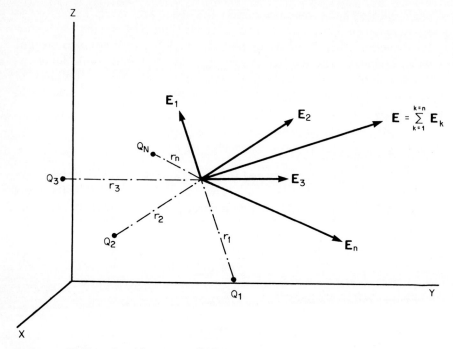

**Figure 5.1** Field produced by $n$-point charges.

ditions. In order to evaluate the field intensity at a point resulting from the cumulative effect of $n$ = point charges for instance, the different relative directions require that each constituent be handled as a vector. In Fig. 5.1 the resultant field at point $p$ can be determined by calculating the vector sum of the field components from each of the $n$ charges on a test charge $+P$. That is,

$$\mathbf{E}_p = \mathbf{E}_t = \sum_{i=1}^{n} E_i \mathbf{r}_i = E_1 \mathbf{r}_1 + E_2 \mathbf{r}_2 + \cdots + E_n \mathbf{r}_n \qquad (5.13)$$

$$\mathbf{E}_p = \frac{Q_1}{4\pi\epsilon r_1^2} \mathbf{r}_1 + \frac{Q_2}{4\pi\epsilon r_2^2} \mathbf{r}_2 + \cdots + \frac{Q_n}{4\pi\epsilon r_n^2} \mathbf{r}_n \qquad (5.14)$$

where $\mathbf{E}_p$ and $\mathbf{E}_t$ indicate the electric-field vector on the test charge placed at $p$, and $\mathbf{r}_1, \mathbf{r}_2, \mathbf{r}_n$ indicate unit vectors in the directions of $\mathbf{r}_i$ for $i = 1$ to $n$.

Most applications in solving static problems in the factory can be accomplished satisfactorily by considering the additive effects of single charge sources.

### 5.2.3 Lines of force

Electric fields are often depicted diagrammatically by lines of force. This technique was first introduced by Faraday and was used exten-

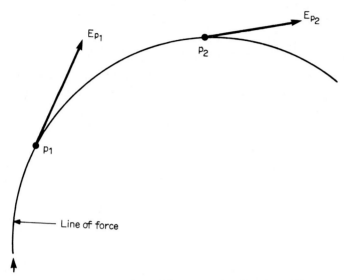

**Figure 5.2** A tangent on a line of force is in the direction of $E$.

sively by Maxwell to facilitate conceptual visualization as reinforcement for theoretical premises. Conventionally the number of lines is proportional to the magnitude of charge present. Common practice is to use one line per unit charge or some other ratio. The directional convention is from positive to negative, as discussed for fields in Sec. 5.2. Since different points in space will experience field influences from charged sources at varying relative directions, force lines are usually curved. A tangent at any point on a line of force is in the direction of $E$ at that location as indicated by Fig. 5.2.

## 5.3 Work to Move Charges

To move a small charge $\delta Q$ from an infinite distance to the surface of a sphere of radius $r$ would require an amount of work $W$ determined by integrating the force over distance.

$$W = - \int_{\infty}^{r} \frac{Q \delta Q}{kr^2} dr \qquad (5.15)$$

$$W = \frac{Q \delta Q}{kr} \qquad (5.16)$$

## 5.4 Conservative Nature of Electric Field

Suppose a point charge is to be moved in the presence of an electric field along an irregular path $p_{ab}$ from point $a$ to point $b$ as indicated in

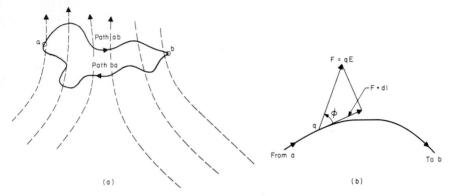

**Figure 5.3** Conservative electric field: (*a*) Movement of a charge around any closed contour in the presence of an electric field, (*b*) for 90 > φ > 270 degrees positive, work is required. For other directions, the charge is moved by the field forces.

Fig. 5.3*a*. The work required is

$$W_{ab} = -\int_a^b qE\cos\phi\, dl = -q\int_a^b E\cdot dl = -\int_b^b f\cdot dl \quad (5.17)$$

Justification for the negative sign can be seen geometrically. Observe in Fig. 5.3*b* that for φ < 90 degrees and φ > 270 degrees, the force along l is in the same direction as the field. Positive cosines for these ranges mean that the work is negative: that is, the point charge is moved by the force of the field. For the range of 90 degrees > φ > 270 degrees, however, positive work must be applied to move the charge along the path at an incremental length $dl$. Since the cosine is negative in this range, the equation appropriately indicates positive work by virtue of the minus sign preceding the integral.

The result of integration of Eq. (5.24) is that the work is dependent upon the starting and ending point of the length traversed. Extend this analysis to the work required to traverse the entire contour, that is from $a$ to $b$ along $p_{ab}$ then from $b$ to $a$ along $p_{ba}$. Since the final position is the same the net work should be zero.

$$W_{\text{closed path}} = W_{ab} + W_{ba} = 0$$

$$= -q\int_a^b E\cdot dl - q\int_a^b E\cdot dl = 0$$

which can be stated in terms of traversing a closed curve back to the point of beginning.

$$\oint_c E\cdot dl = 0 \quad (5.19)$$

Equation (5.19) is sometimes called the conservative property of electrostatic fields.

## 5.5 Potential Difference as a Function of Electric Field

A volt was defined in Chap. 4 as the work to move a unit charge 1 cm in a field of 1 dyn/C. Mechanically, potential energy is proportional to height. The slope of an inclined plane is defined as the increase of height, or potential energy, per horizontal unit of distance. When an item is moved mechanically to a different height, the change in potential energy is independent of the path. This is because the calculation is always made with respect to the direction of the gravitational field and related to 1 lb. The change in height could be defined as

$$\delta H = \frac{\delta P.E.}{lb} \text{ in } \left(\frac{ft \cdot lb}{lb}\right) = \delta H \text{ in } \cdot ft \qquad (5.20)$$

Electrically, the potential energy of a charge is proportional to voltage. Voltage is determined by the field at a given point. The change in voltage acquired by moving a charge in the direction of the field is related to the work involved by

$$\delta V = \frac{\delta W}{\text{unit charge}} \qquad (5.21)$$

$$\delta V = -\frac{f \delta x}{\text{unit charge}} \qquad (5.22)$$

where $x$ is distance in the direction of the field. The minus sign comes about because that the potential increases when the movement of a positive charge is in the direction opposite to the field.

Therefore:

$$\delta V = -E \delta x \qquad (5.23)$$

or

$$E = -\frac{\delta V}{\delta x} = -\frac{dV}{dx} = -\text{grad } V = -\Delta V \qquad (5.24)$$

In the general case, where the direction of $E$ is defined in rectangular coordinates,

$$\mathbf{E} = -\left(\frac{\delta V}{\delta x}\mathbf{x}, \frac{\delta V}{\delta y}\mathbf{y}, \frac{\delta V}{\delta z}\mathbf{z}\right) \qquad (5.25)$$

A convenient result of Eq. (5.25) is that when the field is in a single axial direction such as $r$,

$$V = - \int E \cdot dr \qquad (5.26)$$

### 5.5.1 Potential

To find the potential difference between infinity and a sphere's surface the work per unit charge is

$$\frac{W}{\delta Q} = \frac{Q}{kr} \qquad (5.27)$$

where Eq. (5.16) was used to substitute $Q\delta Q/kr$ for $W$.

Since the potential at infinity can be taken as 0 V, the potential at the surface with respect to infinity $V_{is}$ is equal to the potential $V$ at the surface.

Therefore:

$$V_{is} = V = \frac{Q}{kr} \qquad (5.28)$$

defines the potential, sometimes called absolute potential, on the sphere. In the general case, potential is referred to infinity. This is identical with the common practice of using a ground, or zero volts, reference.

### 5.6 Capacitance

From Eq. (5.28) we can define a term $kr = Q/V$. This geometry-dependent ratio of charge to voltage was defined as capacitance in Chap. 4.

$$C = \text{capacitance} = kr = \frac{Q}{V} \qquad (5.29)$$

### 5.7 Energy

A system containing electrostatic charge has potential energy equal to the work required to bring the charge to the system. In a capacitor one conductor is positively charged while the other is negatively charged to the same magnitude. The work required to move a charge from one plate to the other is

$$\delta W = V \delta Q = \frac{Q \delta Q}{C} \qquad (5.30)$$

The work required to move all charge is obtained by integration.

$$W = \int_0^Q \frac{Q\delta Q}{C} = \frac{Q^2}{2C} \qquad (5.31)$$

$$W = \frac{C^2V^2}{2C} = \frac{1}{2}CV^2 = \frac{1}{2}QV \qquad (5.32)$$

## 5.8 Conductors in a Field

A conductor has no current flow once it has attained its charge. However, a conductor has many free electrons available. Free electrons have negative charge and would be accelerated in the opposite direction of any field present in the conductor. For this to be the case, a current could be detected. It is only logical to deduce that all charge must be on the surface of the conductor.

### 5.8.1 Conductive sphere

If a charge of magnitude $Q$ is placed on a conductive sphere it will reside on the surface and produce an external field that is radial. In other words, the field is the same as that produced by a charge $Q$ located at the center of the sphere. That is, from Eq. (5.11),

$$E = \frac{Q}{kr^2}$$

However, it must be noted that if the charge really were at the center of the sphere, $r$ would be 0 and $E$ would be forced to infinity. In other words, the field intensity *at a point charge,* would be infinite. The point charge is a concept that is useful in determining field intensities *at finite distances* from the point charge(s).

Reiterating and generalizing from the case of a homogeneously conductive sphere, charges on any conductor would be uniformly distributed on the outer surface due to mutual forces of repulsion of like charges. It is also readily deduced that the field must be entirely normal to the surface; otherwise any tangential component would result in a current on the surface.

### 5.8.2 Two separated spheres

Two spheres of opposite charge will have a field pattern such that all lines will leave the positively charged sphere and traverse to the negatively charged sphere. For two spheres of like charge, none of the lines travel from sphere to sphere, but because of repulsive force they are emitted toward *infinity*. The field at a large distance from two such spheres is identical to that produced by a single sphere having charge equal to $2Q$. Figure 5.4 shows field lines for a positively charged sphere, a pair of oppositely charged spheres, and a pair of positively charged spheres. The pattern would be identical for point charges of the polarities indicated.

### 5.8.3  Comparison of fields on different sized spheres

Consider two spheres of radii $R$ and $r$, respectively, where $R > r$. If both spheres are analyzed independently for field intensity when at a given test voltage $V_t$, we get

$$V_t = \frac{Q}{4\pi\epsilon_0 R} \tag{5.33}$$

and

$$V_t = \frac{q}{4\pi\epsilon_0 r} \tag{5.34}$$

Therefore, $Q/q = R/r$, where $Q$ and $q$ are the charges on the large and small spheres, respectively. The ratio of the fields is

$$\frac{E_R}{E_r} = \frac{Q}{4\pi\epsilon_0 R^2} \times \frac{4\pi\epsilon_0 r^2}{q} = \frac{Qr^2}{qR^2} = \frac{r}{R} \tag{5.35}$$

Therefore,

$$E_R = \frac{r}{R} \times E_r \tag{5.36}$$

This means that at equal distances from the surfaces, the field from the larger sphere is smaller. The point is that although the field intensity would be the same at points equidistant from the centers of either sphere, the field is greater at equal distances from the surface of the smaller sphere. This is analogous to the generalization that field intensities are higher with spheres of greater or sharper curvature. This can be shown to hold for geometric shapes other than spherical and is useful in analysis of electrostatic behavior as well as in designing to avoid high field concentrations.

### 5.8.4  Equipotential surfaces

Equation (5.28) gives the voltage on a conductive sphere or the potential difference between the sphere's surface and infinity (which is zero)

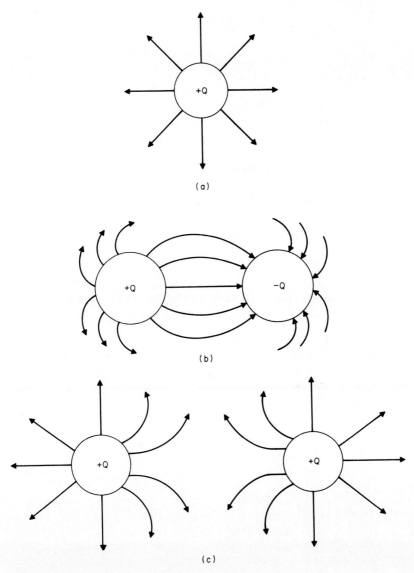

**Figure 5.4** Field lines for (a) positive charged sphere; (b) oppositely charged spheres; (c) two positively charged spheres.

$$V_{is} = \frac{Q}{kr} = \frac{Q}{4\pi\epsilon r} \qquad (5.28)$$

The potential at any other radial distance between $r$ and infinity is given by

$$V_i = \frac{Q}{kr_i} = \frac{Q}{4\pi\epsilon r_i} \qquad (5.37)$$

where $r_i$ is the distance from the center of the sphere.

Any such radius $r_i$ defines a spherical surface that is, by definition, equipotential. That is to say that any point on such a sphere is at the same potential with respect to infinity, or with respect to the conductive sphere's surface. Figure 5.5 shows field lines (solid) and equipotential surfaces (dashed) for a positively charged sphere. Equipotential surfaces exist around a charged conductor of any shape. For an irregular shape the field lines are closer at sites of greater curvature, consistent with higher fields at such locations.

A mechanical analogy exists in topographical contour maps showing elevation levels. Just as each elevation level represents the amount of work required to move a unit weight to that level against

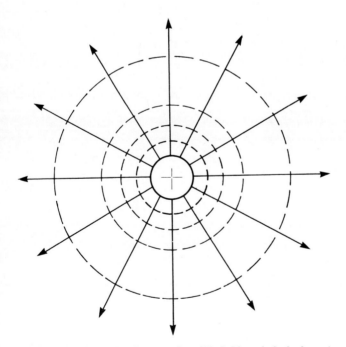

**Figure 5.5** Charged sphere with solid field and dashed equipotential lines.

gravity, equipotential lines represent the work required to move a unit charge from ground potential to that voltage level. Topographical elevation contours also get closer for locations of higher curvature (or steepness).

Considerations based upon behavior of electric fields can help clarify the nature of equipotential surfaces. Any component of the electric field parallel to an equipotential surface would imply a difference of potential which is, by definition, impossible. Therefore equipotential surfaces must be perpendicular to the composite field direction at any given point in space.

Conceptually, equipotential surfaces exist for any electric field although construction and/or visualization is often beyond practical application. This would be the case for complex systems of conductors, and in particular for insulators having inhomogeneous charge distributions in both polarity and magnitude. However for less complex systems that can be approximated by regular shaped conductors, the construction of equipotential surface contours can facilitate the analysis of field effects. Figure 5.6 shows field lines (solid) and equipotential surfaces (dashed) for pairs of spheres with both positive and opposite charges.

### 5.8.5 Concentric conductive spheres

Such an equipotential surface as defined in the preceding paragraph is analogous to having a thin conductive spherical surface with the radius $r_i$. This realization correctly infers that a thin conductive spherical surface of radius $r_i$ could be inserted in the field without altering the effects of the field at points further away than $r_i$. Such an inserted conductive sphere must have an opposite charge on its inner surface since no net charge was added to the system.

#### 5.8.5.1 Potential difference between concentric spheres.
The potential difference between the two spheres is calculated by subtracting Eq. (5.37) from Eq. (5.28).

$$V = V_{is} - V_i = \frac{Q}{4\pi} \left( \frac{1}{r} - \frac{1}{r_i} \right) \qquad (5.38)$$

#### 5.8.5.2 Capacitance of concentric spheres.
The capacitance of these concentric spheres is

$$C = \frac{Q}{V} = 4\pi\epsilon \frac{r r_i}{r_i - r} \qquad (5.39)$$

#### 5.8.5.3 Induction effects of concentric spheres.
If the outside surface of the outer sphere is tied to ground the outer potential becomes zero fol-

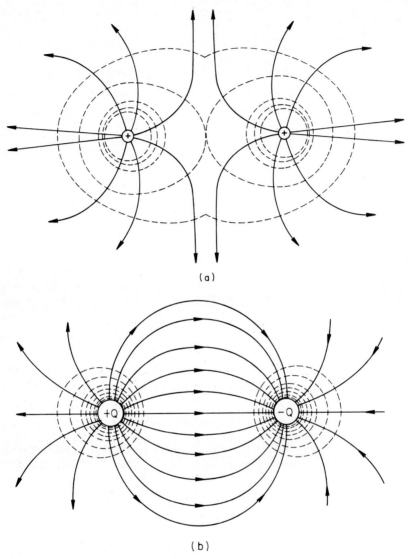

**Figure 5.6** Field and equipotential lines for (a) positively charged sphere pair; (b) oppositely charged sphere pair.

lowing the transient static discharge. However, the inner side of the outer sphere is still at $-Q$, due to the influence of the inner sphere. This means that a net charge of $-Q$ has been placed on the outer sphere by induction.

**5.8.5.4 Capacitance of isolated sphere.** If $r_i$ is extended to infinity, we get

$$C = \frac{4\pi\epsilon r r_i}{r_i - r} = 4\pi\epsilon r \qquad (5.40)$$

which is, appropriately, the capacitance of an isolated sphere.

## 5.9 Displacement Field

The dielectric constant $k$ is a factor that reduces the field $E$. The term $D$ is used to designate the field due to the charge, but independent of the dielectric.

$$D = \frac{Q}{4\pi r^2} = \epsilon E \qquad (5.41)$$

$D$ is called the displacement field.

In a vacuum, $\epsilon = \epsilon_0$, so

$$D = \epsilon_0 E \qquad (5.42)$$

In terms of dielectric constant,

$$D = k\epsilon_0 E \qquad (5.43)$$

The field line concept is easily converted to representation of fields by tubes of flux. An arbitrary-sized tube of flux leaving a charged surface as illustrated in Fig. 5.7 must cover a geometrically symmetrical larger area as the radial distance is increased.

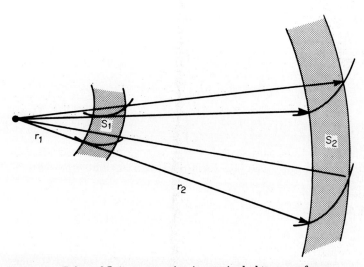

**Figure 5.7** Tubes of flux transversing increasingly larger surfaces.

## 5.9.1 Flux density

If flux lines are cut at two or more radial distances $r_i$ from the sphere in such a way as to describe areas that increase with $r$, the flux remains constant as it passes each area. Since the areas get larger with distance, the flux per unit area (flux density) decreases with radial distance. The relationship is

$$F_1 = F_2 = F_i \tag{5.44}$$

where $F$ = flux = constant

$$A_1 \frac{F_1}{A_1} = A_2 \frac{F_2}{A_2} = A_i \frac{F_i}{A_i} = \text{constant} \tag{5.45}$$

where $A_i$ = area at distance $i$ and $F_i A_i$ = flux density. This relationship is satisfied when

$$dF = D\,dA \tag{5.46}$$

where $dF$ = the incremental amount of flux crossing normal to the surface area element $dA$ and $D$ is the displacement field at the site caused by the charge. The total amount of flux $F$ over the sphere is

$$F = \int_A D\,dA = 4\pi r^2 D \tag{5.47}$$

$$F = 4\pi r^2 \frac{Q}{4\pi r^2} = Q \tag{5.48}$$

## 5.9.2 Gauss' law

The total flux leaving a charged sphere is equal to $Q$ and is constant at any radial distance from the sphere.

$$F = \int_A D \cdot dA = Q \tag{5.49}$$

is known as Gauss' law and is valid where $A$ is any closed surface and the dot product is observed for all $D$ and $A$. The generalized form is

$$\int_A D \cdot dA = Q = \text{total charge enclosed in surface} \tag{5.50}$$

In a practical sense this is not quite true because at large distances both field strength and charge density reduce to zero.

## 5.10 Fields Produced by Selected Charge Configurations

Electric fields in the factory and other industry environments are often the result of one or more charged items of, or approximating, common geometric shapes. The field intensity can be derived by various analytical means, including application of Gauss' law. In the following examples we shall endeavor to illustrate such analysis for

1. An infinite sheet of charge
2. An infinite charged conductive plate of finite thickness
3. A charged conductive cylinder
4. A line of charge
5. A spherical shell of charge
6. A uniform spherical charge distribution
7. The field just outside any charged conductor
8. The field between two charged conductive plates

### 5.10.1 An infinite sheet of homogeneous charge density

Applying Gauss' law, a cylindrical gaussian surface is constructed through the sheet as shown in Fig. 5.8. The cylinder walls are perpendicular to the sheet and the identical ends have area $A$. Such an infinitely large sheet charged homogeneously can be said to have a charge density of $\sigma$ per unit area. If the charge is thus uniformly dispersed over the sheet there can be no field component normal to cylinder walls. Symmetry requires that the field $E$ is identical in each direction normal to the sheet. Therefore at each end of the cylinder the normal component of the field is equal to $E$. Applying Eq. (5.48),

$$\int_A D \cdot dA = Q = \epsilon_0 \int_A E_n dA = 2\epsilon_0 EA = \sigma A$$

Thus,

$$E = \frac{\sigma}{2\epsilon_0} \quad (5.51)$$

### 5.10.2 An infinite charged conductive plate of finite thickness

**5.10.2.1 Field intensity determined by geometrical analysis.** As discussed in Sec. 5.7, charge on a conductor is distributed over the outer

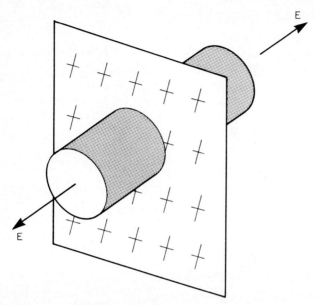

**Figure 5.8** Gaussian surface around sheet of charge.

surface by mutual forces of repulsion. Therefore the two side surfaces of the charged plate would have equal charge densities σ for a unit area. Refer to Fig. 5.9a to follow the determination of the field intensity by adding geometrical considerations to the results of the charged sheet example. The field produced on either side is the net effect produced by the sum of the influence of both sides. Due to the symmetry and factors already considered in the preceding infinite sheet example, the field is normal to the surfaces of the plate. Let $E_1$ equal the component of $E$ due to the sheet of charge on side 1 and $E_2$ be the component due to side 2. Then outside the plate,

$$E = E_1 + E_2$$

From the preceding example,

$$E_1 = E_2 = \frac{\sigma}{2\epsilon_0}$$

Therefore,

$$E = E_1 + E_2 = \frac{\sigma}{\epsilon_0} \tag{5.52}$$

Internal to the plate the equal fields are in opposite directions, so the net field is zero.

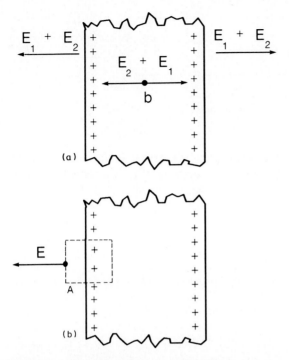

**Figure 5.9** Field determination for a conductive plate by (a) geometrical considerations; (b) Gauss' law.

**5.10.2.2 Field intensity by Gauss' law.** In order to arrive at this result by direct application of Gauss' law, draw a gaussian cylinder with ends having areas $A$ on one side of the plate as shown in Fig. 5.9b. The part of the cylinder lying inside the conductor is in a region between two equally charged surfaces. It is therefore obvious that the field is zero in this region. Thus the only field component is normal to the circular area of the cylinder outer end. Applying Gauss' law,

$$\int_A D \cdot dA = Q = \epsilon_0 \int_A E_n \, dA = \epsilon_0 E A = \sigma A$$

or

$$E = \frac{\sigma}{\epsilon_0} \tag{5.52}$$

### 5.10.3 Conductive cylinders

The geometry of wires used so frequently in electronics is that of cylindrical surfaces. For this reason potential and capacitance is addressed in addition to field intensity.

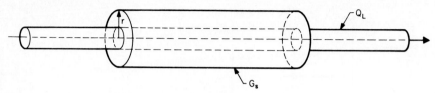

**Figure 5.10** Field of charged conductive cylinder.

**5.10.3.1 Field intensity.** Fields around cylindrical surfaces are similar to those around spheres in many respects. Taking the case of a cylinder with a concentric gaussian cylinder $G_s$ as shown in Fig. 5.10 we can apply Gauss' flux theorem to a unit length:

$$\int_A D \cdot dA = Q_L \quad (5.53)$$

where $Q_L$ is the charge per unit length, and $A$ is the area of a unit length.

The area of a cylinder wall of radius $r$ is

$$A_c = 2\pi r \times \text{length} \quad (5.54)$$

thus, for a unit length,

$$A_c = 2\pi r$$

By symmetry all flux is radially directed, thus there is no component normal to the end faces of $G_s$. Therefore,

$$\int_A D \cdot dA = 2\pi r D = Q_L$$

Since $D = \epsilon E$,

$$E = \frac{Q_L}{2\pi \epsilon r} \quad (5.55)$$

**5.10.3.2 Potential due to a cylinder.** To determine the change in potential at a point due to the presence of a charged cylinder of radius $r_1$, we can arbitrarily place that point on an outer gaussian cylinder of radius $r$. The potential difference can then be calculated from the work required to move a unit charge between the two surfaces, or from Eq. (5.26),

$$V = -\int E\, dr \quad (5.26)$$

$$V_{r_1-r} = -\frac{Q_L}{2\pi\epsilon} \int_{r_1}^{r} \frac{dr}{r} \tag{5.56}$$

where $V_{??}$ is the voltage at $r$ with respect to $r_i$.

$$V_{r_1-r} = V_r - V_{r_1} = -\frac{Q_L}{2\pi\epsilon} \ln\frac{r}{r_1} = \frac{Q_L}{2\pi\epsilon} \ln\frac{r_1}{r} \tag{5.57}$$

This result is logical if one considers that $V_{r_1}$ is more positive than $V_r$.

**5.10.3.3 Capacitance of concentric cylinders.** The preceding results would have been identical if the gaussian cylinder were replaced by an actual physical conductive cylinder. Such an actual concentric cylinder configuration is typical of coaxial cables. Capacitance per unit length is

$$C = \frac{Q_L}{V} = \frac{2\pi\epsilon}{\ln r/r_1} \tag{5.58}$$

Note that in this case, the voltage was taken in the $V_{r_1} - V_r$ sense in order to assure a positive capacitance.

### 5.10.4 A line of charge

Consider the line as being along the $z$ axis, as shown in Fig. 5.11, with its midpoint at the origin. Enclose the line with a cylindrical gaussian surface having radius $r$ and height $2h$. Let $h$ approach infinity. There can be no field component along the $z$ axis because symmetry causes cancellation. The flux is therefore in the radial direction of $G_s$ normal to the cylinder wall. Applying Gauss' law,

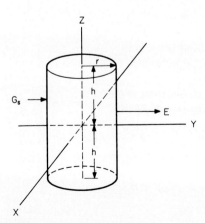

Figure 5.11 Field of line of charge.

$$\int_A D \cdot dA = 2Q_L h$$

$$2\pi r(2h)\epsilon E = 2Q_L h$$

or

$$E = \frac{Q_L}{2\pi \epsilon r} \tag{5.59}$$

Thus, the field from a line charge is identical to that of a charged cylinder. In other words the field effect of a charged conductive cylinder is identical to that produced if the surface charge were concentrated at a line coincident with the axis of the cylinder. This is somewhat analogous to the similar relationship between spheres and central point charges.

### 5.10.5 A spherical shell of charge

Consider a spherical shell of charge having radius $r_s$ and uniform surface charge density of $q_s$. By geometric analysis the shell can be visualized as consisting of an infinite number of point charges, pairs of which are located on opposite ends of a diameter through the center of the shell. Any such pair of point charges would cancel the effects of any inwardly directed field lines. Any component of $E$ in a direction other than outwardly radial would also be canceled out from simple geometry considerations. Figure 5.12 illustrates this effect at points $p_n$ and $p'_n$.

Application of Gauss' law supports these geometrical considerations. To determine the field intensity inside the shell a concentric gaussian sphere is inserted with radius $r_{gi}$ just slightly less than $r_s$. From Gauss' law,

$$\int_A D \cdot dA = Q_{enclosed}$$

but $Q_{enclosed} = 0$. Therefore,

$$D = \epsilon E = 0 \text{ inside the shell} \tag{5.60}$$

To determine the field outside the shell, a concentric gaussian sphere is constructed with $r_{g0} > r_s$. Then,

$$\int_A D \cdot dA = Q_{enclosed} = 4\pi r_{g0}^2 \epsilon_0 E = q_s 4\pi r_s^2$$

or,

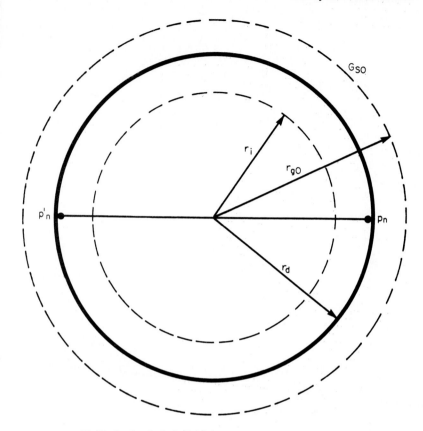

Figure 5.12  Field of spherical shell of charge.

$$E = \frac{q_s r_s^2}{r_{g0}^2 \epsilon_0} \tag{5.61}$$

directed radially outward from the shell.

### 5.10.6  A spherical charge distribution

In this example, we consider a volume distribution of charge rather than the frequently used examples of conductors which have their charge on the outer surface. The distribution is spherical with radius $r_d$ and volume charge density of $\delta_v$. Physically this could be the representation of a spherical cloud assumed to be charged homogeneously. In order to determine the field inside the sphere, a gaussian sphere is constructed with radius $r_g$ smaller than the radius $r_d$ of the charge distribution. Due to the homogeneity of charge distribution and symmetry any resultant $E$ field must be radial. Applying Gauss' law,

$$\int_A D \cdot dA = q_{enclosed}$$

The volume of a sphere $= \frac{4}{3}\pi r^3$

$$Q_{enclosed} = \delta_v \frac{4}{3}\pi r_g^3 = \int D \cdot dA = 4\pi r_g^2 \epsilon_0 E$$

Therefore,

$$E_{inside} = \frac{\delta_v r}{3\epsilon_0} \qquad (5.62)$$

for any $r$ less than $r_d$.

For $E$ external to the distribution, construct a gaussian sphere with $R_g > r_d$. Again, by symmetry, the field is radial. Applying Gauss' law,

$$\int_A D \cdot dA = Q_{enclosed} = 4\pi r_g^2 \epsilon_0 E = \delta_v \frac{4}{3}\pi r_d^3$$

or

$$E = \frac{\delta_v r_d^3}{3\epsilon_0 r_g^2} = \frac{Q}{4\pi \epsilon_0 r_g^2} \qquad (5.63)$$

where $r_g$ can be replaced by $r$ for any $r = r_d$.

The resultant field outside the radius of the spherical distribution is observed to be identical to that found outside a solid conductive sphere, a spherical shell of charge, or a point charge.

### 5.10.7 The field just outside any conductor

Consider the field at a site on an arbitrarily shaped conductor. The total charge is distributed on the outer surface. As discussed in prior examples, if any component of $E$ were in a direction other than normal to the surface, charges would move about on the surface. However, immediately following any transient conditions that might have occurred during charge accumulation, a conductor's surface reaches equilibrium with the charges held stationary by mutually repulsive forces. To estimate the field intensity just outside the conductor (prior to the influence of any other charged items) a gaussian surface is constructed. A cylindrical gaussian surface on a portion of an irregular surface is depicted in Fig. 5.13. The bottom face of the cylinder is inside the conductor, where the field is zero. For reasons already discussed there is no field component parallel to the surface, thus the field normal to the cylinder wall is zero. This leaves only the field nor-

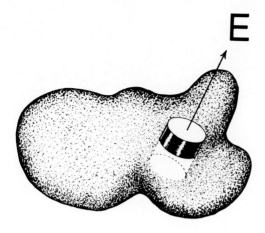

Figure 5.13 Field just outside any conductor.

mal to the cylinder face just outside the conductor. Applying Gauss' law,

$$\int_A D \cdot dA = Q_{\text{enclosed}} = \epsilon_0 E A = \sigma A$$

Therefore,

$$E = \frac{\sigma}{\epsilon_0} \qquad (5.64)$$

which is identical to the field outside the infinitely charged plate. The fields determined for the spherical-shaped conductors can easily be converted to this same form if surface density is used rather than $Q$.

### 5.10.8 Parallel conductive plates

Two parallel conductive plates, as shown in Fig. 5.14a and separated by air or some other dielectric, form a capacitor. The capacitor is charged with each plate at opposite polarities. In actuality most of the charge would reside on the inner opposing faces of the two plates. The small percentage remaining on the opposite outer faces and at the edges result in fringing. Such effects are typically minimal and are neglected in this analysis. In other words the charges are regarded as being entirely and homogeneously distributed on the inner faces of each plate.

#### 5.10.8.1 Geometrical determination of field intensity.
Again we can apply the results from the infinite sheet of charge in Sec. 5.10.1, supplemented with additional geometric considerations. In Fig. 5.14a,

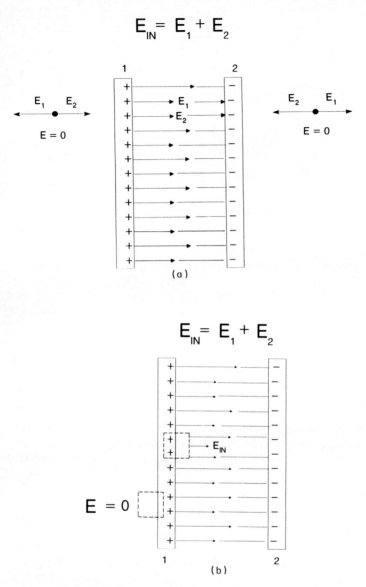

**Figure 5.14** Field of charged parallel conductive plates determined by (a) geometrical considerations; (b) Gauss' law.

just outside of either plate, we can see that $E_1$, the field from positive sheet of charge 1, and $E_2$, the field from negative sheet of charge 2, are of equal magnitudes but in opposite directions and thus cancel. In between the plates, the resultant fields $E_1$ and $E_2$ are again equal but in the same direction. From Eq. (5.51),

$$E = \frac{\sigma}{2\epsilon_0} = E_1 = E_2 \qquad (5.51)$$

$$E_{\text{between}} = E_1 + E_2 = \frac{\sigma}{\epsilon} \qquad (5.65)$$

**5.10.8.2 Field intensity determined by Gauss' law.** If we construct two cylindrical gaussian surfaces, one on the inside and one on the outside of either plate as shown at side 1 in Fig. 5.14b, we get

$$\int_A D \cdot dA = Q = \epsilon_0 \int_A E_n \, dA = \epsilon_0 EA$$

or

$$E = \frac{Q}{A\epsilon} = \frac{\sigma}{\epsilon} \qquad (5.66)$$

directed toward the negative plate and $E = 0$ directed outward since we assumed all charge at the inner surface.

**5.10.8.3 Potential difference between plates.** The potential difference between the plates is

$$V = E \times \text{distance} = Ed = \frac{Qd}{\epsilon A} \qquad (5.67)$$

**5.10.8.4 Capacitance of parallel plates.** Capacitance is

$$C = \frac{Q}{V} = \frac{Q\epsilon A}{Qd} = \frac{\epsilon A}{d} \qquad (5.68)$$

## 5.11 Combinations of Capacitors

If $n$ capacitors are connected in parallel, the combined capacitance is

$$C = c_1 + c_2 + \cdots + c_n \qquad (5.69)$$

If $n$ capacitors are connected in series, the combined capacitance is

$$C = \frac{1}{c_1} + \frac{1}{c_2} + \cdots + \frac{1}{c_n} \qquad (5.70)$$

## 5.12 Green's Theorem of Reciprocity

Green calculated the effects of groups of charged conductors on one another. In a system containing $n$ charged conductors: 1, 2, 3, ..., $n$, the potential at conductor 1 as a result of the charges on all other conductors is

$$V_1 = \frac{Q_2}{4\pi\epsilon r_{21}} + \frac{Q_3}{4\pi\epsilon r_{31}} + \cdots + \frac{Q_n}{4\pi\epsilon r_{n1}} \tag{5.71}$$

The potential on conductor 2 due to charges on all other conductors is

$$V_2 = \frac{Q_1}{4\pi\epsilon r_{12}} + \frac{Q_3}{4\pi\epsilon r_{32}} + \cdots + \frac{Q_n}{4\pi\epsilon r_{n2}} \tag{5.72}$$

and so on to the potential at conductor $n$.

$$V_n = \frac{Q_1}{4\pi\epsilon r_{in}} + \frac{Q_2}{4\pi\epsilon r_{2n}} + \cdots + \frac{Q_{n-1}}{4\pi\epsilon r_{(n-1)n}} \tag{5.73}$$

If these equations are now multiplied by charges $Q'_1, Q'_2, Q'_n$, the series becomes

$$V_1 Q'_1 = \frac{Q_2 Q'_1}{4\pi\epsilon r_{21}} + \frac{Q_3 Q'_1}{4\pi\epsilon r_{31}} + \cdots + \frac{Q_n Q'_1}{4\pi\epsilon r_{n1}} \tag{5.74}$$

$$V_2 Q'_2 = \frac{Q_1 Q'_2}{4\pi\epsilon r_{12}} + \frac{Q_3 Q'_2}{4\pi\epsilon r_{32}} + \cdots + \frac{Q_n Q'_2}{4\pi\epsilon r_{n2}} \tag{5.75}$$

$$V_n Q'_n = \frac{Q_1 Q'_n}{4\pi\epsilon r_{1n}} + \frac{Q_2 Q'_n}{4\pi\epsilon r_{2n}} + \cdots + \frac{Q_{n-1} Q'_n}{4\pi\epsilon r_{(n-1)n}} \tag{5.76}$$

If we now write similar equations for a different set of charges, $Q'_1, Q'_2, \ldots, Q'_n$, we get

$$V'_1 = \frac{Q'_2}{4\pi\epsilon r'_{21}} + \frac{Q'_3}{4\pi\epsilon r_{31}} + \cdots + \frac{Q'_n}{4\pi\epsilon r_{n1}} \tag{5.77}$$

$$V'_2 = \frac{Q'_1}{4\pi\epsilon r_{12}} + \frac{Q'_3}{4\pi\epsilon r_{32}} + \cdots + \frac{Q'_n}{4\pi\epsilon r_{n2}} \tag{5.78}$$

$$V'_n = \frac{Q'_1}{4\pi\epsilon r_{in}} + \frac{Q'_2}{4\pi\epsilon r_{2n}} + \cdots + \frac{Q'_{n-1}}{4\pi\epsilon r_{(n-1)n}} \tag{5.79}$$

If these equations are now multiplied by charges $Q_1, Q_2, Q_n$, the series becomes

$$V'_1 Q_1 = \frac{Q_1 Q'_2}{4\pi\epsilon r_{21}} + \frac{Q_1 Q'_3}{4\pi\epsilon r_{31}} + \cdots + \frac{Q_1 Q'_n}{4\pi\epsilon r_{n1}} \tag{5.80}$$

$$V'_2 Q_2 = \frac{Q_2 Q'_1}{4\pi\epsilon r_{12}} + \frac{Q_2 Q'_3}{4\pi\epsilon r_{32}} + \cdots + \frac{Q_2 Q'_n}{4\pi\epsilon r_{n2}} \tag{5.81}$$

$$V'_n Q_n = \frac{Q_n Q'_1}{4\pi\epsilon r_{1n}} + \frac{Q_n Q'_2}{4\pi\epsilon r_{2n}} + \cdots + \frac{Q_n Q'_{n-1}}{4\pi\epsilon r_{(n-1)n}} \quad (5.82)$$

Upon examination it can be seen that the sums of the $VQ$ products of Eqs. (5.74) through (5.76) are equal to the sums of Eqs. (5.80) through (5.82). Remember that $r_{12} = r_{21}$, etc. This relationship, given in Eq. (5.83), is called Green's reciprocation theorem.

$$\sum V'_n Q_n = \sum V_n Q'_n \quad (5.83)$$

### 5.12.1 Elastance

Elastance is a term often mistakenly interpreted as the reciprocal of capacitance. Although elastance terms are ratios of voltage and charge, there is an important distinction in the definition that must be noted. Elastance is derived from a system of conductors where all are ungrounded and all but one are uncharged. Capacitance terms are derived from a system of conductors where all but one are grounded.

### 5.12.2 Elastance relationship of a system of conductors

Green's reciprocation theorem can be applied to determine the mutual elastance of separated conductors. For instance, in such a system of conductors, a charge $Q_A$ on $A$ will cause a potential $V_B$ on $B$, and a charge $Q'_B$ on $B$ will cause a potential $V'_A$ on $A$. Applying Green's theorem we get

$$V'_A Q_A + V'_B Q_B = V_A Q'_A + V_B Q'_B \quad (5.84)$$

Since it is known that $Q_A$ develops $V_B$ while $Q_B = 0$ and $Q'_B$ develops $V'_A$ while $Q'_A = 0$, we can get the following from Eq. (5.78):

$$V'_A Q_A = V_B Q'_B \quad (5.85)$$

This can be restated as

$$\frac{V'_A}{Q'_B} = \frac{V_B}{Q_A} \quad (5.86)$$

The ratios given in Eq. (5.86) are called mutual elastance and show the effect on the potential at one conductor due to a charge on another. The term is typically written as follows:

$$S_{AB} = \frac{V_B}{Q_A} \quad \text{or} \quad S_{BA} = \frac{V_A}{Q_B} \quad (5.87)$$

Equation (5.86) proves that

$$S_{AB} = S_{BA} \tag{5.88}$$

The potential at any conductor in a group of $n$ charged conductors can be expressed as

$$V_i = S_{1i}Q_1 + S_{2i}Q_2 + S_{3i}Q_3 \cdots + S_{ni}Q_n \tag{5.89}$$

$S_{ii}$ = self elastance which is equal to $V_i Q_i$ only when the charges on the other conductors are all equal to zero.

### 5.12.3 Capacitive relationships between conductors

Just as voltage on a conductor can be written in terms of charge on other conductors, the charge on a conductor can be written in terms of the voltage on other conductors. Green's reciprocation theorem can be applied to determine the mutual capacitance of separated conductors. In the case of conductors $A$ and $B$, a potential $V_A$ on $A$ will cause an induced charge $Q_B$ on $B$ with $V_B = 0$; and a potential $V'_B$ on $B$ will cause an induced charge $Q'_A$ on $A$ with $V'_A = 0$. Applying Green's theorem we get

$$V'_A Q_A + V'_B Q_B = V_A Q'_A + V_B Q'_B \tag{5.90}$$

Since it is known that $V_A$ induces $Q_B$ while $V_B = 0$, and $V'_B$ induces $Q'_A$ while $V'_A = 0$, we get the following from Eq. (5.85):

$$V'_B Q_B = V_A Q'_A \tag{5.91}$$

This can be restated as

$$\frac{Q_B}{V_A} = \frac{Q'_A}{V'_B} \tag{5.92}$$

The ratios given in Eq. (5.87) are called mutual capacitance, which shows the effective induced charge at one conductor due to a potential on another. The term is typically written as follows:

$$C_{AB} = \frac{Q_A}{V_B} \quad \text{or} \quad C_{BA} = \frac{Q_B}{V_A} \tag{5.93}$$

Equation (5.92) shows that

$$C_{in} = C_{ni} \tag{5.94}$$

The charge on any conductor in a group of $n$ conductors having potentials $V_1, V_2, \ldots V_n$, can be written:

$$Q_i = C_{i1}V_1 + C_{i2}V_2 + C_{i3}V_3 + \cdots + C_{in}V_n \quad (5.95)$$

The $C_{ii}$ term that arises is called self-capacitance, whereas all others are mutual capacitance coefficients. Mutual capacitance is $Q_i/V_n$ with the voltages on all conductors other than $n$ set to 0 V. Since induced charges are of opposite sign, all mutual capacitance coefficients are negative, while the self-capacitance coefficient is positive. The self- and mutual-capacitance terms can be derived by inverting the $n \times n$ matrix of equations attained from writing Eq. (5.89) for all $V_i$. Likewise a similar inversion of the matrix from Eq. (5.95) will yield elastance terms from known capacitance terms.

### 5.12.4 Physical interpretation of mutual capacitance

To visualize mutual capacitance in more general terms consider two irregularly shaped conductors in a dielectric medium. Such a configuration makes up a capacitor. Figure 5.15a shows such a capacitor with an applied voltage. In Fig. 5.15b conductor $A$ is referenced to ground, as is a third conductor $C$ positioned nearby. Some of the flux lines from conductor $B$ are now diverted to $C$ inducing a negative charge on the side of $C$ nearest $B$. This effect is termed mutual capacitance of $C$ with respect to $B$ or $C_{CB}$, where

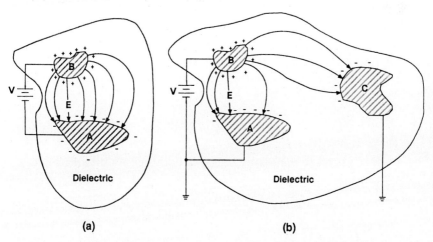

**Figure 5.15** (a) Capacitance of two irregularly shaped conductors and (b) mutual capacitance $C_{CB}$ of $C$ with respect to $B$. $C_{BB}$ is the self-capacitance of $B$ due only to the charge on $B$.

$$C_{CB} = \frac{Q_C}{V_B} = \frac{Q_C}{V}$$

### 5.12.5 Capacitor current

Figure 5.15*b* shows an example of self- and mutual capacitance. Capacitor current was discussed in Chap. 4 in terms of the inductive effect of one plate on the other. Further insight may be gained by considering that in Fig. 5.15

$$Q_B = C_{BB}V_B \quad \text{and} \quad Q_C = C_{CB}V_B \quad (5.96)$$

Differentiating,

$$\frac{dQ_B}{dt} = C_{BB}\frac{dV_B}{dt} \quad \text{and} \quad \frac{dQ_C}{dt} = C_{CB}\frac{dV_B}{dt} \quad (5.97)$$

Current is defined as the derivative of charge.

$$I = \frac{dQ}{dt} \quad (5.98)$$

Therefore, substituting into Eq. (5.97):

$$I_B = C_{BB}\frac{dV_B}{dt} \quad \text{and} \quad I_C = C_{CB}\frac{dV_B}{dt} \quad (5.99)$$

### 5.12.6 Energy

A system containing electrostatic charge has potential energy equal to the work required to bring the charge to the system. In a capacitor, one conductor is positively charged while the other is negatively charged to the same magnitude. The work required to move a charge from one plate to the other is

$$\sigma W = V\sigma Q = \frac{Q\sigma Q}{C} \quad (5.100)$$

The work required to move all charge is obtained by integration.

$$W = \int_0^Q \frac{Q\,dQ}{C} = -\frac{Q^2}{2C} \quad (5.101)$$

$$W = \frac{(CV)^2}{2C} = \frac{CV^2}{2} = \frac{QV}{2} \quad (5.102)$$

For a system of $n$ conductors,

$$W = \frac{1}{2}\sum_n V_n Q_n \qquad (5.103)$$

which is equivalent to

$$W = \frac{1}{2}\sum_n Q_n \times \sum_n V_n \qquad (5.104)$$

Substituting Eq. (5.89) into Eq. (5.103):

$$W = \frac{1}{2}\sum_i Q_i(S_{1i}Q_1 + S_{2i}Q_2 + S_{3i}Q_3 + \cdots S_{in}Q_n) + \cdots$$

$$= \frac{1}{2}(S_{11}Q_1^2 + S_{22}Q_2^2 + \cdots + S_{ii}Q_i^2 + \cdots + 2S_{12}Q_1Q_2 + 2S_{13}Q_1Q_3 + \cdots$$

$$+ 2S_{in}Q_iQ_n) \qquad (5.105)$$

Likewise substituting Eq. (5.95) into Eq. (5.103):

$$W = \frac{1}{2}\sum_i V_i(C_{1i}V_1 + C_{2i}V_2 + C_{3i}Q_3 + \cdots + C_{in}V_n)$$

$$= \frac{1}{2}(C_{11}V_1^2 + C_{22}V_2^2 \cdots + C_{ii}Q_i^2 + 2C_{12}V_1V_2 + 2C_{13}V_1V_3 + \cdots$$

$$+ 2C_{in}V_iV_n) \qquad (5.106)$$

## 5.13 Boundary Conditions of Insulators

Properties of insulators are quite different from conductors. Of particular importance as far as fields are concerned are the facts that insulators can have interior electric fields, and the surface need not be equipotential on charged insulators.

To analyze an interface of two insulators consider a gaussian surface in the shape of a box as shown in Fig. 5.16a. If there is free surface charge at the boundary,

$$\oint_G D \cdot ds = Q = \rho_s \Delta s \qquad (5.107)$$

Allow the height of the box $h$ to approach zero. This for the moment eliminates any tangential component of $D$, leaving

$$D_{n2}\Delta s - D_{n1}\Delta s = \rho_s \Delta s \qquad (5.108)$$

dividing by $\Delta s$,

$$D_{n2} - D_{n1} = \rho_s \qquad (5.109)$$

where directions are assumed as in Fig. 5.16b and $\rho_s$ is positive.

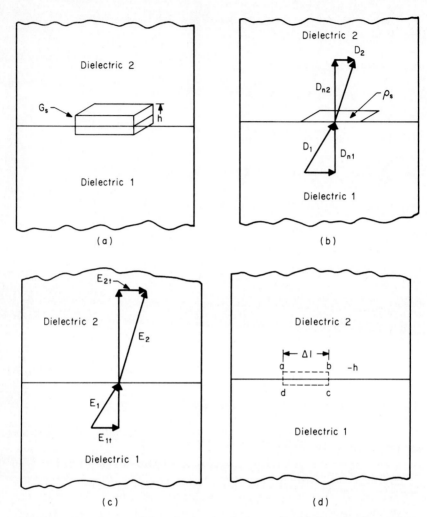

**Figure 5.16** Boundary conditions for two insulative media. (*a*) Gaussian box surface enclosing boundary with height $h$ approaching zero; (*b*) flux through boundary; (*c*) fields through boundary; (*d*) hypothetical rectangular path with height $h$ approaching zero.

To revisit the tangential components consider the fields shown in Fig. 5.16*c* and the path *abcd* shown in Fig. 5.16*d*.

$$\oint E \cdot dl = \int_a^b E_{2t} \cdot dl + \int_b^c E_{bc} \cdot dl + \int_c^d E_{1t} \cdot dl + \int_d^a E_{da} \cdot dl \quad (5.110)$$

If $h$ is allowed to approach zero, the contributions along paths *bc* and $da$ are zero. Therefore, using the conservative field relationship,

$$\oint E \cdot dl = 0 = \int_a^b E_{2t} \cdot dl + \int_c^d E_{1t} \cdot dl \qquad (5.111)$$

therefore,

$$E_{2t} \Delta l = E_{1t} \Delta l \qquad (5.112)$$

or $E_{2t} = E_{1t}$ = tangential components of electric field of the two surfaces.

The conservative property of the field $E$ is independent of the dielectric constant of the media. At an interface of two media the tangential component of the field has been shown to be continuous. The normal component of the field is affected by the dielectric constant transition. From Eq. (5.109):

$$\epsilon_2 E_{n2} - \epsilon_1 E_{n1} = \rho_s \qquad (5.113)$$

For the case where no free charge has been placed on either insulator surface by rubbing or some other means $\rho_s$ would be 0, and

$$D_{n2} = D_{n1} = \epsilon_2 E_{n2} = \epsilon_1 E_{n1} \qquad (5.114)$$

## 5.14 Maxwell's Equations

In 1864, Maxwell introduced four equations that define all electrostatic and electromagnetic phenomena. The first two of these equations listed in Table 5.1 could be used as the basis for analysis of electrostatic problems. In effect by utilizing Gauss' law and the conservative-field relationship we have used Maxwell's equations 1 and 2, respectively, in integral form.

TABLE 5.1 Maxwell's Equations Arranged with Electrostatic Relationships as First and Second

|  | Integral form | Point, or differential form |
|---|---|---|
| First | $\oint E \cdot dl = 0$ | $\nabla \times E = 0$ |
| Second | $\oint D \cdot ds = \int \rho dv$ | $\nabla \cdot D = \rho$ |
| Third | $\oint H \cdot dl = \int J \cdot ds$ | $\nabla \cdot H = J$ |
| Fourth | $\oint B \cdot ds = 0$ | $\nabla \cdot B = 0$ |

## References

1. James Clerk Maxwell, *A Treatise on Electricity and Magnetism*, Dover Publications, Inc., New York, 1954.
2. Oliver Dimon Kellogg, *Foundation of Potential Theory*, Dover Publications, Inc., New York, 1953.
3. James R. Wait, *Electromagnetic Wave Theory*, Harper & Row, Publishers, Incorporated, New York, 1985.
4. Ralph Morrison, *Grounding and Shielding Techniques in Instrumentation*, 3d ed., John Wiley & Sons, Inc., New York, 1986.
5. W. R. Smythe, *Static and Dynamic Electricity*, 2d ed., McGraw-Hill Book Company, New York, 1950.
6. G. G. Skitek and S. V. Marshall, *Electromagnetic Concepts and Applications*, Prentice-Hall, Inc., Englewood Cliffs, N.J., 1982.
7. Allen Nussbaum, *Electromagnetic Theory for Engineers and Scientists*, Prentice-Hall, Inc., Englewood Cliffs, N.J., 1965.
8. David K. Cheng, *Field and Wave Electromagnetics*, Addison-Wesley Publishing Company, Menlo Park, Calif., 1983.
9. Francis W. Sears and Mark W. Zemansky, *University Physics*, Addison-Wesley Publishing Company, Cambridge, Mass., 1957.
10. Clayton R. Paul and Syed A. Nasar, *Introduction to Electromagnetic Fields*, McGraw-Hill Book Company, New York, 1987.
11. William D. Greason, *Electrostatic Damage in Electronics: Devices and Systems*, John Wiley & Sons, Inc., New York, 1987.

Chapter

# 6

# Physical Concepts of Static Electrification

This chapter explores physical phenomena believed to be involved with static electrification and important related variables. The concepts are somewhat scientific in nature and are often overlooked by the practitioner of ESD control. Fortunately, the complex physical nature of static electricity can often be disregarded in the solution of practical problems. Occasionally, however, oversimplification can lead to erroneous conclusions. Static accumulation processes can involve contact electrification, electrolytic phenomena, spray electrification, and piezo- and pyroelectrification. Brief discussions of molecular interfaces, contact potential, Fermi levels, and work function are presented briefly to facilitate understanding of these different types of electrification. Further reading of the listed references for more thorough coverage of the topics is encouraged.

## 6.1 Static Electrification

Loeb[1] defined static electrification as follows: "Static Electrification covers all processes for producing segregation of positive and negative electrical charges by mechanical actions which operate by contact or impact between solid surfaces, between solid and liquid surfaces, or in the rupture or separation of solid or liquid surfaces by gases or otherwise, including also ionized gases. These involve such processes as frictional, contact, or triboelectrification, spray electrification and electrification in dust, snow or in thunderstorms."

## 6.2 Molecular Interfaces

Since static electrification involves segregation of charges, the types of bonds restricting such separations are addressed briefly. Van der

Waal's forces are induced dipole moments between atoms caused by electrons rotating around their nuclei. The resulting dipole at an atom causes an attractive force with other atoms. Since the electrons are in constant motion the attraction is a fluctuating momentary effect. Electrovalent bonds are said to exist when two oppositely charged ions are formed through electron transfer between a metal and a nonmetal. The immediate result is a mutual attraction termed ionic or electrovalent bond. Covalent bonds are those based on shared electrons. For example, nonconductors cannot transfer electrons readily, but sometimes can share electrons at the molecular level.

Energy transfer is involved when an atom acquires or loses an electron. Ionization energy is defined as the energy required to remove one electron from an atom. Electron affinity is the energy released when a neutral atom acquires an electron to become a negative ion.

## 6.3 Band Structure of Solids

Quantum mechanics describes electrons in a solid as being grouped into quanta of energy within defined bands. Between bands are ranges of energy not permitted for electrons of the material. These are called forbidden bands or forbidden gaps. The band formation comes about because solids are made up of combined atoms. When such atoms are brought together interatomic forces are of large magnitude. The nuclei, including their completed inner shells, are closer to one another than the radii of their outer valence electrons in the isolated atomic state. This factor and the Pauli exclusion principle demands that the narrow energy levels of the free atom state broaden and become diffused in bands with each electron at different energy levels. There may be many electrons at distinct levels until all valence electrons are included.

### 6.3.1 Insulators

A material can conduct only if it has outer electrons in a partially filled energy band. If the band is full, an electron cannot acquire an increase in energy or be accelerated by an electric field because there are no allowed spaces. Such a material is an insulator. The forbidden band for insulators is sufficiently wide (about 12 eV) that electrons cannot be excited enough to cross the gap to the conduction band, even at rather high temperatures.

### 6.3.2 Metals

Materials that are electrically and thermally conductive at normal temperatures are called metals. The outer electrons of metals are either in a valence band that is partially filled or one that overlaps the levels of the conduction band. Such a band configuration allows these outer electrons unfilled spaces at higher energy levels, and is therefore a conductor.

### 6.3.3 Semiconductors

Semiconductors have band structures similar to insulators except the forbidden gap is about one-tenth as wide. Therefore at moderate temperatures a few electrons are likely to be sufficiently thermally energized to cross the gap to the conduction band. Figure 6.1a depicts the configuration for metallic conductors, Fig. 6.1b the band structure for insulators, and Fig. 6.1c for semiconductors.

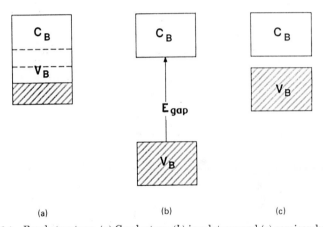

Figure 6.1 Band structure. (a) Conductors; (b) insulators; and (c) semiconductors.

## 6.4 Fermi Levels

The Fermi-Dirac distribution function gives the probability of an electron state of energy $E$ being occupied by a free electron.

$$f(E) = \left[1 + \exp\left(\frac{E - E_f}{kT}\right)\right]^{-1} \tag{6.1}$$

The Fermi level ($E_f$) is defined as the highest filled energy level at absolute zero. At absolute zero, free electrons will occupy the lowest available energy state up to the Fermi level, as depicted by the energy density diagram of Fig. 6.2a. At higher temperatures, thermal energy permits some electrons to occupy states above the Fermi level, leaving vacancies at states slightly below the Fermi level as indicated by Fig. 6.2b. In relation to the Fermi-Dirac distribution, room temperature is

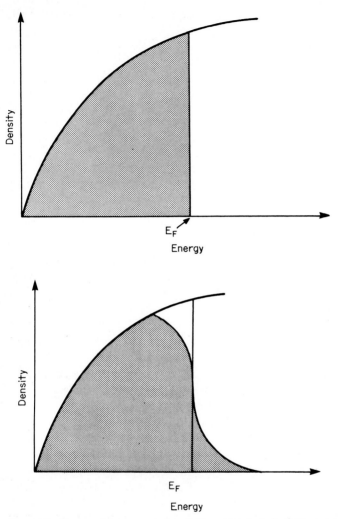

**Figure 6.2** Energy states occupied up to Fermi level at absolute zero and above Fermi level at higher temperatures.

considered low. Thus the absolute zero condition can be used as a good approximation for most ambient conditions.

## 6.5 Work Function

Figure 6.3 depicts the energies of conduction electrons in a metal. Assuming the absolute zero approximation for room ambient, the highest energy level is $E_f$. The lowest allowable energy level is zero, as was indicated in Fig. 6.2. The energy required to remove an electron from

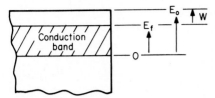

**Figure 6.3** Energies of conduction electrons in a metal.

this lowest level to just outside the surface is $E_o$. The work required to remove an electron completely outside the material from the top of the filled Fermi band is called the work function $W$, where

$$W = E_o - E_f \qquad (6.2)$$

For metals typical work function values range from 2 to 6.5 eV. Variables that influence the effective work function are surface contamination, internal impurities, temperature, lattice orientation with respect to the surface and external fields.

## 6.6 Contact Potential Difference or Volta Potential

Volta, as early as 1801, believed there was an intrinsic potential difference between dissimilar *dry* metals in contact. He distinguished this concept as being different from the electrolytic phenomena occurring with metals in solutions. In 1898, Lord Kelvin presented his studies to the Royal Institution which proved the existence of contact potentials between different metals. R. A. Millikan further clarified the understanding of this phenomenon in 1915, when he showed the relationship of the potential to the difference in the work functions for free electrons in the two metals.

Figure 6.4 shows the free electron levels of a metal $A$ contacting the inside of a U-shaped dissimilar metal $B$. The Fermi levels adjust to equalize as shown along $cd$. The work functions $W_A$ and $W_B$ are now at different levels as indicated. While work function is the work to remove an electron from the Fermi level, $\phi$ is generally used to symbolize the potential opposing the work. Thus,

$$W = \phi \times e \qquad (6.3)$$

where $e$ is the charge of one electron = $4.803 \times 10^{-10}$ absolute electrostatic units.

The potential $\phi$ is also called the thermionic work function. The potential at a metal's surface can be obtained by dividing $E_o$ by $e$. Since

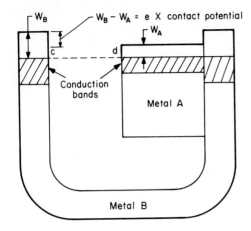

**Figure 6.4** Metals $A$ and $B$ in contact result in contact potential $V_c = \phi_A - \phi_B$

the Fermi levels have equalized, we can state

$$E_{fA} = E_{fB} \tag{6.4}$$

or

$$E_{oA} - W_A = E_{oB} - W_B \tag{6.5}$$

dividing by $e$,

$$V_A - V_B = \frac{W_B - W_A}{e} = \phi_A - \phi_B \tag{6.6}$$

This potential difference is called contact or Volta potential and is equal to the difference in thermionic work functions of the two metals.

## 6.7 Contact Electrification

When two solids of geometrical surface area $A$ *contact* one another, actual contact is made by many smaller areas $a$ rather than the entire continuous surface. Thus many contact points within several angstroms of one another actually transfer charge. The resultant charge transfer is believed due to a contact potential phenomenon, even when insulators are involved.

Contact electrification between solids was described in Chap. 4 as the simple charge sharing between a previously charged and an uncharged conductor. In the usual context, contact electrification also includes Volta electrification or triboelectric generation. Volta electrification is the result of electron transfer caused by different energy levels in dry metal-to-metal or metal-to-insulator contact. Triboelectric generation does not require rubbing as the name implies but is caused by contacts of solid surfaces and subsequent separation. Usually dissimilar surfaces are involved such as two insulating solids or an insulator and a metal or two solids containing the same ion-indifferent concentrations. Electron or ionic transfer is responsible.

### 6.7.1 General variables

In Chap. 4 it was mentioned that contact charging could not always be accurately predicted by a simple triboelectric series. Solid-to-solid electrification is subject to many misleading variables, such as films of moisture, Helmholtz double layers, electrolytic effects, impurities on surfaces or in solids, external electric fields, temperature and humidity variations, and the type of contact. Type of contact could be simply touching, rubbing, rolling, sliding, hammering, or scraping. Another important factor is proper measurement circuitry, with due accounting of resistivities, groundings, capacitances, and measurement apparatus error. Complete prediction of static generation results for any two materials must account for these and other variables. For instance, one must realize that part of the perceived voltage increase upon separation of two materials is due to the reduction of mutual capacitance between them.

### 6.7.2 Back discharge

All electrostatic contact charging is limited by recombination to a conductive material or an electrolyte by gas discharge or field emission. Even insulators possess some residual conductivity and charges will be slowly transferred or recombined. Thus the amount transferred in contact and separation must exceed the recombined amount for any given two materials. One of the materials being insulative tends to increase the static potential accumulated, largely because of minimal backflow. During separation, the charge that is recombined can exceed the residual charge on the material. This could be a frequently unseen factor relating to ESD in the electronics industry. Some variables that affect recombination are the materials in contact, conductivity, surface conditions, speed of separation and contact, surface areas and geometries in contact, temperature, ambient gas, external fields, pressure, and capacitance. Gasses adsorbed on the surfaces or in the gap will affect back discharge.

### 6.7.3 Adsorbed gasses

The best of conditioning or sample preparation techniques is incapable of removing all contamination from a surface. The surface work function can be altered by just a monolayer of adsorbed gasses. Adsorption refers to surface layers as opposed to absorption involving penetration to greater depths. Monolayers are often in the form of chemical bonds with surface atoms of the solid material. An oxide is a classical example. The same type of layer can occur on liquids.

### 6.7.4 Effect of work functions

It has been established that when two materials of different work functions are separated, the one with the greater work function will take on a negative polarity. Experimental works[2,3] have demonstrated a definite relationship between effective work function of polymers and the resultant polarity and magnitude from triboelectric contact and separation with metals.

### 6.7.5 The effects of other variables on contact electrification

Numerous other variables affect the magnitude and polarity attained through contact electrification. Fine particles of contamination are next to impossible to dislodge. Such matter can affect electrification, especially at reduced contact pressures. The degree of contact itself is varied. Contact can be touching, sliding, rubbing, and rolling, with other variations, such as speed of contact and separation, frequency, contact time, pressure, surface area, and more. Material characteristics, surface finish and condition, and sample preparation and environment are also major considerations. Triboelectric charge generation through friction brings about several problems: abraded particulate and material transfer, hot spots, and surface reactions with the environment.

Additional environmental factors can be involved. Generally, an increase in temperature means more available energy and increased conductivity of insulating materials. Radiation can have an ionizing effect, and at high intensities can even alter insulators into conductors. Recovery time is in terms of half-life of the charge carriers and can be from fractions of a second to days. A permanent degradation that is usually much less than the temporary effects remains after *recovery*. Integrated dosage degradation reported in the literature normally refers to the permanent degradation in relation to the intensity × time product. Polarization occurs frequently, and is an often overlooked variable in evaluation of static control materials. A common polarization effect is strong apparent charging or *surface* charging due to polarization in some insulators. Insulators of greater than

$10^{11}$ $\Omega \cdot$ cm volume resistivity tend to have this propensity for polarization.

## 6.8 Other Electrification Processes

Electrolytic phenomena such as take place in galvanic cells also occur at liquid-solid interfaces. Helmholtz double layers result at metal surfaces and other materials in contact with liquids which normally have high dielectric constants. Spray electrification occurs in aerating, atomizing, or bubbling of liquids, and in separations in high-velocity jets by collisions with solid surfaces. Some crystalline structures will become charged due to the mechanical stresses and possible molecular reorientation caused by applied pressure or heat. These somewhat similar means of static generation are called piezo- and pyroelectrification. When a solid is deformed or fractured, charge generation can also result from reorientation of polar molecules or ionic displacement.

### 6.8.1  Pyro- and piezoelectrification

Certain asymmetric crystals develop a charge due to polarization when exposed to elevated temperature. This pyroelectric effect is present as long as the temperature is maintained. Crystals exposed to pressure can also exhibit an electrostatic potential. The term piezoelectrification is used to denote this type of charge generated as a result of molecular displacement. Exposure to temperature gradients can cause similar internal stresses and electrostatic charging that is identical to the piezoelectric effect. The distinction is that true pyroelectrification is not dependent on a temperature gradient, but simply on elevated temperature.

### 6.8.2  Helmholtz layers

The interface of a liquid and solid generally involves ionization effects called Helmholtz layers. Metal ions of one polarity are believed to be released into the liquid at the interface, and in like manner, some dissociated soluble ions in the liquid, of the opposite polarity, are adsorbed by the solid. Some of the liquid's ionic molecules will become aligned with the field around the metal ions at the interface. The two layers of aligned molecules as shown in Fig. 6.5 are called Helmholtz layers. In this example the polar liquid molecules have aligned themselves around positive ions emitted from the metal, forming the outer layer. The potential across the double layer is called the Helmholtz or

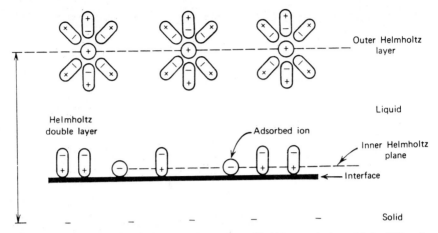

**Figure 6.5** Helmholtz double layers. (*From Moore,[14] with permission of John Wiley & Sons, Inc.*)

zeta potential. Helmoltz layering is very important in the electrification of solids whereby fast removal of the outer layer by moving liquid leaves a charged surface behind. The liquid will carry away an equal amount of opposite charge.

### 6.8.3 Spray electrification

Spray electrification was first investigated by studies of waterfalls. Early studies indicated that fine waterfall sprays were negative whereas the larger droplets near the fall were positively charged. The amount of charging was reduced as salt content increased. Further investigations verified that static generation from disturbances in water occurs only when there is spraying, bubbling, aerating, or otherwise breaking into small droplets. Spray electrification is a result of electrical dipoles inherent in certain liquids at the interface with gases. Such dipoles usually have negative outward polarity and positive inward polarity, although reverse polarities are believed possible.

Symmetrical or homogeneous charge separation can also occur when small solid particles or liquid droplets are separated from larger pieces of the same material. This brings about particles of equal and opposite charges with zero net charge. The amount and polarity of charge is determined by the statistics of available random charges on both sides of the ruptured boundary. This same mechanism is the predominant cause of dust-cloud charging.

### 6.8.4 Metal in solution

A metal immersed in a liquid, such as water, with a high-dielectric constant will tend to go into solution in the form of ions. The nature of dissolved substances present in the liquid, such as acids, bases, or salts, and the characteristics of the metal in solution will either result in positive ions from the metal or complex negative ions in reaction with the solution. This solution continues until the metal acquires a fixed potential. Equilibrium is maintained by continual ionic bombardment on the surface of the metal and the processes of solution. Metals such as zinc, or the alkali atoms, tend to emit positive ions into solution. Other metals react differently depending upon their characteristics and the nature of salts, acids, and/or bases in the solvent. Ions of either polarity, or none at all, may be given off. If the conditions of such a solution are closely controlled, the potentials acquired for specific metals can be accurately defined. The normal calomel electrode is used to standardize dimensions of the metal in solution. Relative rankings of the voltages acquired are called electromotive-force series. An example is given in Table 6.1.

**TABLE 6.1 Electromotive Force Series of Selected Elements (referenced to $H_2$)**

| | | | |
|---|---|---|---|
| $Li_+$ | +3.045 | $Sn^{++}$ | +0.136 |
| $K^+$; $Rb^+$ | +2.925 | $Pb^{++}$ | +0.126 |
| $Ba^{++}$ | +2.90 | $H^+$ | 0 |
| $Ca^{++}$ | +2.87 | $Cu^{++}$ | −0.337 |
| $Na^+$ | +2.714 | $Cu^+$ | −0.521 |
| $Y^{+++}$; $Mg^{++}$ | +2.37 | $Hg^{++}$ | −0.789 |
| $Al^{+++}$ | +1.66 | $Ag^+$ | −0.7991 |
| $Mn^{++}$ | +1.18 | $Pd^{++}$ | −0.987 |
| $S^{--}$ | +0.92 | $Cl^-$ | −1.36 |
| $Zn^{++}$ | +0.763 | $Au^{+++}$ | −1.50 |
| $Cr^{+++}$ | +0.74 | $Au^+$ | −1.68 |
| $Fe^{++}$ | +0.44 | $F^-$ | −2.85 |
| $Ni^{++}$ | +0.25 | | |

Two dissimilar metals in solution will acquire different potentials creating a galvanic cell. When electrical contact is made through a circuit between the two metals, a current results. The current provides more ions to be released at each metal in solution and the battery will function until various degradation mechanisms take place. The process and potentials are distinct from contact or Volta potentials discussed in Sec. 6.6.

### 6.8.5 Flow electrification

Electrification has been a problem with liquids flowing through metal and insulated pipes. The degree is dependent upon the nature of the pipe material, its conditioning, the resistivity and the amount of contamination in the liquid. Two often quoted contributing factors are ungrounded sections of conductive pipe and moisture ($H_2O$) content in the liquid. Fires have reportedly been started in this manner in dry-cleaning establishments. Soap solutions added to the cleaning solvents have been used to increase conductivity and to alleviate the problem. Flow electrification seems to be limited to liquids with volume resistivity greater than $10^9\ \Omega \cdot$ cm, and it becomes a serious concern at $>10^{11}\ \Omega \cdot$ cm.

## 6.9 Polarization

The change in the positioning of charges that occurs when insulating materials are influenced by electric fields is called polarization. Sometimes these displacements due to polarization transpire very slowly, possibly over a period of weeks. At other times it can occur in microseconds or less. The tendency toward depolarization after removal of the field influence can also be variably time-consuming.

Static electrification or polarization of dielectrics in electric fields takes place at the atomic level for elemental dielectrics and at the molecular level for compounds. Sometimes an entire dielectric volume is polarized in the presence of an electric field. The mechanisms of polarization are complex and not well-understood. Very little experimentation supports the speculative descriptions of the process. The relative effect of such changes is often referred to as dielectric absorption.

### 6.9.1 Induced dipoles

Simple inductive charge separation for conductors as discussed in Chap. 4 is the result of movement of free electrons within the conductor. An insulator does not have such free electrons, yet a similar induction effect takes place in the presence of a field. This phenomenon of atomic electrons being displaced toward the anode in an electric field and the nucleus toward the cathode is the result of induced dipoles. The term "induced dipole" infers correctly that the atom will return to its equilibrium position when the field is removed.

A dielectric over a charged surface will become polarized internally and the upper surface will appear to be charged to the same polarity as the underlying surface. If the ambient air has sufficient ionic content to bleed off this charge the surface will eventually appear to be neutral. However, when the positively charged surface below is neutralized or removed, the dielectric will take on a charge of opposite polarity.

#### 6.9.1.1 Alignment of polar and nonpolar molecules.
Certain molecules that are electrically symmetrical with their nucleus, such as hydrogen

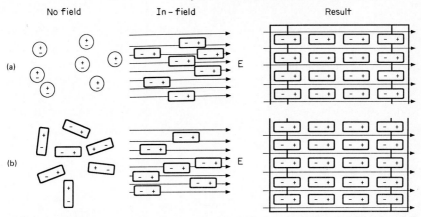

**Figure 6.6** The macroscopic effect of an electric field on (a) nonpolar and (b) polar molecules is identical.

and neon, are nonpolar in the absence of an external electric field. Other molecular structures are such that the electron and nucleus centers are offset. For instance, $H_2O$ has both hydrogen atoms on one side of the oxygen atom. A similar case exists in $N_2O$ with two nitrogen atoms on one side. These molecules are thus polar even in the absence of an electric field. The presence of an electric field has a polarizing effect on nonpolar molecules and aligns both polar and nonpolar molecules in the direction of the field. Figure 6.6 is a representation of nonpolar and polar molecules in the absence and presence of an electric field. From a macroscopic viewpoint the effect is the same: an excess of charges of opposite polarity on either side of the insulator.

**6.9.1.2 Electronic polarization.** A shift in the position of an atom's charges due to an electric field is called electronic ionization. Forces of the applied field on the oppositely charged electrons and protons reach an equilibrium state with a residual polarized effect. The dipoles are the result of shifted electron clouds with respect to the position of the nucleus as indicated in Fig. 6.7a.

**6.9.1.3 Ionic polarization.** The field-induced separation of molecular charge groups of ionic molecules is called ionic polarization. Sodium chloride, with resultant Na(+) and Cl(−) separation, is a good example. The altered separation of bond length between two atoms is sometimes referred to as atomic polarization. Another form occurs in the case of an altered bond angle among three atoms. Ionic polarization is depicted in Fig. 6.7b.

**6.9.1.4 Orientation polarization.** Preexistent dipoles in a polar material that are not in alignment with a field have a torque applied to them. The nucleus shift in these molecules is only partial due to the inertial mass as compared to that of an electron. This results in im-

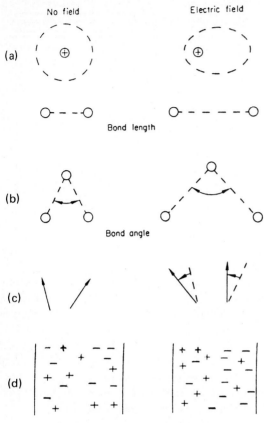

**Figure 6.7** (a) Atomic polarization; (b) ionic polarization; (c) orientation polarization; (d) space charge polarization. (From Jowett,[4] with permission of John Wiley & Sons, Inc.)

perfect alignment with the field lines and the angular displacement is called orientation polarization as shown in Fig. 6.7c. The reaction time can be very slow, not only because of the inertial mass involved but also the crowding effects of neighboring molecules.

**6.9.1.5 Space-charge polarization.** Randomly located residual charges in an insulative material after forming are referred to as space charges. These random charges can also result from radiation or thermal degradation. Polarization of a space charge is said to have occurred when a field causes a displacement of the residual charges as depicted in Fig. 6.7d.

### 6.9.2 Polarization effects

The response times of polarization can vary from nanoseconds to years. This relates directly to the elapsed time required to realize the full effects on some of the parameters mentioned in this section. The sum total of such charges in a dielectric as a result of an applied electric field is sometimes referred to as the applied electric field "dielectric absorption" characteristic $D_A$.

#### 6.9.2.1 Dielectric constant variations.
The frequency response due to dipoles in a capacitor's dielectric can be interpreted as changes in the dielectric constant and dissipation factors. This results from dipole movement in response to the alternating signal, especially at low frequencies. The result is manifested as a high-dielectric constant. If the frequency is gradually increased, a point of resonance is reached where maximum energy is absorbed. This relates to a decreased dielectric constant and a peak in the dissipation factor. Above the resonance rate there is no further contribution from the dipoles.

#### 6.9.2.2 Effects on resistivity.
Perhaps the most important effect of polarization as it relates to ESD control are changes in volume resistivity. In the case of an initially neutralized insulator, the effect will result in an asymptotic increase as shown in Fig. 6.8. The value and the rate of increase is determined by such things as the dielectric absorption constant of the material, temperature, and moisture content, and possibly other material characteristics. Note that an initially low volume resistivity material can exceed an initially higher resistivity material with time if the dielectric absorption characteristic is excessive. Since time is required for depolarization, it is hypothesized that a previously polarized insulator might behave as indicated in Fig. 6.9. This

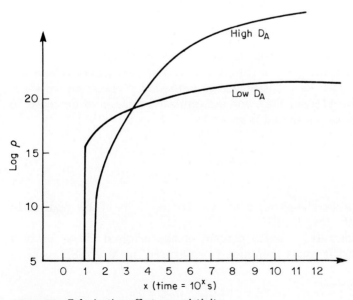

**Figure 6.8** Polarization effect on resistivity.

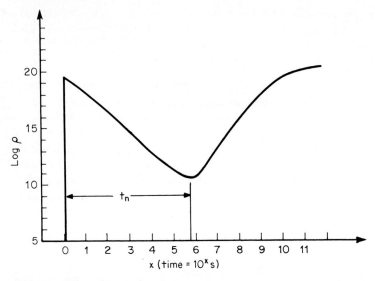

**Figure 6.9** Effect of prepolarized material on resistivity.

points out the importance of long duration measurements to get actual resistivity, which is seldom done. Short-term resistivity measurements can be very misleading.

### 6.10 Relaxation Time and Decay Time

Just as important as the accumulation rate or magnitude of charge during electrification is the rate of decay. The variable of relaxation time is often used to quantify charge reduction. Relaxation time $\tau$ is defined as the elapsed time for accumulated charge to decrease to $1/e \times$ its original value and is given by

$$\tau = \frac{\varepsilon}{\sigma} \qquad (6.7)$$

where $\varepsilon$ and $\sigma$ are the permittivity and conductivity of the medium in the decay path.

Static voltage decay to 10 percent of the original value (or to a fixed voltage) is more commonly measured in the ESD control field.

## 6.11 Discharging Effects of Points

If a highly charged conductive cone is distorted until it becomes very pointed, the point will eventually begin to ionize the air. An ion is an atom or molecule having a net deficiency or excess of electrons. In the case of the charged point, the air or gas molecules will begin to lose or gain electrons to the point depending on the polarity of the charge on the point. These gas ions are now opposite in polarity to the point and are repelled away from the point by the electric field of the charged point. The movement of such ions causes collisions and consequential movement of gas molecules as depicted in Fig. 6.10. This movement is called an "electrical wind" or "ion wind" even though the overall effect is largely due to bulk movement of the gas involved in addition to the ions. A result of the continued electron transfer is a rapid loss of charge or neutralization of the point.

**Figure 6.10**  Electrical or ion wind.

Bringing a grounded needle near a charged electroscope, as shown in Fig. 6.11, will illustrate this phenomenon. The needle point initially takes on a polarity opposite that of the electroscope by induction. The oppositely charged point now causes electron transfer until the electroscope is neutralized.

### 6.11.1 Corona

A highly charged metal point will readily exhibit corona breakdown if placed near a large conductive surface. Corona can be considered a low-power arc, or ionization that does not result in complete breakdown of the insulative part of the path. Voltage breakdown, on the other hand, would be considered a high-power arc. Corona can initiate degradation mechanisms that start with high resistance and degrade to low resistance and finally breakdown. This often occurs in voids or by corona cutting action on the surface of insulators.

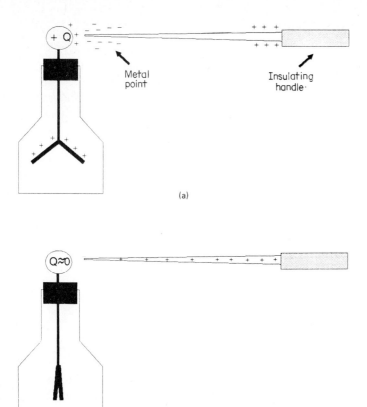

**Figure 6.11** Neutralization of an electroscope by electron wind. (*a*) An uncharged sharp needle point brought near a positively charged electroscope takes on negative charge through inductive charge separation; (*b*) a subsequent electron wind results in neutralization of the electroscope and a residual positive charge on the needle.

#### 6.11.1.1 Corona from insulative materials.
Brush discharges or corona discharges from insulative materials sometimes occurs. In particular, the likelihood of such discharges is enhanced by the presence of sharp conductive points, especially if grounded.

### 6.12 Sparks

Complete electrical breakdown of a gas across a gap between electrodes occurs at potentials exceeding those which bring about the less abrupt corona effect. Generally, a spark consumes much energy and is associated with an audible *crack*.

### 6.12.1 Sparkover of conductive points

Gas ions near a charged sharp conductive geometrical point are subject to strong forces of repulsion or attraction depending upon the polarities involved. Such ions can acquire velocities sufficiently large that impact with other gas molecules can cause electron separation and thus new ions. Two ions are formed with each such collision. The newly created ions are subject to the same field forces and cause new ions upon further collisions. In actuality the process occurs very rapidly and the gas around the sharp point becomes conductive because of the many ions present. Electrons then move rapidly through this ionized path neutralizing the charge on the pointed object in less than a microsecond. This electron movement is seen as spark. Air is heated and expanded, then cooled and contracted, which results in a crack similar to a whiplash.

### 6.12.2 Sparks from insulative materials

Experiments with cotton, terylene and cotton blends, and nylon have shown that all these materials can cause a spark discharge sufficient to ignite certain gases at low humidities. Nylon can cause such ignition at higher humidities. Simple resistivity tests are inferred as a reasonable safety guide for homogeneous materials. Jowett[4] suggests $10^{11}$ Ω per square as the limit. Where personnel safety is an issue, there should be no compromise. So many variables are involved with using marginally safe materials that only approved well-documented spark-free materials should be used.

### 6.12.3 Air breakdown on insulator surfaces

An insulator in the presence of air presents a special case worthy of discussion that frequently occurs in the world of electronics such as depicted in Fig. 6.12. An electric field in the longitudinal direction means that the air is a parallel insulator. Consider the fact that a certain amount of moisture will be absorbed onto the solid insulator dependent upon humidity and characteristics of the material. The moisture film provides a somewhat semiconductive layer also in parallel. Leakage current in this film might cause sufficient heat to cause local evaporation leaving dry spots with good insulating properties. When electric potential across these spots exceeds the breakdown potential of air, an arc will occur. This increases the dry spot length

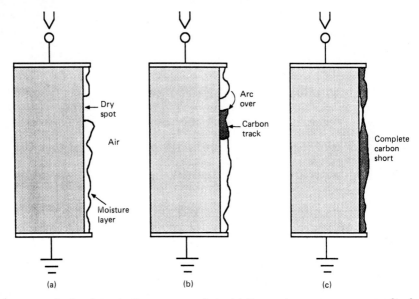

**Figure 6.12** An insulator in the presence of air. (*a*) Dry spots can occur as a result of leakage current heating in the absorbed moisture layer; (*b*) arcs across a dry spot can increase its length and deposit carbon; (*c*) continual arcing can result in a carbon track short across the insulator.

and continued arcs can leave a conductive carbon track along the length of the insulator. Some materials do not produce carbon byproducts and are better for high-voltage applications.

### 6.12.4 Paschen's law

Paschen's law states that the potential required to cause sparkover in a gas is determined by the product of the pressure and the spark length in a uniform field. The relationship for air is shown in Fig. 6.13. Other gasses show similar curves with varying minimum locations and slopes. Pressure is in torr · cm. Gaseous discharge involves electron release, collisions, further release, recombinations, and ionization.

## 6.13 Breakdown in Fluids

Contaminants are a major factor in voltage breakdown of liquids. Field forces tend to draw contaminants to the highest stress areas where they contribute to arcing. The arcing in turn tends to decompose the contaminants and the liquid into other constituents that

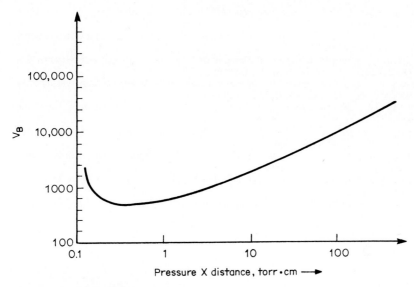

**Figure 6.13** Paschen's law.

could be construed as an increase of contaminants leading to further deterioration.

## 6.14 Lightning

Lightning flashes occur between clouds as well as between clouds and earth, in a manner similar to that described in the preceding paragraph. Grounded lightning rods become charged inductively to a charge opposite the clouds above, then attract a lightning bolt to them which is conveyed to the ground through an intended path bypassing the remainder of the building protected in this manner.

## References

1. Leonard B. Loeb, *Static Electrification,* Springer-Verlag, Berlin-Gottingen-Heidelberg, 1958.
2. D. K. Davies, "Charge Generation on Dielectric Surfaces," *Journal of Applied Physics,* vol. 2, 1969, pp. 1533–1537.
3. W. D. Greason and I. I. Inculet, "Insulator Work Function Determination from Contact Charging with Metals," *Proceedings of IEEE IAS Annual Meeting,* 1975, pp. 428–435.
4. Charles E. Jowett, *Electrostatics in the Electronics Environment,* John Wiley & Sons, Inc., New York, 1976.
5. A. S. Grove, *Physics and Technology of Semiconductor Devices,* John Wiley & Sons, Inc., New York, 1967.

6. J. A. Cross, *Electrostatics: Principles, Problems, and Applications*, Adam Hilger, Bristol, England, 1987.
7. Charles Kittel, *Elementary Solid State Physics: A Short Course*, John Wiley & Sons, Inc., New York, 1962.
8. William D. Greason, *Electrostatic Damage in Electronics: Devices and Systems*, Research Studies Press, Latchworth, England, 1987.
9. Francis W. Sears and Mark W. Zemansky, *University Physics*, Addison-Wesley Publishing Company, Cambridge, Mass., 1957.
10. Arthur W. Adamson, *Physical Chemistry of Surfaces*, John Wiley & Sons, Inc., New York, 1976.
11. Harry C. Gatos, *The Surface Chemistry of Metals and Semiconductors*, John Wiley & Sons, Inc., New York, 1959.
12. Henry Semat, *Introduction to Atomic and Nuclear Physics*, Holt, Rinehart and Winston, Inc., New York, 1962.
13. Charles Kittel, *Introduction to Solid State Physics*, John Wiley & Sons, Inc., New York, 1967.
14. A. D. Moore, *Electrostatics and its Applications*, John Wiley & Sons, Inc., New York, 1973.

Chapter

# 7

# Nature of Static Damage

## 7.1 Characteristic Traits of ESD Failures

In a 1970 technical paper[1] entitled "Characteristic Traits of Semiconductor Failures," coauthor W. J. Lytle and I discussed the resultant physical features on failed electronic devices. The topic was the burnout and diffusion results of thermal and electrothermal stresses on devices. The theme was that by observance of the microscopic residual traits on a failed device, the failure analyst might deduce the nature of any transient overstress involved was the theme. The reader is cautioned that interpretation of these traits is an acquired skill and that other factors such as failure symptoms and circumstantial evidence of transient sources must be taken into consideration. Nevertheless, certain features can be used to help distinguish ESD failures from other causes. The predominate ones are discussed in this chapter along with selected examples of static damaged parts. This chapter is not intended to be inclusive of all aspects of ESD damage as certain topics are more appropriately addressed in Chaps. 8 through 13.

## 7.2 Mathematical Models of Failure Mechanisms

Models have been proposed for many of the mechanisms involved with semiconductor, oxide, and metallization structures. Researchers continually question and modify the models as new data is obtained. Selected models supported by experimental data are included in this chapter for the primary mechanisms associated with ESD.

## 7.3 MOS Damage

In thermal oxide a field strength of about 8 to $10 \times 10^6$ V/cm brings about breakdown. This is about 80 to 100 V for the typical 1000

A gate thickness. Often this breakdown will occur at lesser field strengths due to imperfections or defects in the oxide. The breakdown of silicon nitride, $Si_3N_4$, is about the same, around $9 \times 10^6$ V/cm. However, a higher dielectric constant means that thicker gates are used for silicon nitride resulting in higher breakdown voltages. Silicon nitride is also more conductive allowing a part of the transient to bleed off rather than contribute to breakdown.

### 7.3.1  Adiabatic nature of oxide ruptures

Oxide failure, like most mechanisms of interest herein, are basically thermal in nature. This fact is often overlooked because of the extremely low energy levels required for damage. The damaging energy level of much less than 1 µJ coupled with the extremely short transient times to failures are believed to be due to the poor thermal conductivity of the oxide. Thus thermal behavior is reported[2] to be adiabatic (with no appreciable heat loss). The relationship supported by data[3] is

$$\text{Power} = P = K_1 t^{-1} \tag{7.1}$$

where $K_1$ depends upon the oxide material, thickness, breakdown potential, and damage energy level.

### 7.3.2  Characteristic traits of ESD failures of MOS devices

Typically, an MOS failure from ESD results in a fracture of the gate oxide due to exceeding the breakdown voltage. High currents through the breakdown site causes localized heating and usually results in a metal-silicon alloy through the gate-fracture site. The end result is a resistive short across the gate. The short can be gate to source, drain, or substrate, depending on structure and imperfections present in the oxide.

The cross-sectional structure of a simple unprotected discrete MOS transistor destroyed by ESD is shown in Fig. 7.1. Depending upon the polarity of the transient and the biasing of the device, either the source or drain will be at maximum potential difference with respect to the gate. In many cases the substrate is tied in common with the drain, but the source and drain regions are also of lower resistance. Therefore the most likely site of ESD damage would be expected at the source or drain region. In some cases ESD damage will occur from gate to substrate. This is indicative of preexistent oxide defects, or geometrical or dopant irregularities resulting in a preferential weaker path.

Nature of Static Damage    127

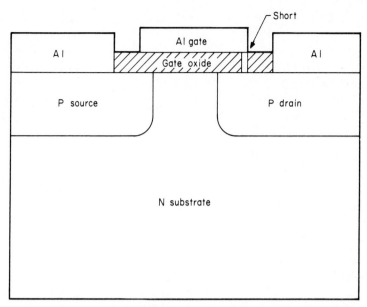

**Figure 7.1** The cross-sectional structure of a simple unprotected discrete-MOS transistor destroyed by ESD.

Gate oxide shorts are most likely to occur in defect-free oxides at the corners (1), or at least at the edges (2), of the source or drain regions under the gate metal as shown in Fig. 7.2. The electric field would be expected to be higher at such locations. In practice, occurrences at the metal edges are also common. Typically, P-channel devices have thinner oxides so they are more sensitive than N-channel devices.

Human body simulators and other models are discussed in detail in Chap. 8. The present standard model used for part susceptibility clas-

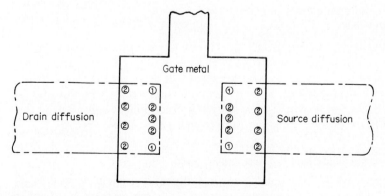

**Figure 7.2** In defect-free oxides gate shorts are most likely to occur at the corners (1), or edges (2), of the source or drain regions.

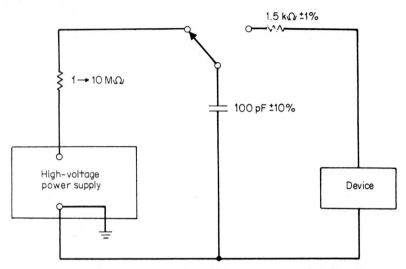

**Figure 7.3** The present standard model used for part susceptibility classification.

sification is shown in Fig. 7.3. Figure 7.4 shows a typical ESD-damaged discrete MOS device, in this case a 3N128, exposed to 150 V from a standard human body simulator circuit. This device had a gate to source leakage of 24 μA, well in excess of the 50 pA limit. Upon microscopic examination after opening the device TO-5 can, no visible trait is detected until the gate metallization is removed. The tiny defect site is located over the source region. Note the slight hump of what appears to have been molten material. This is an aluminum silicon alloy resulting from the eutectic temperature reached by $i^2R$ heating during the transient.

### 7.3.3 Characteristic traits of EOS failures of MOS devices

An electrical overstress (EOS) failure other than ESD is usually of a longer time duration, generally greater than 50 μs. Typical EOS sources are 110 V ac at 60 cycles, accidental short to a dc or ac potential exceeding the gate-oxide breakdown or system transients of varying durations. In these cases the heating is normally sustained longer than during the typical ESD exposure, thus resulting in more extensive damage.

### 7.3.4 Characteristic traits of processing faults

Oxide defects from poor process controls by device manufacturers can result in pinhole failures of the type shown in Fig. 7.5. Such a defect can produce an immediate resistive short, but sometimes occurs dur-

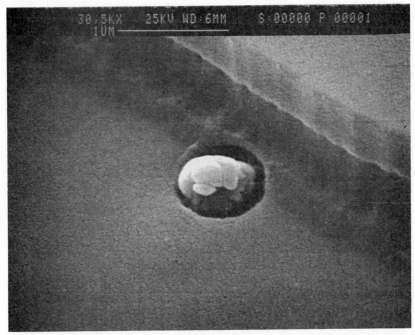

**Figure 7.4** A typical ESD-damaged unprotected discrete MOS device, in this case a 3N128, exposed to 150 V from a standard human body simulator circuit.

ing normal bias conditions. Severe defects will fail early but some are time-dependent decreasing reliability in device user's end product. Clues to this defect are the absence of the alloy at the defect site associated with ESD and EOS failures.

### 7.3.5 Other MOS failure mechanisms

There are other more subtle failure mechanisms in MOS devices that can result from ESD. Typically the symptom is a shift in the gate turn-on or turn-off threshold voltage. Microscopically visible traits are either nonexistent or very slight source or drain to substrate junction damage. Charge trapping in the oxide can also be involved in threshold shifts.

### 7.3.6 Latchup

A notorious problem with complementary MOS (CMOS) circuitry is a latchup behavior due to inherent PNPN (or reverse) type paths. Resultant high-current induced damage or degradation depends upon the degree of supply current limiting. Typical results include damaged junction sites and/or open metallization lines. Figure 7.6 shows the cross section of a CMOS circuit with the associated parasitic bipolar

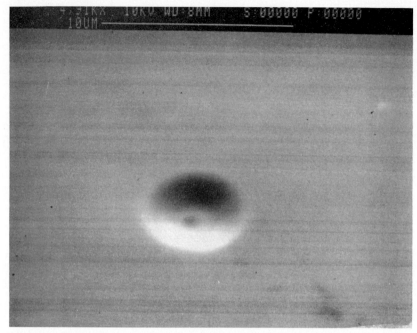

**Figure 7.5** Oxide defects resulting from poor process controls by device manufacturers can result in pinhole failures of the type shown.

circuitry. When the gain product of the lateral and vertical parasitic transistors exceeds unity, latchup occurs. The vertical P+-to-N-to-P transistor and the lateral N+-to-P-to-N transistor act as a P-to-N-to-P-to-N structure. Variables affecting latchup tendency include physical layout and injection current or applied voltage. Increased complexities of Very Large-Scale Integration (VLSI) have an increased tendency toward latchup.

## 7.4  Junction Damage

### 7.4.1  Characteristic traits of ESD-caused junction failures

The most prevalent PN junction damage from ESD occurs in the reverse biased condition and is evidenced by a degraded I-V characteristic. Degradation can be from a nearly negligible shift of the curve to a short circuit. Physical damage is invisible upon microscopic exami-

Nature of Static Damage    131

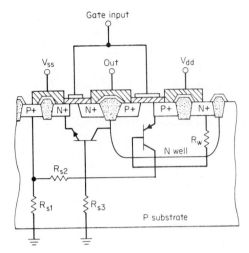

**Figure 7.6** Cross section of a CMOS circuit with the associated parasitic bipolar circuitry involved in latchup.

nation of the chip surface, except in the more severe cases. Figure 7.7 shows a typical severely ESD-damaged PN junction. This example is an input base emitter junction of an LM148 operational amplifier after exposure to 4000 V from a standard human body simulator. The fine *cracked-glass* feature across the junction has become widely ac-

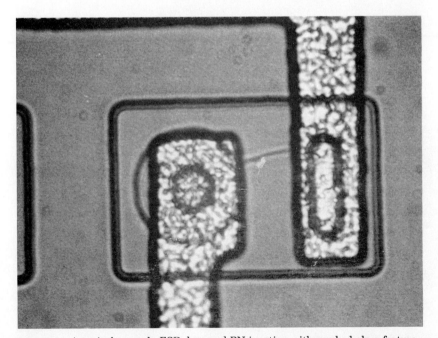

**Figure 7.7** A typical severely ESD-damaged PN junction with cracked-glass feature.

cepted as a common result of ESD. The trait is often observed at inputs of operational amplifiers as well as other PN junctions. Other more subtle damage to bipolar junctions are discussed in Chap. 10.

### 7.4.2 Characteristic traits of EOS-caused junction failures

Ordinarily EOS transients are of longer duration than ESD as discussed in Sec. 7.3.3. EOS-caused failures can be of either forward or reverse bias. The "white spear",[1] which is an aluminum-silicon alloy across the surface of the junction as shown in Fig. 7.8, is usually indicative of reverse-bias EOS. The transistor transistor logic (TTL), hex gate input in this example was exposed to 17 V for 200 µs. The peak current was 300 mA.

Forward bias EOS is typically evidenced by resultant damage of high currents such as melting and/or vaporization of intraconnects. An uncommon occurrence is the *white-spear* feature produced by high-voltage EOS transients in the forward-bias direction. Associated signs of high-current melting of intraconnects as seen in Fig. 7.9 are indicative of a

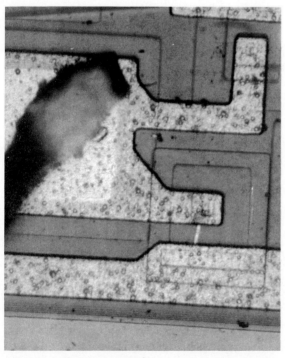

**Figure 7.8** The white spear,[1] an aluminum-silicon alloy across the surface of the junction, is usually indicative of reverse-bias EOS.

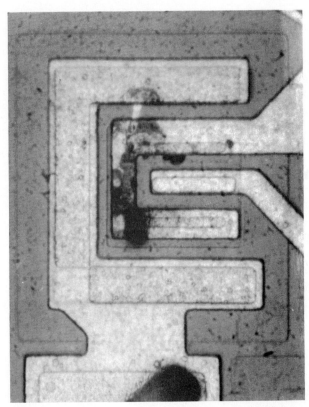

**Figure 7.9** Associated signs of high-current melting of intraconnects are indicative of a forward-biased transient.

forward-biased transient. In this example the TTL hex 2 input gate was exposed to 14 V for 200 μs with a peak current of 500 mA.

### 7.4.3 Wunsch-Bell model

Damage occurs because of localized temperature increase during avalanche breakdown. During this condition the conductivity of the least-doped polar side of the junction is believed to become intrinsic and a negative resistance region occurs. Resistance continues to decrease with increased temperature, so current is crowded into the localized hot-spot area resulting in thermal runaway. The tiny cross section involved will finally reach the melting point of silicon if sufficient energy is available. This mechanism has been modeled by Wunsch and Bell[4] as follows:

$$\frac{P}{A} = \sqrt{\pi k C_p}(T_m - T_i)t^{-1/2} \approx 1.81 \times 10^3 t^{-1/2} \quad (7.2)$$

where P/A = power density or power per unit junction area
  $k$ = thermal conductivity
  $\rho$ = density
  $C_p$ = specific heat
  $T_m$ = melting point of silicon
  $T_i$ = starting temperature
  $t$ = time in seconds

The above equation is sometimes written as

$$\frac{P}{A} = kt^{-1/2}$$

The equation holds for materials in a quasiadiabatic region. For silicon, this occurs at about 100 ns. A plot of power density versus time to raise silicone from 25°C to 1415°C is shown in Fig. 7.10. Another factor that can be derived from the equation is the starting temperature effect. Obviously, the higher the $T_i$ the lower the damage threshold. This fact has been observed experimentally as well.

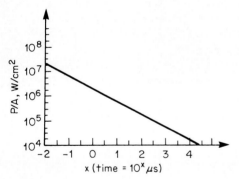

Figure 7.10 Wunsch-Bell plot of power density versus time to raise silicone from 25°C to 1415°C.

Generally reverse-bias damage occurs more easily than forward-bias damage because of the order of magnitude difference in voltage or correspondingly higher ratio of current required to reach damaging power densities. Emitter-base junctions are the smallest and subject to the highest power densities so they are most likely to be damaged.

The extended equation is

$$\frac{P}{A} = k_1 t_m^{-1} + k_2 t_m^{1/2} + k \qquad (7.3)$$

The first term is for short pulse widths in the adiabatic range. The second term is the intermediate duration transients where some heat dis-

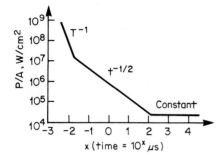

**Figure 7.11** Wunsch-Bell generalized plot of these conditions for silicon with three regions.

sipation occurs. The last term $k$ is for steady-state conditions where failure becomes independent of pulse width. Figure 7.11 shows a plot of these conditions for silicon.

### 7.4.4 Speakman model

Electrostatic discharge through a bipolar junction consists of a charged capacitor discharging through a resistance. For a human body discharge the resistance is the body resistance plus contact resistance plus resistance of the device. Usually any parasitic $R$, $L$, or $C$ components present are insignificant and can be ignored. Figure 7.12 shows the circuit diagram for a human body static discharge through a PN junction in the reverse-bias direction. $C_b$ and $R_b$ are the human body capacitance and resistance. $V_d$ is the voltage across the device, in this example the reverse-breakdown voltage. $R_d$ is resistance across the junction. $R_c$ is the contact resistance to ground. For purposes of illustration, $R_c$ is considered negligible. The parameters of $V_d$ and particularly $R_d$ are complex and dynamic during a transient condition. For simplification they are considered as constant in illustrating the Speakman[5] model.

The discharge current waveform for this $RC$ circuit is a decaying exponential as shown in Fig. 7.13. The current $i(t)$ is

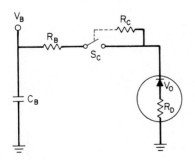

**Figure 7.12** Circuit diagram for a human body static discharge through a PN junction in the reverse-bias direction.

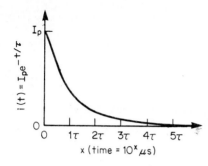

**Figure 7.13** Discharge current waveform for this $RC$ circuit is a decaying exponential.

$$i(t) = I_p e^{-t/\tau} \tag{7.4}$$

where

$$I_p = \text{peak current} = \frac{V_b - V_d}{R_b + R_d + R_c} \tag{7.5}$$

and

$$\tau = \text{total resistance} \times \text{total capacitance}$$
$$= (R_b + R_d + R_c)C_b \tag{7.6}$$

The instantaneous power is

$$P(t) = V_d \times i(t) + R_b \times i^2(t) \tag{7.7}$$

Substituting Eq. (7.4) for $i(t)$,

$$P(t) = V_d \times I_p e^{-t/\tau} + R_b I_p^2 \times e^{-2t/\tau} \tag{7.8}$$

Approximately 99 percent of the power is dissipated in five time constants. The Speakman model is based upon the average power during that period calculated by integrating $P(t)$ over five time constants.

$$P_{AV} = \frac{1}{5\tau} \int_0^{5\tau} V_d I_p e^{-t/\tau} dt + \frac{1}{5\tau} \int_0^{5\tau} R_b I_p^2 e^{-2t/\tau} dt \tag{7.9}$$

$$P_{AV} = \frac{V_D I_P}{5}(1 - e^{-1/2}) + \frac{R_b I_p}{10}(1 - e^{-10}) \tag{7.10}$$

where $P_{AV}$ is the average power dissipated during five time constants. Since $e^{-5}$ and $e^{-10}$ are negligible,

$$P_{AV} = \frac{V_d I_p}{5} + \frac{R_b I_p}{10} \tag{7.11}$$

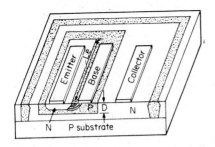

**Figure 7.14** ESD transient current is perpendicular to the area defined by the depth of the base region $D_b$ times the emitter length $L_e$.

In order to investigate correlation with the Wunsch-Bell model a conversion to power density must be made. The ESD transient current is perpendicular to the area defined by the depth of the base region ($D_b$) times the emitter length ($L_e$) depicted in Fig. 7.14.

$$A = D_b \times L_e \tag{7.12}$$

Power density can then be calculated by dividing Eq. (7.11) by Eq. (7.12).

$$P_d = \frac{P_{AV}}{A} = (5 \times D_b \times L_e)^{-1}\left(V_d I_p + R_b \frac{I_p^2}{2}\right) \tag{7.13}$$

**7.4.4.1 Sample calculation.** A sample calculation for discharge of a person at a 5000-V potential through a representative PN junction shows that high power densities are likely during ESD. The emitter length and base junction depth are assumed to be $6.1 \times 10^{-3}$ cm and $0.26 \times 10^{-3}$ cm, respectively. Thus the area is

$$A = 6.1 \times 10^{-3} \text{ cm} \times 0.26 \times 10^{-3} \text{ cm} = 1.6 \times 10^{-6} \text{ cm}^2$$

$V_d$ is 10 V. The base contact is $0.76 \times 10^{-3}$ cm from the emitter junction. The device resistance is the base-region resistance calculated from

$$R_d = \text{base-sheet resistance} = (\Omega \text{ per square}) \times \text{number of squares} \tag{7.14}$$

$$R_d = \text{base-sheet resistance} \times \frac{b - e \text{ separation}}{\text{emitter length}} \tag{7.15}$$

For our example,

$$R_d = 200 \times \frac{\Omega}{\text{square}} \times \frac{0.76}{6.1} \text{squares} = 25 \text{ } \Omega$$

The human body capacitance and resistance are assumed to be 100 pF and 1.5 k$\Omega$, respectively. The following can be derived:

$$\tau = C_b(R_b + R_d) = 100 \times 10^{-12} \times 1.525 \times 10^3 = 152.5 \text{ ns}$$

$$5\tau = 762.5 \text{ ns}$$

$$I_p = \frac{V_0 - V_d}{R_b + R_d} = \frac{5000 - 10}{1525} = 3.27 \text{ A}$$

$$P_{AV} = \frac{V_d I_p}{5} + \frac{R_d I_p^2 2}{10} = \frac{10 \times 3.27}{5} + \frac{25(3.27)^2}{10}$$

$$= 6.54 + 26.73 = 33.27 \text{ W}$$

$$P_d = \frac{P_{AV}}{A} = \frac{33.27 \text{ W}}{1.6 \times 10^{-6}} \text{ cm}^2 = 20.79 \frac{\text{MW}}{\text{cm}^2}$$

With this degree of power density there should be little wonder that a PN junction can be damaged by ESD.

**7.4.4.2 Correlation with Wunsch-Bell model.** Table 7.1 lists some of Speakman's[5] experimental results for different conditions that fit the Wunsch-Bell plot as shown in Fig. 7.15. The beauty of this fit is that with the appropriate constants known, the static threshold to failure should be predictable. One needs to know

1. $V_d$, the reverse breakdown voltage
2. $R_d$, the junction resistance, which usually can be measured via slope of the I-V characteristic or calculated as in Eq. (7.13)
3. $A$, the cross-sectional area of the current path
4. The $C_b$ and $R_b$ parameters of the human body simulator circuit

Then, one can

**TABLE 7.1 Selected Speakman Experimental Results that Were Found to Fit the Wunsch-Bell Curve Extended (Fig. 7.15).**
The damaged junction of device A had an area of $1.6 \times 10^{-6}$ cm$^2$, a reverse breakdown of 7.5 V, and a bulk resistance of 25 Ω. The damaged junction of device B had an area of $0.95 \times 10^{-6}$ cm$^2$, a reverse breakdown of 7.5 V, and a bulk resistance of 70 Ω.

| Device | Experimental model | | | Calculated parameters | | |
|---|---|---|---|---|---|---|
| | $C$ (pF) | $R$ (Ω) | $V$ (volts) | $5\tau$ (µs) | $I_p$ (A) | $P/A$ (kW/cm$^2$) |
| A | 220 | 300 | 350 | 0.357 | 1.05 | 2700 |
| " | 100 | 300 | 525 | 0.163 | 1.59 | 5400 |
| " | 30 | 330 | 500 | 0.053 | 1.39 | 4300 |
| " | 27 | 300 | 575 | 0.044 | 1.75 | 6300 |
| " | 27 | 150 | 525 | 0.024 | 2.87 | 16000 |
| B | 1000 | 1000 | 250 | 5.35 | 0.23 | 750 |
| " | 30 | 710 | 500 | 0.117 | 0.63 | 3900 |
| " | 30 | 300 | 350 | 0.056 | 0.93 | 7600 |

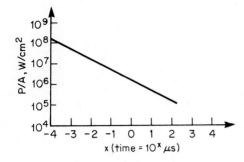

**Figure 7.15** Wunsch-Bell plot extended to include subnanosecond times.

1. Calculate $5\tau$ using Eq. (7.4)
2. Refer to the Wunsch-Bell plot for the damage power density $P_d$
3. Multiply $P_d$ by the area $A$ to determine $P_{AV}$
4. Calculate $I_p$ using Eq. (7.9)
5. Then calculate the threshold voltage $V_t$ that would produce $I_p$ using Eq. (7.3)

## 7.5 Metallization Damage

Integrated-circuit intraconnect metallization damage usually means an open circuit from melting and vaporization. In this chapter the meaning is extended to include intermetallic shorts and for discussion, melted and reflowed metallization stripes.

### 7.5.1 Open metallization

#### 7.5.1.1 Experimental proof of adiabatic conditions during short pulses.
From past research on metallization stripes, I have reported[6] data verifying adiabatic behavior during square-pulse transients as long as 824 ns. That is, for such short transient conditions there is no heat transfer loss. When a metal stripe heats under such conditions the following relationships hold:

$$J^2 t_m = K_m \qquad (7.16)$$

and

$$J^2 t_0 = K_0 \qquad (7.17)$$

where $J$ = current density
$t_m$ = time to melt
$t_0$ = time to open

and both $K_m$ and $K_0$ are constants.

TABLE 7.2  Resistivity of Aluminum

| Temperature, °C | $\rho$, $\mu\Omega \cdot$ cm |
|---|---|
| 0 | 2.42 |
| 20 | 2.65 |
| 100 | 3.50 |
| 200 | 4.62 |
| 300 | 5.81 |
| 400 | 7.05 |
| 500 | 8.36 |
| 600 | 9.77 |
| 660(Solid) | 10.95 |
| 660(Liquid) | 24.2 |

The time to melt or open a defect-free metal stripe can be determined by careful analysis of the voltage waveform across the stripe during the transient condition. Table 7.2 lists the volume resistivity of aluminum as a function of temperature. The voltage waveform will increase as the stripe heats up under constant current conditions. A discontinuity exists as the phase changes from solid to liquid, and at the phase change from liquid to vapor. A typical voltage waveform is shown in Fig. 7.16. The first discontinuity, where the metal changes from solid to liquid, is taken as the time to melt. The second discontinuity, where the metal changes from liquid to vapor, is taken as the time to open. There is an indeterminate period of conduction in the vapor state prior to open circuit, but once vaporization takes place the stripe is destined to become open circuit.

Pulse data taken for 140 constant area (step-free) stripes of 10.7 kÅ thickness and varying widths is given in Table 7.3. The stripes were "second metal" test patterns over silicon dioxide and thermal oxide as shown in Fig. 7.17. Analysis of the data resulted in calculated values of $5.18 \times 10^8 \pm 7.3$ percent and $1.03 \times 10^9 \pm 12.6$ percent for $K_m$ and $K_0$, respectively. Thus the pulse conditions proved to produce adia-

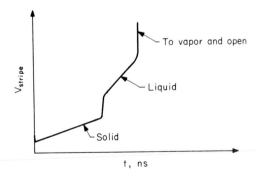

Figure 7.16  A typical voltage waveform across an adiabatically pulsed metal stripe indicates the solid-to-liquid and liquid-to-vapor phase changes.

**TABLE 7.3** Pulse data taken for 140 constant area (step-free) stripes of 10.7 kA thickness and varying widths.

| Nominal line width ($\mu m$) | Measured width ($\mu m$) | Area ($cm^2$) | $I_{melt}$ (A) | $J_{melt}$ (A/cm$^2$) | $t_m$ (nsec) | $t_o$ (nsec) | $K_m$ (Constant) | $I_{(open)}$ (A) | $K_o$ (Constant) |
|---|---|---|---|---|---|---|---|---|---|
| 5 | 2.9$^{(2)}$ | 3.1 × 10$^{-8}$ | 2.76 | 8.9 × 10$^7$ | 70.00 | 160.00 | 5.55 × 10$^8$ | 2.46 | 1.01 × 10$^8$ |
| 7 | 5.0$^{(2)}$ | 5.35 × 10$^{-8}$ | 2.81 | 5.2 × 10$^7$ | 189.60 | 432.14 | 5.23 × 10$^8$ | 2.68 | 1.08 × 10$^8$ |
| 8 | 6.0$^{(2)}$ | 6.42 × 10$^{-8}$ | 2.82 | 4.39 × 10$^7$ | 269.69 | 577.50 | 5.2 × 10$^8$ | 2.69 | 1.01 × 10$^8$ |
| 9 | 7.1$^{(1)}$ | 7.6 × 10$^{-8}$ | 2.82 | 3.7 × 10$^7$ | 353.12 | 660.00* | 4.87 × 10$^8$ | 2.72 | 0.845 × 10$^{9*}$ |
| 10 | 8.05$^{(2)}$ | 8.6 × 10$^{-8}$ | 2.82 | 3.28 × 10$^7$ | 455.83 | N/A | 4.92 × 10$^8$ | N/A | N/A |
| 13 | 11.6$^{(2)}$ | 12.4 × 10$^{-8}$ | 2.83 | 2.28 × 10$^7$ | > 500 | N/A | N/A 10$^8$ | N/A | N/A |
| 5 | 2.9$^{(3)}$ | 3.1 × 10$^{-8}$ | 4.13 | 1.33 × 10$^7$ | 26.9 | 67.0 | 4.76 × 10$^8$ | 3.65 | 9.34 × 10$^9$ |
| 7 | 5.0$^{(3)}$ | 5.35 × 10$^{-8}$ | 4.19 | 7.8 × 10$^7$ | 86.0 | 200.5 | 5.27 × 10$^8$ | 3.90 | 1.06 × 10$^9$ |
| 8 | 6.0$^{(3)}$ | 6.42 × 10$^{-8}$ | 4.23 | 6.59 × 10$^7$ | 129.5 | 296.0 | 5.62 × 10$^8$ | 4.04 | 1.17 × 10$^9$ |
| 9 | 7.1$^{(3)}$ | 7.6 × 10$^{-8}$ | 4.23 | 7.6 × 10$^7$ | 181.0 | 279.5 | 5.61 × 10$^8$ | 4.06 | 1.06 × 10$^9$ |
| 10 | 8.05$^{(3)}$ | 8.6 × 10$^{-8}$ | 4.24 | 4.93 × 10$^7$ | 215.5 | 470.5 | 5.24 × 10$^8$ | 4.09 | 1.06 × 10$^9$ |
| 13 | 11.6$^{(3)}$ | 12.4 × 10$^{-8}$ | 4.24 | 3.42 × 10$^7$ | 414.5 | 824.0 | 4.85 × 10$^8$ | 4.12 | 0.906 × 10$^9$ |

(1) Single data point
(2) Data averaged from 16 stripes of listed width
(3) Data averaged from 10 stripes of listed width

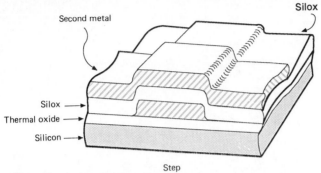

**Figure 7.17** The stripes were second metal test patterns over silicon dioxide and thermal oxide as shown.

batic behavior as anticipated. Similar pulsing in the range of $t_m \geq 1.5$ μs resulted in nonadiabatic conditions; that is, a constant $K_m$ could not be obtained from the data representing 60 stripes.

**7.5.1.2 Theoretical calculation of $K_m$ for comparison.** In order to calculate an approximate value of $K_m$ consider that[7]

$$\text{Energy density} = \delta T \times \text{heat capacity/cm}^3 + \text{heat of fusion/cm}^3 \quad (7.18)$$

$$0.239 J^2 t_m = \delta T H_c + H_f \quad (7.19)$$

$$K_m = J^2 t_m = \frac{\delta T H_c + H_f}{0.239 \rho} \quad (7.20)$$

where
$\rho$ = volume resistivity of aluminum in $\Omega\text{cm}$
$\delta T = 660°C - 25°C = 635°C$
$H_c = \dfrac{\text{Heat capacity}}{\text{Volume}} \text{ cal/(cm}^3 \cdot °\text{C)}$
$H_f = \dfrac{\text{Heat of fusion}}{\text{cm}^3} = 248.5 \text{ cal/cm}^3$

To derive the proper $\rho$ and $H_c$ consider that

- All temperature rise is from $\rho J^2$ heating
- $J = I/A$ is constant for constant $I$, therefore $J^2$ is constant

The rate of temperature rise is

$$\text{Rate} = \frac{\text{Rise}}{\text{Unit time}} \sim \rho$$

Therefore

$$\frac{\text{Unit time}}{\text{Rise}} \sim \rho^{-1} \qquad (7.21)$$

Since ρ is not linear as a function of temperature, Eq. (7.19) is solved by iteration of the values from Table 7.2 to determine ρ at 0.5t. The calculated result is

$$\rho = 5.1 \times 10^{-6}\ \Omega \cdot \text{cm}$$

which corresponds to the value at 240°C.

$$H_c \text{ at } 240°C \text{ is } 0.637 \frac{\text{cal}}{\text{cm}^3 \cdot °C}$$

therefore

$$K_m = \frac{635 \times 0.637 + 248.5}{0.239 \times 5.1 \times 10^{-6}} = 5.36 \times 10^8$$

Thus the theoretical value is in close agreement with our experimentally determined value of $5.18 \times 10^8$.

### 7.5.2 Metallization open is a characteristic trait of EOS

As in preceding discussions of EOS failures, high current and longer duration effects are characteristic. Therefore a vaporized metal line to the point of open circuit is generally a good indication of EOS as opposed to ESD. Again the reason being the usual existence of energy-absorbing elements in the circuit path would dissipate most of the ESD transient.

### 7.5.3 ESD opens at metallization steps

The importance of the preceding experimental and theoretical investigation of adiabatic heating of integrated metal stripes in Sec. 7.5.1 relates to possible damage mechanisms. Pulses in the adiabatic heating range of sufficient current and duration to reach vaporization will result in open circuit at sites of constriction, such as oxide steps. ESD pulses are likely to fall in the adiabatic range for most metal configurations particularly with the trend toward ever decreasing geometries. Thus, a frequent concern for ESD causing such opens at reduced-area oxide steps exists. Figure 7.18 shows a second metal step of questionable integrity. There are two reasons why ESD failures of such steps are infrequent.

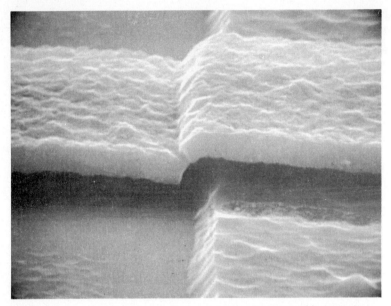

Figure 7.18 A second metal over first metal step of questionable integrity.

1. Usually there are other energy absorbing elements in the path, greatly reducing the transient current through the metal to below damaging current densities.
2. Modern processing has greatly improved metal step coverage.

**7.5.3.1 Hypothetical metallization opens due to ESD.** Data was presented in Table 7.3 and other sources on the current and time conditions to bring about failure. Consider an ESD across a defect-free nonstepped metallization stripe. In order for the metal to melt, assuming the worst case of adiabatic conditions, the following must hold:

$$J^2 t_m = K_m \quad (7.16)$$

Using the standard human body model,

$$\tau = 100 \times 10^{-12} \times 1500 = 150 \text{ ns} \quad (7.22)$$

For melt to occur the longest available time is

$$t_m(\max) = 5\tau = 750 \text{ ns} \quad (7.23)$$

Thus,

$$J^2 = \frac{I^2}{A^2} = \frac{K_m}{t_m} = \frac{5.36 \times 10^8}{750 \times 10^{-12}} = 7.15 \times 10^{17}$$

$$I_{AV}^2 = 0.715 \times 10^{18} \times A^2 \quad (7.24)$$

$$I_{AV} = 0.846 \times 10^9 \times A \qquad (7.25)$$

Take the example of 3 μm width and 10,000-Å thickness with negligible resistance in series.

$$I_{AV} = 0.846 \times 10^9 \times 3 \times 10^{-6} \times 10,000 \times 10^{-8} = 2.538 \times 10^{-1}$$

To calculate the peak current required, consider Eq. (7.11)

$$P_{AV} = \frac{V_d I_p}{5} + \frac{R_b I_p^2}{10} \qquad (7.11)$$

except that in this example only the second term applies with $R_b = R_{AV}$. Thus,

$$I_p^2 = \frac{10 I_{AV} R_{AV(circuit)}}{R_{AV}} = 10 I_{AV} \qquad (7.26)$$

$$I_p = 3.164 \sqrt{I_{AV}}$$

In our example,

$$I_p = 1.595 \text{ A}$$

which indicates that melting could be expected for a standard human body voltage of

$$V = I_p R_{HB} = 1.595 \times 1500 = 2392.5 \text{ V}$$

If we were to take the same example with 500 Ω in series, Eq. (7.26) takes on a different light.

$$I_p^2 = \frac{10 I_{AV} R_{AV(circuit)}}{R_{AV(stripe)}} \qquad (7.26)$$

Since the resistance terms are no longer equal we must assign values to them. For illustration, stripe resistance will be set at a purposely high value of 5 Ω. The average circuit resistance then is 505 Ω. Therefore,

$$I_p^2 = \frac{10 I_{AV} \times 505}{5} = 1010 I_{AV}$$

$$I_p = 31.78 I_{AV}^{1/2} = 31.78(0.2538)^{1/2} = 31.78 \times 0.504 = 16.0 \text{ A}$$

Thus it would require $V = 16.0 \times 2005 = 32,080$ V to melt the stripe. For a 6-μm stripe the voltage required to melt would be double or 64,160 V. This does not equate to open-circuit failure, but simply a reflow of the metal that did not reach vaporization. It is well-documented that melted metallization lines can actually improve cross-sectional coverage at constricted sites due to the reflow process. In general, a human body ESD is unlikely to cause an open circuit in a metallization stripe.

## 7.6 Confusion Traits between ESD and EOS

### 7.6.1 Short duration EOS transient

**7.6.1.1 MOS devices.** A short-duration transient of a few microseconds that exceeds the gate-oxide breakdown potential will cause a rupture in the oxide and a resistive short circuit. This damage cannot be distinguished from ESD damage by microscopic examination of the residual physical traits, which are identical.

**7.6.1.2 Bipolar devices.** The input base emitter junction of an LM148 operational amplifier in Fig. 7.19 was damaged by a 1-μs pulse of 150 V amplitude. This *cracked glass* was described[1] in the characteristic traits paper as resulting from a 1-μs transient. This paper was written prior to any ESD experience on my part. A short EOS transient produces a trait identical to that of ESD. This is because the applied electrical stress is virtually identical to an ESD transient. (See Fig. 7.7.)

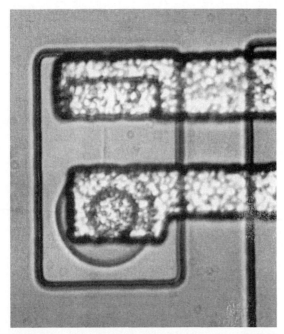

**Figure 7.19** Input-base emitter junction of an LM148 operational amplifier damaged by a 1-μs pulse of 150 V.

### 7.6.1.3 Differentiation of short EOS from ESD damage.

Sometimes it is difficult to distinguish between ESD and short EOS-transient damage, whether to MOS or bipolar circuitry. The investigator must rely heavily on other failure analysis factors such as evidence supporting either case found at the scene of the failure occurrence. Such investigations are discussed in Chap. 9.

### 7.6.2 Masking of ESD failures

A proposition often encountered is the contention that ESD might be masked by subsequent $i^2R$ heating during normal bias conditions. The resultant failure trait then might look like it was produced by longer transient conditions such as EOS. It is the author's opinion that such an occurrence would be extremely rare. Sufficient current limiting would be present in other elements of most circuits to prevent severe EOS-like damage. A clue to such an occurrence would be the existence of an open circuit in the path. For if normal biasing produced virtually unlimited current across the stress site it would be expected to continually reduce resistance and increase current until open circuit resulted.

### 7.6.3 Exception to the rule: ESD-caused open metallization

Figure 7.20 shows a rather completely vaporized metal line. At first encounter an experienced failure analyst would recognize this as evidence of electrical overstress of the high-current long-duration variety. This particular failure was caused by an ESD of 6000 V from the standard human body simulator circuit. At point $A$ in Fig. 7.21 the second (upper) metal crossed over a first (lower) metal stripe with a separation dielectric of several thousand angstroms of silicon dioxide. The electrostatic discharge between the output and $V$-pins broke down this oxide and this left only the metal line in the path. With no other circuit elements to absorb energy, the ESD pulse became a severe overstress to the metal line causing the violent destruction shown. This illustrates the importance of careful analysis with due consideration of all pertinent facts.

## 7.7 Wire Damage

Wires or "interconnects" open as a result of electrothermal stresses under EOS conditions. The combination of $I^2R$ heating and electromigration, and even final arcing can be involved. A minimum current density of about $10^7$ A/cm$^2$ is required to bring about fail-

**Figure 7.20** Vaporized metal line at 400x.

ure in any transient condition. A gold interconnect must reach 1063°C for melting to occur. Taking the case of a 0.7-mil diameter gold wire with 0.5 Ω resistance,

$$A = \pi r^2 = \frac{22}{7} \times \left(\frac{0.7}{2} \times 10^3 \text{ in}\right)^2 = \frac{22}{7} \times \left(\frac{0.35}{0.39} \times 10^{-3}\right)^2$$

$$= \frac{22}{7} \times 8.97^2 \times 10^{-8} = \frac{22}{7} \times 80.5 \times 10^{-8} = 2.53 \times 10^{-6} \text{cm}^2$$

$$\frac{I}{A} = 10^7$$

$$I = 10^7 \times 2.53 \times 10^{-6} = 25.3 \text{ A}$$

Even to attain a peak rather than average current of 25.3 A from the standard human body model would require over 37,950 V. To attain an average current of 25.3 A from the human body model would require $P_{AV} = I_{AV}^2 R_B = 640 \times 1500 = 960{,}000$ W.

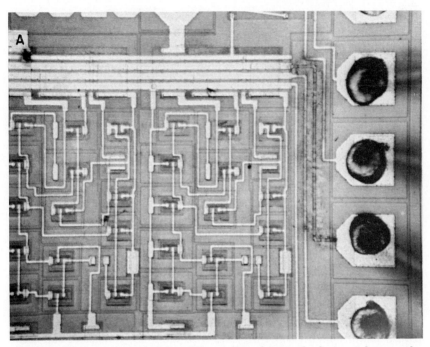

**Figure 7.21** The same vaporized metal line at 100x showing the first metal crossunder location at point $A$.

$$P_{AV} = \frac{R_B I_P^2}{10} = 150{,}000 \text{ W}$$

$$I_p^2 = 96{,}000 \times \frac{10}{1500} = 6400$$

$$I_P = 80 \text{ A}$$

which would require

$$V_b = 31.6 \times 1500 = 47{,}400 \text{ V}$$

which proves that wires do not open from human body ESD.

## 7.8 Passive Device Damage

Thin-film resistors of various materials are susceptible to ESD. Included are the commonly used discrete nichrome as well as tantalum-nichrome chips used in hybrid assemblies. Thick-film resistors have had a history of ESD susceptibility, but material and processing improvements have rendered such devices relatively immune to ESD. Anodic tantalum capacitors are susceptible to ESD levels of about 1000 V from the human body.

### 7.8.1 Thin-film resistors

Thin-film resistors are known to be affected by exposure to ESD. Experiments have revealed that both increases and decreases in resistance can be brought about by ESD. The levels at which such degradation occurs is dependent upon the materials and geometries involved. Typical resistance changes are in the range of 0.5 to 2 percent. Thus the most critical problem is in applications requiring high-stability film resistors of less than one percent tolerance.

**7.8.1.1 Thin-film ESD failure mechanisms.** The two degradation effects of increased and decreased resistance are indicative of two failure mechanisms. Experience has shown that physical ESD damage traits to thin-film resistors are extremely subtle and generally not discernible even with high-magnification scanning electron microscope (SEM) examination. Some discrete film resistors consist of the resistive material deposited in spirals around an insulative ceramic cylinder. Arcing across adjacent spirals or across voids in the film are believed to result in added current paths thus decreasing overall resistance slightly. Increase in resistance is most likely the result of thermal burnout or material changes as a result of sites of high-current concentration.

### 7.8.2 Thick-film resistors

Thick-film resistive elements are made from semiconductive metal oxides of either palladium, iridium, rhodium, or ruthenium. The resistive element is fired into a glass frit with a silver or gold additive and trimmed to desired resistances. The resistive element may be doped to improve electrical characteristics. Electrostatic or high-voltage *rf* trimming has been reported as successful by some custom manufacturers, especially with palladium oxide, which was commonly used in early thin-film resistors.

A comparative study[8] of resistors made from pastes from five different commercial sources showed varying results after printing and firing according to the manufacturer's instructions. The test consisted of an estimated 10-kV exposure achieved by placing each test substrate in a polyethylene bag and rubbing with latex finger cots. Measurements before and after exposure showed that materials of resistivities ≈ 1 kΩ per square changed very little and the change was an increase in resistance. Materials of higher (10 to 100 kΩ per square) resistivities suffered decreases in resistance that was somewhat proportional to their resistivities. Typical changes for 1-kΩ, 10-kΩ, and 100-kΩ resistivities were less than 1 percent, less than 10 percent, and about one-fifth of the 100-kΩ population exceeded 10 percent. A perhaps more deleterious effect was the fact that static exposure also changed the temperature coefficient of resistance in an unpredictable fashion as indicated in Table 7.4.

TABLE 7.4  Changes in Temperature Coefficients of Resistance After ESD

| Resistivity (Kilohms/square) | Coded commercially available manufactures paste | | | | | |
|---|---|---|---|---|---|---|
| | A | B | C | D | E | F |
| 1 | 0 | −10 | −10 | −5 | +10 | +10 |
| 10 | +5 | −20 | −10 | −5 | +20 | +25 |
| 100 | +10 | −25 | −10 | 0 | +10 | +30 |

**7.8.2.1 Failure mechanisms for thick-film resistors.** The glass frit is the vehicle that supports the resistive element, and provides adhesion to the substrate. The relative concentrations of the glass and the semiconductive oxide control the resistivity. From a microscopic view the resistor is made up of islands of the oxide and the noble metal in nonuniform states of contact with one another. These include noncontact as well as oxide-to-oxide, oxide-to-metal, and metal-to-metal contacts of varying integrity. Negative resistance change can be explained by the hypotheses of breakdowns of isolated pockets of the glass dielectric. Such breakdowns would result in continuity between more islands of the metal, or even the resistive oxide, resulting in greater conductivity. A preponderance of increased metal in the conductive path would also account for positive increases in the temperature coefficients, which is characteristic of the metal. If the added islands were of certain oxide types the temperature coefficient could be negative.

Several theories have been proposed to explain positive resistance change. These include the formation of additional metal oxides which have lower resistivity than the parent metal, the displacement of noble metal ions from the resistive oxides thus increasing their resistivities, and burning away of conductive particulate. All of these mechanisms could result in either positive or negative changes in temperature coefficient depending on the types of materials and the physical changes involved.

## 7.9 Effects on Very Thin Oxides

VLSI and very high-speed integrated circuit (VHSIC) devices are being fabricated with oxides in the range of 50 Å to 300 Å. Y. Fong and C. Hu[9] studied the effects of high-electric field transients on thin oxides. The studies were conducted in order to better predict future needs in ESD protection circuits.

### 7.9.1 Electric field strength across thin oxides

Time to breakdown was measured as a function of square-pulse voltage for oxides of various thicknesses between 55 Å and 200 Å. Figure 7.22 shows that thinner oxides broke down in shorter times at a given oxide potential than did thicker oxides, as was expected. Figure 7.23 of the same data replotted in terms of electric fields across the oxides shows that thinner oxides can withstand greater field strengths.

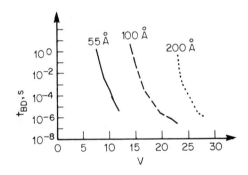

**Figure 7.22** Thinner oxides broke down in shorter times at a given oxide potential than did thicker oxides, as was expected.

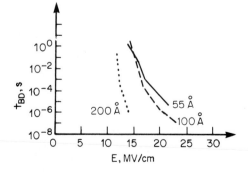

**Figure 7.23** The same data of Fig. 7.22 replotted in terms of electric fields across the oxides shows that thinner oxides can withstand greater field strengths.

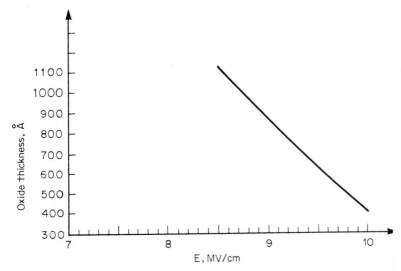

**Figure 7.24** Similar data reported by J. M. Soden[10] replotted in terms of oxide thickness versus field strength.

Figure 7.24 shows similar data reported by J. M. Soden.[10] Variations in anticipated thresholds can occur due to imperfections in the oxide causing earlier breakdown. These can usually be distinguished by their central rather than peripheral topographical locations.

### 7.9.2 Threshold shifts

Threshold shifts were observed in thin-oxide MOSFETs after exposures to square pulses for times shorter than the time to breakdown $t_{BD}$. Applications in some circuitry would result in system failures for such threshold shifts. Data in Fig. 7.25 shows that thicker oxides are more susceptible to threshold shifts for given oxide-electric fields.

### 7.9.3 Future protective circuits need

In spite of the increased field strength capability of thinner oxides in the range studied, improvements in future protective networks are necessary. This is due to the fact that thinner oxides can still sustain only smaller voltages, in spite of the greater relative field strengths..

## 7.10 Reversible Damage

### 7.10.1 Reversible charge-induced failures

R. E. McKeighen et al.[11] report an example of a CMOS matrix switch with a charge-induced failure mode that could be healed by exposure

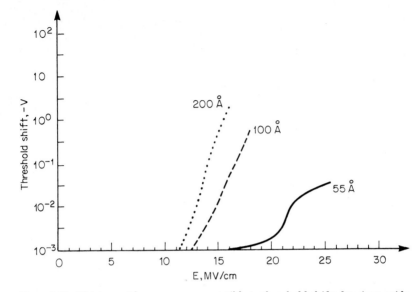

**Figure 7.25** Thicker oxides are more susceptible to threshold shifts for given oxide-electric fields.

to x-ray or ultraviolet (UV) light. The devices in question were made in both metal-lidded and ceramic-lidded packages. Failures occurred only in the ceramic-lidded packages. Switch failures always occurred in adjacent pairs in the devices and were "cured" when the ceramic lids were removed during failure analysis. Further investigation showed that these devices could be cured by exposing the packaged device to x-ray or by exposing the die to UV light.

These observed symptoms were similar to those caused by the surface inversion mechanism reported in 1978.[12] Figure 7.26a depicts the package with a positively charged lid. Polarization in the ceramic is shown in Fig. 7.26b. Figure 7.26c shows that corona discharge can occur when the field strength inside the package exceeds the breakdown limit. This limit is dependent upon the ambient pressure inside the package. In this particular example the atmosphere was 80 percent nitrogen and 20 percent oxygen at 0.4 atm at 25°C. At this pressure the breakdown limit is 24 kV/cm. The spacing of the package was about 1 mm, which relates to about 2.4 kV required to bring about corona discharge. Such a discharge would bring about transfer of positive ions to the die surface and negative ions to the ceramic lid. The most likely conclusion is that the positive ions would be electron-deficient nitrogen and the positive ions would be electron-rich oxygen. Figure 7.26d shows the residual charge on the integrated circuit chip leading to a surface inversion failure.

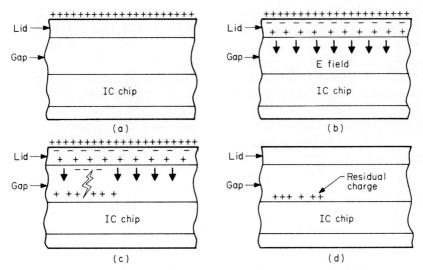

**Figure 7.26** Static caused surface inversion. (*a*) Package with a positively charged lid; (*b*) polarization in the ceramic; (*c*) corona discharge; (*d*) residual charge on the chip's surface.

Calculations showed that 2.1 nC/cm$^2$ surface charge was required to bring about the breakdown field strength. A 1-V threshold on a 1000-Å thick oxide requires a field strength of 100 kV/cm to bring about a conductive inversion channel in the P well of the N-channel device depicted in Fig. 7.27. Further calculations showed that once the die surface acquires 17.7 nC/cm$^2$ charge density a parasitic surface inversion channel is formed causing leakage.

Charge measurement with a Faraday cup showed that 1.3 nC could be obtained by rubbing with Q-tips or chemwipes; freeze spray could generate 17 nC as could a zerostat gun. Failures were produced by applying 3000 V to the ceramic lids with a charged 132-μm needle probe. Positive voltage reproduced disabled output failures, and negative

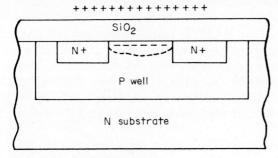

**Figure 7.27** Surface inversion.

voltage reproduced both disabled output and cross-talk failures. Both of these were representative of actual failure experience. Application of x-ray for 2 to 5 min cured the failures just as in the actual cases. UV curing was verified in a dark room.

Experiments with 12-μm diameter probes at 100 to 150 V disclosed several locations of the device layout that were vulnerable to this purposely induced inversion. The field strength from this needle was estimated to be about $20 \times 10^6$ V/m. The reason for paired output failures was found to be the proximity of the matched 3 input NOR gates. This led to overlapping of N-type transistors in two neighboring channels by patches of surface charge.

### 7.10.2 Recoveries of ESD failures on subsequent exposures

In past experiments associated with failure duplication attempts with both bipolar and MOS types of devices I have often observed apparent failure reversal. Although functionally recovered, I-V characteristics usually are improved rather than fully recovered. This effect was more predominate in MOS circuitry. I have theorized that many of these reversals were due to burnout of tiny cross section filamentary resistive shorts.

R. G. Taylor and J. Woodhouse[13] are in agreement with the burnout of filaments in oxides or dielectrics, but offer a different outlook on junction damage. The results of their work is summarized in Secs. 7.10.2.1 and 7.10.2.2.

**7.10.2.1 Shorted and degraded junction recovery.** Figure 7.28 shows observed behavior of 64K N-channel MOS dynamic random access memory (NMOS DRAM) after successive 2-kV exposures from a human body simulator circuit in accordance with Mil Std 883C method 3015.2. Failure analysis showed the filamentary junction short characteristic after removal of glassivation, aluminum, and field oxide.

**7.10.2.1.1 Degraded junctions.** Junction damage occurs when joule heating from second breakdown conditions leads to temperatures above the melting point of silicon (1420°C). Initial damage at low ESD levels occurs at sites of high-field concentration which is typically at the corners. Figure 7.29 shows a cross-sectional sketch of such damage. The damage was initially invisible even after removal of glassivation, aluminum, and field oxide. An additional polysilicon etch of 27$M$ hydrofluoric acid and 16$M$ nitric acid revealed the defect site.

Elemental analysis by wavelength dispersive x-ray with the sample rotated 30 degrees to maximize response provided information on

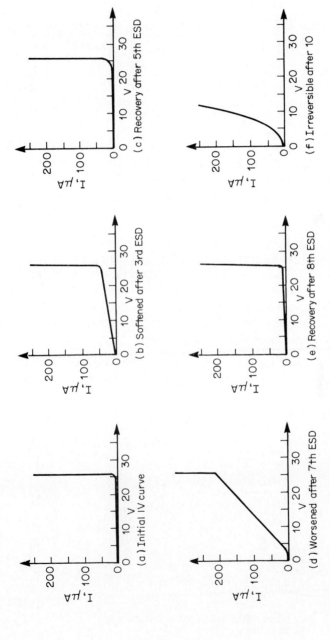

**Figure 7.28** Observed behavior of 64K NMOS DRAM after successive 2 kV exposures from a human body simulator circuit.

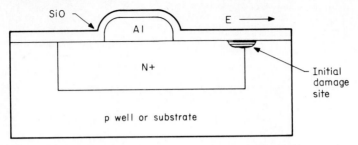

**Figure 7.29** Cross-sectional sketch of slightly damaged PN junction with damage only at the junction edge, usually at a corner.

the elemental nature of the damage site. A phosphorus distribution showed the expected high levels of concentration in the N+ region with significant levels in the $B$ region of Fig. 7.30. No phosphorus was detected in the $C$ region, and just a few isolated low-level responses in the $A$ region. Similar analysis of additional samples exposed to multiple ESD exposures of 4 kV from the human body model provided further insight into the nature of the failure mechanism. The phosphorus in the $A$ region is apparently a "front" preceding extension of the $B$ region further into the $A$ or P-doped region as shown in Fig. 7.31. No aluminum was detected in any of the regions associated with the damage sight. Analysis by laser Raman spectroscopy showed the $C$ region to consist of polycrystalline silicon, which explains the etching requirement to highlight the damaged area.

**7.10.2.1.2 Shorted junctions.** At higher ESD-induced energy levels joule heating is increased due to higher currents. When heating is suf-

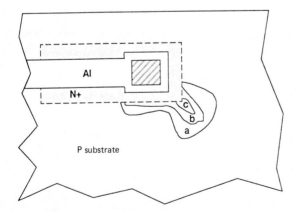

**Figure 7.30** A phosphorus distribution by wavelength dispersive x-ray analysis indicated significant levels in the $B$ region, no phosphorus in the $C$ region, and just a few isolated low-level responses in the $A$ region.

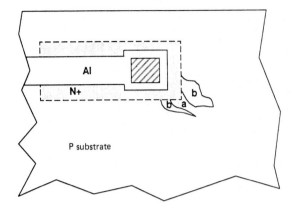

**Figure 7.31** The phosphorous in the *A* region is apparently a "front" preceding extension of the *B* region further into the substrate region.

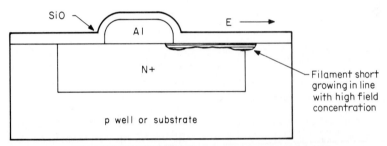

**Figure 7.32** When heating is sufficient to cause the aluminum-silicon contact to exceed the eutectic temperature of 580°C, a filament begins which can short the junction as indicated.

ficient to cause the aluminum-silicon contact to exceed the eutectic temperature of 580°C, a filament begins and shorts the junction as indicated in Fig. 7.32.

**7.10.2.1.3 Recovery mechanism.** The polycrystalline silicon resulting from low-level ESD exposure increases electrical resistance. Junction breakdown during a subsequent low-level exposure is likely to occur adjacent to the original damage site. This would result in a remelt and resolidification moving the polycrystalline location slightly. In addition the N region would continue its extension into the P region by diffusion and redistribution. Electron-beam induced current (EBIC) analysis after successive ESD exposures verified that at a certain point the extended N-dopant moves away from the junction as a wavefront, resulting in recovery. This is illustrated in Fig. 7.33. Infrared microscopy was used to verify movement of the damaged region as

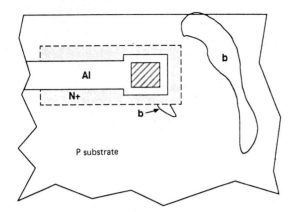

**Figure 7.33** EBIC analysis after successive ESD exposures verified that the extended N dopant moves away from the junction as a wavefront, eventually resulting in recovery.

theorized. In this study only degraded junctions showed recovery. In no case was a shorted junction recovered by subsequent ESD exposure.

**7.10.2.2 Recoveries of dielectric shorts.** Voltage waveforms monitored during ESD exposures of 200 V and 2000 V to an unprotected MOSFET are shown in Fig. 7.34. In Fig. 7.34a, during the 200 V exposure, the gate oxide breaks down at the anticipated level of 100 V and the voltage waveform then decays approximately exponentially through the low-resistance short formed. This short is formed by metallization melted during the transient and transported through the damaged oxide. In Fig. 7.34b, the 2000-V transient is shown to breakdown at 100 V as expected, drop to about 50 V, recover, then repeat this cycle. Finally, the voltage does not quite reach the breakdown level of 100 V, and then gradually decays. This indicates the oxide ended up in an open-circuit condition.

Two possible explanations are proposed for the behavior shown in Fig. 7.34b.

1. The gate to substrate capacitance discharges during breakdown, then recharges and discharges repeatedly until the breakdown potential is no longer reached.

2. Repeated cycles of a metal filament forming and fusing occur until finally left in an open-circuit condition.

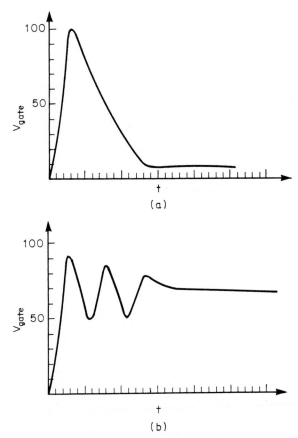

**Figure 7.34** Voltage waveforms during ESD exposure to an unprotected MOSFET. (a) 200 V; (b) 2000 V.

In either case if the waveform remains high an open circuit condition exists and the device would be expected to function. The same type of recovery has been observed by subsequent ESD exposures to previously shorted oxides.

## 7.11 Schottky Diode Damage

Schottky diode junctions consist of a barrier metal in contact with silicon as shown in Fig. 7.35. These junctions are often susceptible to human body model voltage levels of 1000 V or less. Damage that is visible after metal removal usually occurs at the periphery of the junction.

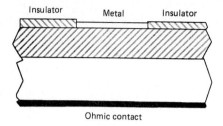

**Figure 7.35** Schottky diode junctions consist of a barrier metal in contact with silicon.

## 7.12 ESD Damage to Gallium-Arsenide Devices

The use of gallium-arsenide devices is in a tremendous growth state that is expected to continue. The primary advantage is high speed due to high-carrier mobility. A. L. Rubalcava et al.[14] reported results of extensive testing of selected elements of gallium-arsenide integrated circuits as summarized in this section. The elements characterized were test patterns fabricated on special evaluation wafers. The susceptibility categorizations were made with the standard human body model in accordance with Mil Std 883 Method 3015.2, except actual thresholds were recorded. Each test step consisted of five pulses with 5 s minimum between pulses. Step stress testing was done until failure or 2000 V, which was the limit of the simulator. Tests were generally repeated starting at higher levels to reduce cumulative effects.

### 7.12.1 Metallization burnout

First metal line patterns consisted of 2-μm wide, 4400-Å thick Ti/Pd/Au in a serpentine run of about 200-μm length. Nominal resistance was 10 Ω. Five of these lines failed by open-circuit melts at 1325 V ± 2 percent regardless of the number of pulses.

Second metal patterns consisted of electroplated gold that was 4-μm wide, 19,000-Å thick in a 1200-μm long serpentine run. Nominal resistance was 5 Ω. After finally reaching 90 pulses at 2000 V, tests were discontinued with no failures. Slight resistance changes of less than 1 percent resulted.

### 7.12.2 Thin-film resistors

Nichrome thin-film resistors of various values ranging from 51 to 459 Ω were evaluated. The lengths and widths of the resistors varied with a total size range of 125 μm² to 6875 μm². After testing 24 resistors the failure thresholds corresponding to a change of greater than 4 percent in resistance were seen to be related to current density. All devices failed at current densities of 24 to 29 mA per micrometer of width. The corresponding voltage levels varied from 220 to 1800 V. Physical damage sites were either near the center of the resistor length or at the nichrome to first metal contact interface.

### 7.12.3 Implanted resistors

N+ resistor patterns with a nominal resistivity of 125 Ω per square were formed by silicon ion implantation in the gallium arsenide. Seventy pulses at 2000 V were applied to 100-μm wide 56 Ω N+ resistors with no observed or measured degradation.

### 7.12.4 MIM capacitors

Metal-insulator-metal (MIM) capacitor patterns were fabricated by sandwiching a 2000-Å dielectric of silicon nitride between first and second metal plates. Thirty-seven capacitors of three physical size types between 2500 μm$^2$ and 128,000 μm$^2$ were evaluated. Most became leaky after pulses of 300 to 500 V.

### 7.12.5 MESFETs

Metal semiconductor field-effect transistors (MESFETs) tested had gate-to-drain spacing that ranged from 1 to 3 μm. Gate lengths ranged from 0.5 to 2 μm, and gate widths were from 6 to 300 μm. The most sensitive polarity was determined to be minus on the gate with the drain grounded. This biasing, which reverse biased the Schottky gate, brought about failures between 50 V and 300 V for the range of sizes. Failures from the same combination with opposite polarity ranged from 400 to 900 V. For a given size the polarity-threshold ratio was usually about three to one.

### 7.12.6 Digital counter

Seventeen 4-bit ripple-counter integrated circuits were evaluated for a total of 1455 pulses. Unprotected outputs were the most sensitive at 500 V. Some inputs and power supply combinations started to degrade at 700 V.

## 7.13 Schottky and Advanced Schottky Device Susceptibility Levels

The introduction of Schottky diodes at TTL inputs for clamping purposes resulted in lower thresholds to failure from ESD. The advanced

Schottky TTL devices became even more susceptible. For example data taken at Westinghouse[15] showed human body thresholds of 100 V and 450 V for the 74F04 and 74F175 devices, respectively. Early versions of this family of devices are particularly susceptible to cumulative multiple subthreshold exposures. Data taken during the 1981 NASA-funded study showed the 74F04 failing after as little as 10 exposures at 60 V, and the 75F175 after 30 exposures at 150 V.

In recent years, integrated circuit manufacturers have greatly improved ESD immunity to varying degrees on these families of devices. The devices are mentioned herein as a caution on the importance of sensitivity testing and classification to avoid such supersensitive parts, and to select those from vendors designing in appropriate levels of immunity.

## 7.14 Cumulative Damage

In personal experiments with a large variety of part types from different technological families I have observed occurrences of cumulative ESD damage. The cumulative damage effect is best illustrated by considering numerous exposures at levels below the damage threshold. Table 7.5 lists representative data on several 1N5711 Schottky diodes. The damage threshold voltage as usually defined is the level for which damage occurs in one ESD exposure from the standard human body model. In this case the threshold is 600 V, for after a single exposure the reverse breakdown voltage $B_V$ and reverse current $I_R$ become degraded. When a 500-V exposure level was applied to a second device, both $B_V$ and $I_R$ were unaffected. After four exposures a sudden catastrophic failure occurred and $B_V$ and $I_R$ are out of specification limits. At 450 V and 375 V the number of exposures to failure were 10 and 37, respectively. At 300 V the device withstood 200 exposures with no apparent degradation.

If one were to repeat the experiment, a cumulative trend would be expected although the number of exposures would likely differ because of part-to-part variations as well as subtle simulator circuit difference. Cumulative damage is discussed in more detail in Chap. 11

TABLE 7.5 Cumulative Damage on Several IN5711 Schottky Diodes (Degraded $I_R$ and $B_V$)

| Level | Number of pulses to degrade |
|---|---|
| 600 V | 1 |
| 500 V | 4 |
| 450 V | 10 |
| 375 V | 37 |
| 300 V | 200 Pulses (no degradation) |

and is reported extensively in the literature. The failure mechanisms involved are subtle and further research is required to bring about understanding beyond theoretical speculation.

## 7.15 Latent Effects

The term latent has two common interpretations that must be clarified to avoid confusion when referring to the literature or otherwise researching the subject. The reason for making so much fuss about the meaning of this term is the possible importance of latent failure considerations on reliability and ESD control expenditures. *Webster's Ninth New Collegiate Dictionary* defines latent as meaning "to lie hidden; to escape notice; and present...but not now visible or active." Therefore the most common inference is that of a hidden or subtle defect.

Synonyms listed in the same reference and in *Webster's Thesaurus* as well suggest a different meaning. Given synonyms include latent, abeyant, dormant, lurking, potential, prepatent, and quiescent. All of which, including latent, are then expanded to show that the meaning is something about to come into effect. For instance, latent is said to "apply to a power or quality that has not yet come into sight or action but may at any time."

An additional note is that the thesaurus lists the terms denoting subtlety such as concealed, hidden, idle, inactive, inert, and related terms. The end result is that much care must be taken to define one's meaning when using the terms latent and latency. In this regard the following is an attempt to separate the two connotations for the purposes of usage in this book.

### 7.15.1 Latent damage

When the literature refers to latent damage meaning subtle or hidden defects, the connotation can be regarded as grammatically correct. If one uses the term latent failure with the same meaning of subtlety intended, and no implication of time-dependent degradation, then the usage can become misinterpreted. In my many years of failure analysis subtle and/or hidden defects often were found to cause failures. For instance, many immediately catastrophic ESD failures leave hidden characteristic failure traits. For the purposes of this book, and elsewhere, we will attempt to use words other than latent to define any concealed or subtle defect.

### 7.15.2 Latent ESD failures

The term latent ESD failures is meant to define failures that occur in time, but are due to some earlier exposure to ESD. Research results in latent failure investigations are covered extensively in Chap. 11. These results show that latent ESD failures can occur, and places a higher importance on ESD matters as far as end usage reliability is concerned.

## 7.16 Walking Wounded

The term "walking wounded"[16] is often used to describe devices where a junction has been degraded by a low-level ESD stress similar to the before and after curves presented in Figs. 2.3 and 2.4. It is important to note however that the parts shown in the referenced figures were catastrophic failures and would not walk if tested at the part level. Part-acceptance test criteria should certainly be specified tight enough to reject a junction with such severe damage.

It may be possible however for such a degraded junction to occur at a higher assembly level where acceptance tests may not detect the degraded rise-time characteristic. Such degradation might be particularly problematic in high-frequency applications.

## References

1. W. J. Lytle and Owen J. McAteer, "Characteristic Traits of Semiconductor Failures," Annual Symposium on Reliability, 1970.
2. D. G. Pierce and K. L. Durgin, "An Overview of Electrical Overstress Effects on Semiconductor Devices," *Electrical Overstress/Electrostatic Discharge Symposium Proceedings*, 1981.
3. T. M. Mazdy, "FET Circuit Destruction Caused by ESD" *IEEE Transactions on Electron Devices*, Sept. 1976.
4. D. C. Wunsch and R. R. Bell, "Determination of Threshold Failure Levels of Semiconductor Diodes and Transistors Due to Pulse Voltages," *IEEE Transactions on Electron Devices*, vol. NS-15, no. 6, December 1968, pp. 244–259.
5. Thomas S. Speakman, "A Model for the Failure of Bipolar Silicon Integrated Circuits Subjected to Electrostatic Discharge," International Reliability Physics Symposium Proceedings, 1974.
6. O. J. McAteer, "Pulse Evaluation of Integrated Circuit Metallization as an Alternate to SEM," *International Reliability Physics Symposium Proceedings*, 1977.
7. J. R. Black, "Short Pulse Fusion Current for Aluminum Film Conductors," unpublished.
8. R. P. Hummel, "The Effect of Static Electricity on Thick Film Resistors," *Insulation/Circuits*, September 1972.
9. Y. Fong and C. Hu "The Effects of High Electric Field Transients on Thin Gate Oxide MOSFETs," *Electrical Overstress/Electrostatic Discharge Symposium Proceedings*, 1987.
10. J. M. Soden "The Dielectric Strength of $SiO_2$ in a CMOS Transistor Structure," *Electrical Overstress/Electrostatic Discharge Symposium Proceedings*, 1979.
11. R. E. McKeighen, W. Daily, T. Pang, P. Huynh, W. Wittman, and B. Wiley, "Reversible Charge Induced Failure Mode of CMOS Matrix Switch," *Electrical Overstress/Electrostatic Discharge Symposium Proceedings*, 1986.
12. M. H. Wood and G. Gear, "A New Electrostatic Discharge Failure Mode," *International Reliability Physics Symposium Proceedings*, 1978.
13. R. G. Taylor and J. Woodhouse, "Junction Degradation and Dielectric Shorting:

Two Mechanisms for ESD Recovery," *Electrical Overstress/Electrostatic Discharge Symposium Proceedings*, 1986.
14. Anthony L. Rubalcava, Douglas Stunkard, and William J. Roesch, "Electrostatic Discharge Effects on Gallium Arsenide Integrated Circuits," *Electrical Overstress/Electrostatic Discharge Symposium Proceedings*, 1986.
15. NASA Funded Study under RADC Contract Number F30602-78-C-0281.
16. D. E. Frank, "The Perfect '10'—Can You Really Have One?" *Electrical Overstress/Electrostatic Discharge Symposium Proceedings*, 1981.

# Chapter 8

# ESD Damage Models

## 8.1 Introduction

The electrostatic discharge problem has been elusive to investigators in more ways than just the subtle nature of the damage involved. Sometimes more puzzling are questions concerning the path of the transient and the original source of the energy or voltage involved. Initial realizations of the possibility of damage to susceptible parts from ESD were restricted to the human body as the source. Thus the human body model (HBM) has been addressed longer than any other. In this author's opinion there are six significant models for ESD damage worthy of serious consideration in attempts to simulate conditions that bring about failure. The primary ones are the human body, charged-device, *field-induced*, field-enhanced, machine (or Japanese) models, and the capacitive-coupled model (CCM).

## 8.2 Human Body Model

Most information available in the literature is based upon the human body ESD model. The human body obviously has capacitance as it can store charge to a degree dependent upon how well it is insulated from ground. The body has a resistance value dependent upon muscle tone, moisture content, contact resistance, and other factors. Most researchers have agreed that a reasonable electrical model consists of a capacitor in series with a resistor. There has been much past research in attempts to determine appropriate values for these components of a typical body. A small amount of inductance, probably a fraction of a microhenry, is believed to have a negligible effect on damage results in most cases.

### 8.2.1 Development of the human body model

A review of selected advances in development of the human body model is included herein. The reason is to give a better perspective as to the reasons for existing differences in models used to derive susceptibility data, and to make the reader aware of the implications of such differences.

**8.2.1.1 Bureau of Mines data.** In the early 1970s the idea of modeling the human body by a capacitor discharging through a combined body-contact resistance had been widely accepted. There was much controversy however, on the proper values to assign these circuit elements. National Bureau of Mines Bulletin 520, 1962[1] reported 22 tests showing an average human body capacitance of 204 pF with a range of 95 to 398 pF. Hand-to-hand body resistance was determined by 100 tests to average 4000 Ω. These values gave a good start at approximating the human body and were used with some modification by many in the electronics industry in establishing early simulation circuits. Several questions remained. What variables affected human body capacitance, and what were representative values of these parameters? The same questions applied to human body resistance as well as the most likely resistive path through the human body. This was readily determined to be from the floor to the hand.

### 8.2.2 Early electronics industry models

Those in the industry used parametric measurements, waveform analysis, and actual part-destruction comparisons between actual human beings and varying simulation circuits to derive a human body model. Figure 8.1 shows such a method reported by Kirk et al.[2] for which values of $C_B$ = 132 to 190 pF and $R_B$ = 87 to 190 Ω were derived. The peak currents produced by a human subject and a $C_B$ = 270 pF, $R$ = 0 model were compared to derive values for the human body elements. The peak current through the $R$ = 0 model is

$$I_0 = \frac{V}{1000}$$

and for the human subject the peak current is

$$I_p = \frac{V}{1000 + R_B}$$

Therefore,

$$R_B = \frac{V - 1000 I_p}{I_p} = \frac{1000(I_0 - I_p)}{I_p}$$

where $I_0$ and $I_p$ are determined from the current waveform through the 1000-Ω resistor. Since the time constant $\tau$ is the total resistance times the total capacitance,

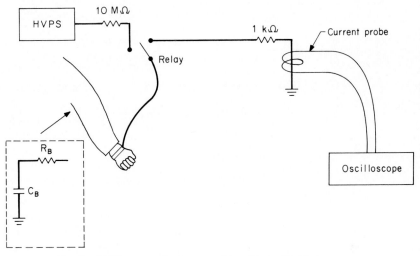

**Figure 8.1** Bendix HBM test method reported in 1976 by W. Kirk.

$$C_B = \frac{\tau}{R_B + 1000}$$

Human body derivations were typically based upon very small samples without due accounting for the many variables to be encountered. This brought about significant differences in models used by different companies with capacitance values ranging from about 50 to 300 pF, and resistances ranging from 0 Ω to 4 kΩ.

### 8.2.3 Standard human body model

After much review of industry-wide determinations of human body models, the Naval Sea Systems Command arrived at a standard model for inclusion in DOD Standard 1686,[3] released in May of 1980. This standard model of 100 pF in series with 1.5-kΩ resistance has been widely accepted, more for the sake of uniformity than accuracy. Other problems associated with implementation of this model will be discussed later.

### 8.2.4 Calculated human body capacitance

Human body capacitance is reportedly[4] made up to two components, $C_S$ and $C_g$, where $C_S$ is calculated considering the body as an isolated sphere and $C_g$ is the parallel plate capacitance of body to floor through the soles.

$$C_S = 4\pi\varepsilon_0 \times \text{area} \tag{8.1}$$

$$C_S = \frac{4\pi\varepsilon_0 H}{2 \times 10^{-12} \times 10^2} = 0.55H \quad \text{pF} \tag{8.2}$$

where $H$ = person's height in centimeters

$$C_g = \frac{k\varepsilon_0 A}{t \times 10^{-12} \times 10^2} = 0.088\frac{kA}{t} \quad \text{pF} \tag{8.3}$$

where $k$ = dielectric constant of shoe soles
$A$ = area of both shoe soles in cm$^2$
$t$ = thickness of shoe soles in cm

As an example, a person 5-ft 8-in (173 cm) tall would have $C_S$ of about 95 pF. Assuming the person has shoe contact area of about 360 cm$^2$ with a sole having dielectric constant of 5 and thickness of 1 cm, $C_g$ turns out to be about 158 pF. This gives a total capacitance of 253 pF. It seems that this method gives approximate capacitance values that tend to be high when compared with reported measured values.

### 8.2.5 Enoch-Shaw human body measurements[5]

Figure 8.2 shows values for human body capacitance as a function of distance above a concrete floor, covered with tile and carpet. The values were derived from measurements of the voltage on a 500 pF capacitor after discharging 200 V from the human subject to the capacitor. The measurement process consisted of

1. Charging a bare-footed human being standing on an insulative platform to 200 V at the prescribed distance from the floor
2. After 5-s stabilization a mercury-wetted relay discharged the person to a known (500 pF) capacitor
3. After 1 μs the voltage on the known (500 pF) capacitor was measured

The human body capacitance was then calculated by

$$C_B = \frac{V_{\text{cap}}}{V_B - V_{\text{cap}}} C \tag{8.4}$$

where $C_B$ and $V_B$ = the human body capacitance and initial voltage (200 V), respectively
$V_{\text{cap}}$ = the voltage on the known capacitance after 1 μs
$C$ = the known capacitance, in this case 500 pF

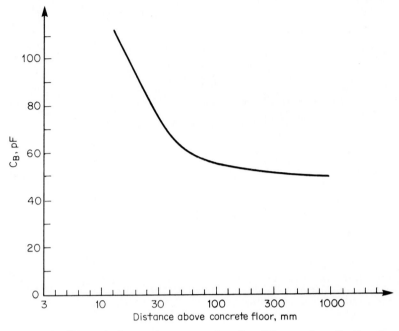

**Figure 8.2** Human body capacitances as a function of distance from the floor from Enoch-Shaw tests.

The data in Fig. 8.2 approaches a minimum $C_B$ when the distance from the floor exceeds a certain value. This minimum value of about 50 pF corresponds to the free-space value $C_s$ of the body considered as a sphere. This equates to a sphere of 378-mm diameter, which seems realistic.

## 8.3 Charged-Device Model

In 1974, Speakman[6] proposed the possibility of destroying a part, such as an integrated circuit, by rapid discharge of accumulated static on the part's own body. This type of failure has since been called the charged-device model (CDM) failure. The charged-device model was addressed in much greater detail in 1980[7] as it became a predominate failure mode at AT&T. Test results showed that triboelectric generation through simulated dual-in-line package (DIP) tube handling resulted in most of the accumulated charge residing on the lead frame. Representative levels were 3 nC on the leads with less than 0.2 nC on the plastic package. This indicated that the bulk of charge was mobile, as it rested on conductors. The device in the general case can be an integrated circuit, hybrid, or other assembly which has a capacitance with respect to ground. The schematic for the CDM is given in Fig.

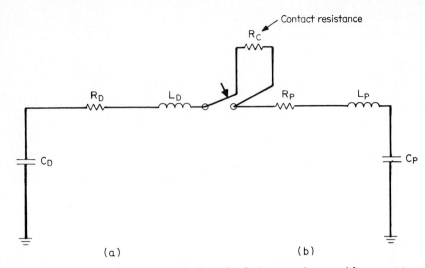

**Figure 8.3** (a) Charged-device model: $C_d$ is the device capacitance with respect to ground, $R_d$ is the resistance of the chip element(s) dissipating the power of the transient, and $L_d$ is the inductance of the leads. (b) $R_p$, $L_p$, and $C_p$ are resistance, inductance, and capacitance in the discharge path to ground.

8.3a. $C_d$ is the device capacitance with respect to ground, $R_d$ is the resistance of the portion of the die dissipating the power of the transient, and $L_d$ is the inductance of the leads. The discharge path to ground contains similar elements as shown in Fig. 8.3b. $R_p$ is the resistance is the path to ground, $C_p$ is the capacitance to ground, and $L_p$ is any inductance to ground. In most real-world conditions, $L_p$ is believed to be negligible and a low impedance path to ground is provided by either a very low $R_p$ or a significantly large $C_p$. For this reason a low resistance is considered to suffice for simplification of the model and analysis. The authors[7] take the case of a 16-pin device with $R_{total} = 10 \, \Omega$, $L_d = 10$ nH, and $C_{total} = 3.6$ pF. The current has the form of a damped sinusoid of

$$I(t) = \frac{V}{\omega L} e^{-\alpha t} \sin \omega t \qquad (8.5)$$

where $\alpha = R/2L$
$\omega = 2\pi f$
$V =$ the voltage on the capacitor at $t = 0$

Both the current and energy for $V_0 = 500$ V are given in Figs. 8.4 and 8.5.

The average power density is equal to

$$P(t) = \frac{1}{tA} \int_0^t Ri^2(t) \, dt \qquad (8.6)$$

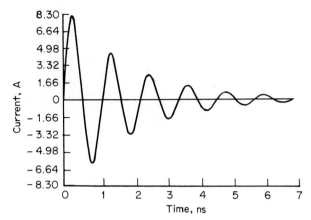

**Figure 8.4** The CDM current waveform for a 16-pin device with $R_{total}$ of 10 Ω $L_d$ of 10 nH, $C_{total}$ of 3.6 pF, and a 500-V potential. *(From Bossard,[7] with permission of ITTRI/RAC.)*

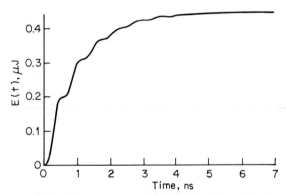

**Figure 8.5** The energy consumed by the device during the CDM transient shown in Fig. 8.4. *(From Bossard,[7] with permission of ITTRI/RAC.)*

The Wunsch-Bell[8] power-density region required to reach the melting point of silicon in the quasiadiabatic region is

$$P_d(t)_{WB} = \sqrt{\pi k C_p}\, (T_M - T_i)t^{-1/2} \tag{8.7}$$

The results were plotted for a junction area of 160 μm² as shown in Fig. 8.6, which indicates that sufficient energy to cause damage is present in the 500-V example cited earlier.

### 8.3.1 Factors affecting CDM failures

Factors affecting charged-device failures are interfacing materials, speed of movement, package configuration, material makeup, and ori-

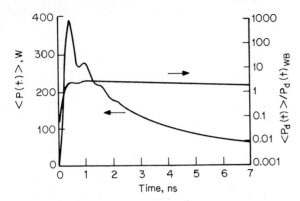

**Figure 8.6** The average power density for a junction area of 160 μm² indicates sufficient energy to cause damage when compared to the extended Wunsh-Bell model as shown in Fig. 7.15. (*From Bossard,[7] with permission of ITTRI/RAC.*)

entation. Of course the interfacing materials, speed of movement, and material makeup of the device affect the triboelectric generation propensity. The package configuration including carriers and fixtures affects triboelectric generation, capacity-to-store energy, and induction-to-conductive parts of the device, such as leads. Device orientation during movement affects these same properties. Orientation during discharge affects the voltage and energy levels because of capacitance dependency as shown in Fig. 8.7.

| Position | 16-pin device capacitance |
|---|---|
| Leads up | Highest ~ 3 pF |
| Leads down | Lower ~ 2 pF |
| Tilted (at contact) | Lowest ~ 1 to 1.5 pF |

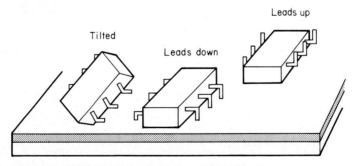

**Figure 8.7** CDM orientation/capacitance dependency.

Consideration of two other important factors might mean that the likelihood of charged-device failures may be less than indicated in Fig. 8.6. First, multiple internal paths could provide a spreading of the available energy thus reducing stresses at individual nodes. Second, the fact that the transient is so fast may mean that the damage model would be more closely approximated by the adiabatic region of the Wunsch-Bell model, requiring more power to reach failure.

The Wunsch-Bell curve given in Fig. 7.11 shows a change in slope from the quasiadiabatic to the adiabatic range at 10 ns. In actuality, the exact point of crossover is unknown for a given structure. Figure 8.8 shows the predicted Wunsch-Bell power to failure for a 160-$\mu m^2$ junction area as a function of crossover points. The failure power would be exceeded only if the quasi-adiabatic range extended down to 1 ns. At the nominal 10-ns crossover the power required to bring about failure is 586 W.

The order of magnitude of power to failure is consistent with actual charge-device data if the adiabatic crossover is at 10 ns or less. It is possible that refinements of the damage models may be required at the high frequencies involved. Regardless, the literature leaves no doubt that charged-device failures are a real problem. In particular, the incidence has been significant during automated handling processes, such as final or incoming tests.

### 8.3.2 CDM waveform profile

A typical CDM failure can occur when a few nanocoulombs accumulate on the device through triboelectric charging in normal processing.

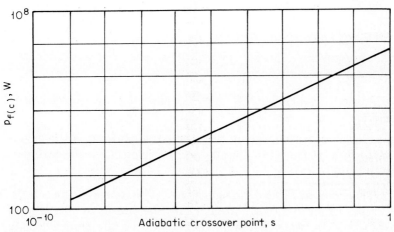

**Figure 8.8** Predicted Wunsch-Bell power to failure for a 160-$\mu m^2$ junction area when the adiabatic region's change of slope is considered.

The rapid discharge that occurs when a lead touches a conductive structure or the test socket can result in damage. The typical discharge waveform is a very fast damped sinusoid with a rise time of about 0.5 ns and total duration of less than 10 ns as shown in Fig. 8.4.

### 8.3.3 CDM damage characteristics

The damped sinusoid transient of less than 10 ns is too fast for input-protection circuitry to turn on, so damage results typically at internal nodes of the chip and often in the gate or field-oxide regions. The resultant damage from a charged-device model failure is typically similar to that produced by the human body model at a much higher voltage.

## 8.4 Field-Induced Model

An electric field can induce a charge on a device, assembly, or isolated conductor of sufficient magnitude to cause ESD damage upon discharge of the induced potential. A double jeopardy exists in that the susceptible item is left with a charge of opposite polarity after discharge and subsequent removal from the field influence. This leaves the possibility for a second ESD event when the item is later grounded. Figure 8.9 shows a printed-circuit board during the sequence of (a) field induction, (b) grounding in the presence of the field, (c) removal from field (or removal of field), and (d) subsequent ESD event upon grounding. It is important to note that the so-called field-induced model (FIM) as described required discharge in addition to the presence of the field. The term field-induction/discharge (FID) is introduced to more accurately describe the model.

### 8.4.1 Part-level field-induced damage

When dealing with the part level we will address two additional aspects of the field-induced model:

1. The induced potential across a part from a highly charged body nearby
2. The induced mobile charge on the conductive portions of the device from the field produced by immobile charges on the insulative portions

**8.4.1.1 Failure caused by field strength alone.** The concern is for a field-sensitive part such as MOS damaged by the presence of a highly charged body. The worst possible case would be the charged body practically touching a sensitive lead with the other lead tied to

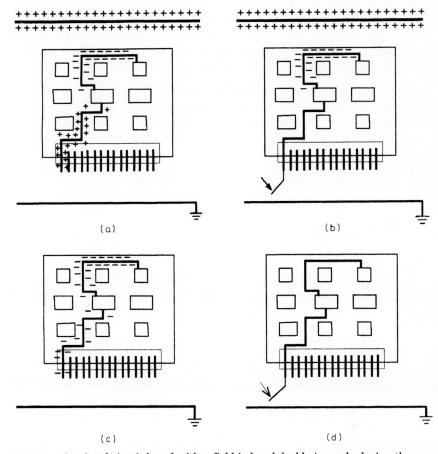

**Figure 8.9** A printed-circuit board with a field-induced double jeopardy during the sequence of: (a) field induction; (b) grounding in the presence of the field; (c) removal from field (or removal of field); and (d) subsequent ESD event upon grounding.

ground. A common real-world example is a device in a DIP tube with one side touching a grounded surface. Unger et al.[9] showed that only devices with ≤100-V sensitivity could be damaged with a body charged to 5000 V touching the outside of the tube. This value was based on a geometrically estimated coupling capacitance across the tube to the top lead of about 2 percent of the capacitance across the device. The voltage divides across the capacitors by

$$V_{device} = \frac{C_{coup}}{C_{device} + C_{coup}} V_{body} \qquad (8.8)$$

From Eq. (8.8) one can infer that direct field-induced failures across a part are likely only when the charged object is in extremely close proximity. For even the most susceptible geometry of an unprotected MOS device with source, drain, and substrate grounded the charged-field source must be within 100 mils of the gate lead.

**8.4.1.2 Field induction from insulative portion of part.** Figure 8.10 depicts the hypothetical sequence of (a) an initially uncharged device, (b) induced charge separation on lead by charged insulative portion of device, and (c) residual charge on leads after discharge to ground. The fact that such a sequence can occur does not mean that a real-world problem is represented. The induced mobile charge on the conductive portions of a device from the immobile charge on the insulative package is much smaller than the mobile charge. Tests show that typical immobile charge is around 0.02 nC with a maximum of about 0.08 nC. The maximum value corresponds to a range of 80 V to as low as 53 V on a 16-pin device in its minimum capacitance orientation of 1 to 1.5 pF as shown in Fig. 8.7. Therefore this type of field induction is felt to be academic and an insignificant contributor to part damage. The case of a part damaged by the sudden discharge of an isolated conductor

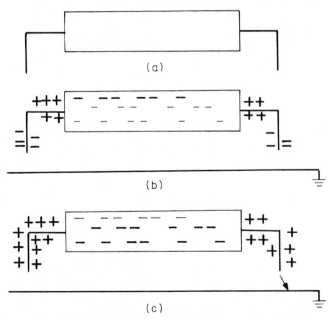

**Figure 8.10** Field induction from the insulative portion of a device: (a) uncharged device with insulative package; (b) charge separation induced on leads by charged package; (c) residual charge of opposite polarity on leads after discharge to ground.

having an induced potential is a hybrid that befits our FID model and is related to the machine model.

### 8.4.2 Validation of assembly-level field-induction model discharge

Enoch and Shaw[5] reported experimental results validating the seriousness of the field-induced* model. The experimental specimen consisted of 74,373 octal-latch circuits of low-power Schottky and high-speed CMOS technologies (HCT) mounted on printed wiring boards. The boards were placed on a charged 6-mm-thick acrylic insulator was shown in Fig. 8.11. A 1.6-mm-thick metal ground plane was placed under the insulator to simulate a static-controlled work surface. The potential on the board was measured by a digital electrostatic fieldmeter monitoring a metal plate attached to the board. The experimental field was generated by rubbing the insulator with a paper-covered cardboard cylinder. The potential on the insulator was uniform within 10 percent. The induced-board potential was measured

---

*Although the referenced authors used the term "field-induced model," the process described is felt to better fit the field induction-discharge model as discussed in Sec. 8.4.

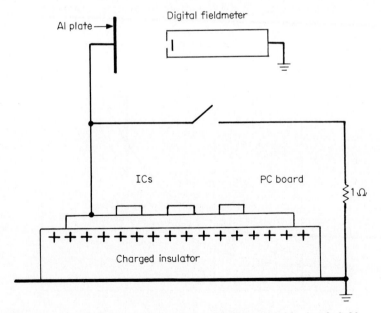

Figure 8.11 Enoch-Shaw experiment to validate assembly level field-induced model.

after placement on the insulator. An edge connector pin connected directly to an input pin of the 74,373 was then grounded by a vacuum relay.

**8.4.2.1 Study results.** Their experiments showed resultant ESD thresholds that varied as a function of series resistance similar to that shown in Fig. 8.12. The dashed line represents catastrophic functional failures whereas the solid line represents parametric specification failures. The curves in Fig. 8.12 were recalculated from a mathematical treatment of the data as explained in Sec. 8.4.2.2. Values of average dynamic part resistance ($R_p$) of 1.993 Ω and 13.833 Ω were derived to fit the data for parametric and functional failures, respectively. The authors[5] had reported similar curve-fit values of $R_p$ of 2 Ω and 15 Ω. The circuit is assumed to consist of the board capacitance $C_{bd}$ discharged through the dynamic resistance of the part $R_p$ plus any series resistance $R_t$ provided by the test circuit. The total available energy $E_a$ is $0.5 C_{bd} V^2$. The energy dissipated in the part is

$$E_p = \frac{R_p}{R_p + R_t} E_a \qquad (8.9)$$

If the damage threshold $V_d$ were determined experimentally with statistical significance for a single series-resistance value of $R_t$, the threshold could then be predicted for other $R_t$ by

$$E_d = \frac{R_p}{R_p + R_t} \left(\frac{1}{2} C_{bd} V_d^2\right) \qquad (8.10)$$

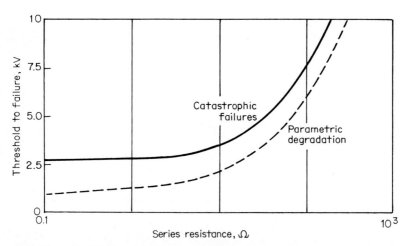

**Figure 8.12** Field-induced board catastrophic (dashed) and parametric (solid) thresholds versus series resistance.

or

$$V_d = \left[\frac{2E_d(R_p + R_t)}{R_p C_{bd}}\right]^{1/2} \tag{8.11}$$

The authors reported no significant difference between lead lengths of 0.2 to 2.2 m which indicated that ordinarily encountered stray inductances did not affect susceptibility.

### 8.4.2.2 Mathematical analysis of FIM board tests.
The analytical method contained herein differs somewhat from that described in the preceding reference, in hopes of some simplification. By using two tests with $R_t = 0$ and $1\ \Omega$ (or some other resistance) respectively, $R_p$ can easily be derived. The reader is cautioned to use an adequate number of samples for statistical significance in deriving such experimental threshold data. With $R_t = 0$,

$$E_p = \frac{1}{2} C_{bd} V_{d0}^2 \tag{8.12}$$

where $E_p$ is the energy dissipated in the part with $R_t = 0$ or the part energy required to cause failure and $V_{d0}$ is the threshold to degrade the part with $R_t = 0$.

With $R_t = 1$,

$$E_p = \frac{1}{2} C_{bd} V_{d1}^2 \frac{R_p}{R_p + 1} \tag{8.13}$$

from which we can get $R_p$.

$$\frac{R_p}{R_p + 1} = \frac{2E_p}{C_{bd} V_{d1}^2} \tag{8.14}$$

$$R_p = \frac{2E_p}{C_{bd} V_{d1}^2}(R_p + 1)$$

$$R_p = \frac{C_{bd} V_{d1}^2 - 2E_p}{C_{bd} V_{d1}^2} = \frac{2E_p}{C_{bd} V_{d1}^2}$$

$$R_p = \frac{2E_p}{C_{bd} V_{d1}^2 - 2E_p} \tag{8.15}$$

Suppose that $C_{bd}$ was measured to be 65 pF and with $0\ \Omega$ for $R_t$, $V_{d0}$ was found to be 816 V. Then $E_p = 0.5 \times 65 \times 10^{-12} \times 816^2 = 21.64$ µJ. If with $1\ \Omega$ for $R_t$, $V_{d1}$ was found to be 1000 V, then to better handle Eq (8.15), calculate the common product $C_{bd}V_{d1}^2 = 65 \times 10^{-12} \times 1000^2 = 65$ µJ.

Now solving for $R_p$,

$$R_p = \frac{2 \times 21.64 \times 10^{-6}}{65 \times 10^{-6} - (43.28 \times 10^{-6})} = \frac{43.28}{21.72} = 1.993 \, \Omega$$

A similar calculation can be made for catastrophic degradation. Suppose that with the same type of board degradation occurs at 2675 V and 2770 V with $R_t = 0 \, \Omega$ and $1 \, \Omega$, respectively. We have

$$E_p = 0.5 \times 65 \times 10^{-12} \times 2675^2 = 232.56 \text{ µJ}$$

and with $V = 2770$ V,

$$C_{bd} \times V_{d1}^2 = 65 \times 10^{-12} \times 2770^2 = 498.74 \text{ µJ}$$

$$R_p = \frac{2 \times 232.56 \times 10^{-6}}{498.74 \times 10^{-6} - (465.12 \times 10^{-6})} = 13.835 \, \Omega$$

### 8.4.3 Induced potentials on isolated conductors

Enoch and Shaw described experimental measurements of the induced potential on an isolated panel from a person at a dc potential $V_b$ of 2000 V. The test setup and equivalent circuit are shown in Fig. 8.13.

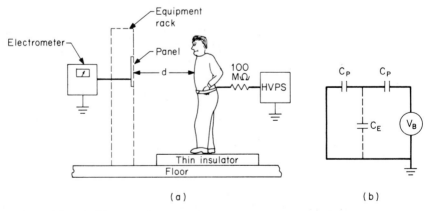

**Figure 8.13** Enoch-Shaw experiment of induced potentials on an isolated panel from a charged human subject: (a) test arrangement, (b) equivalent circuit where $C_{bp}$ is the mutual capacitance between the body and panel: $C_p$ is the capacitance of the panel to ground, and $C_E$ is the capacitance of the electrometer.

$C_{bp}$ is the mutual capacitance between the body and panel; $C_p$ is the capacitance of the panel to ground. The panel voltage is

$$V_p = \frac{C_{bp}}{C_{bp} + (C_p + C_e)} V_b \tag{8.16}$$

where $C_e$ is the electrometer capacitance.

$C_e$ and $C_p$ are known or can be measured; however, $C_{bp}$ depends upon $d$ and the effective area

$$C_{bp} = A_e \frac{\epsilon_0}{d} \tag{8.17}$$

where $\epsilon_0$ is the absolute permittivity = 8.85 pF/m.

Equation (8.16) can be rewritten

$$C_{bp} = \frac{V_p}{V_b}[C_{bp} + (C_p + C_e)] \tag{8.18}$$

$$C_{bp}\left(1 - \frac{V_p}{V_b}\right) = V_p \frac{C_p + C_e}{V_b} \tag{8.19}$$

$$C_{bp} = \frac{V_p(C_p + C_e)}{V_b} \frac{V_b}{V_b - V_p} \tag{8.20}$$

$$C_{bp} = \frac{V_p(C_p + C_e)}{V_b - V_p} \tag{8.21}$$

substituting into Eq. (8.17),

$$A_e = \frac{V_p d(C_p + C_e)}{\epsilon_0(V_b - V_p)} \tag{8.22}$$

The authors calculated an average $A_e$ of 0.126 m² for a panel size of about 8.75 in by 19 in or 0.107 m². Figure 8.14 shows predicted panel voltages with corrections for the effects of the electrometer; that is, $C_e = 0$.

## 8.5 Machine Model

The machine model, often referred to as the Japanese model because of extensive use in that country, consists of 200 pF with 0 Ω in place of the normal series resistance. The machine model is quite representative of common static sources such as charged-isolated panels, carts,

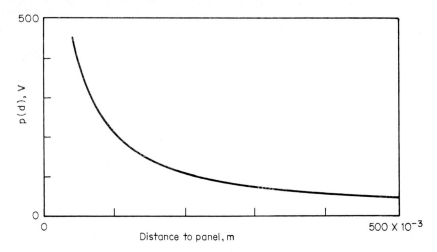

**Figure 8.14** Predicted induced panel voltages for a human charged to 2000 V as a function of distance.

vehicles, conductive tote boxes, or any isolated conductor. The waveform of a machine model (MM) discharge is similar to a CDM waveform as might be expected, since the only difference is a larger charged capacitance. Figure 8.15 shows a typical machine-model discharge waveform into a small (< 10 Ω) resistance. Peak currents can

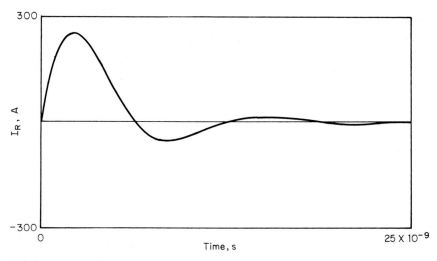

**Figure 8.15** Typical machine model discharge waveform into a small (< 10 Ω) resistance.

reach several hundred amperes, and the duration can be as much as several hundred nanoseconds, depending on the inductance and resistance in the discharge path.

## 8.6 Field-Enhanced Model or Body-Metallic Model

The terms field-enhanced model (FEM) and/or body-metallic model (BMM) are used to described the discharge of a human body through a hand held conductive tool such as tweezers. The result is that the field is concentrated at the sites of sharpest curvature of the tool and effective discharge resistance is reduced. Hyatt et al.[10] describe the model as consisting of a resistance of 350 to 500 $\Omega$ with the same human body capacitance as the HBM. Waveforms of HBM and BMM discharges are shown in Fig. 8.16 for comparison. Primary differences are the faster rise and decay times associated with the lower resistance of the BMM. Since the time constant is proportional to resistance, the transient durations are likewise proportional.

## 8.7 Capacitive-Coupled Model

My personal work experience[11] includes several serious static failures of a type that differed from all the preceding models. For this reason I herein define a new model called the capacitive-coupled model (CCM). This model encompasses the ESD failures that occur from capacitive coupling, whether to ground or from the charged source to a sensitive node of a susceptible item. Regardless of the actual stray capacitance position, the effect is to add a capacitive element in series with the discharge path. For test standardization purposes, a value of about 5 pF is proposed as reasonable and realistic value. The circuit for this model must take on several forms that are dependent upon the source of charge. Figure 8.17 shows (a) the human body CCM, (b) the body metallic CCM, (c) the charged-device CCM which is identical to the field-induced CCM, and (d) the machine model CCM.

The importance of the above CCM varieties is the subtle manner in which failure can occur when the path through the device is not obvious. It is important to note that a CCM discharge does not occur simply because of the presence of stray capacitance. A discharge must occur to a point electrically connected to one plate of the stray capacitance as depicted in Fig. 8.17 before the path can include the susceptible item. The fact that a capacitor projects an infinite impedance to dc and lower impedance with increasing frequency is helpful in visualizing capacitive-coupling behavior.

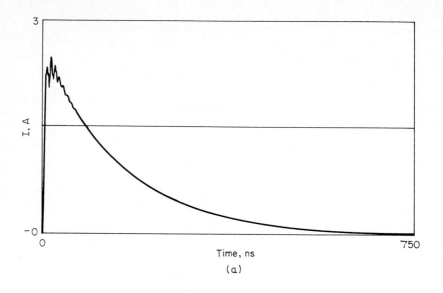

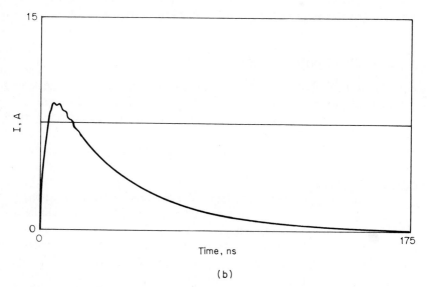

**Figure 8.16** Waveforms of (*a*) HBM and (*b*) body-metallic model (BMM) discharges.

## 8.8 Other Models

The preceding paragraphs are believed to include the major models for ESD failures. Certain variations and possible refinements of these models may be found in the literature. Just two additional models of a somewhat secondary nature but worthy of mention are discussed in this section.

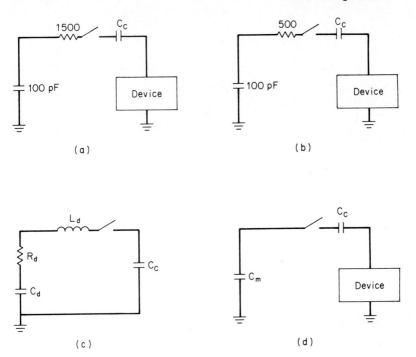

**Figure 8.17** Capacitive coupling models: (*a*) the human-body CCM, (*b*) the body-metallic CCM, (*c*) the charged-device CCM, (*d*) the machine model CCM which is identical to the field-induced–discharge CCM.

### 8.8.1 Floating-device model

Occasionally one will encounter references to the floating device model. This model has been used in studies of ESD test methods to include a device in an ungrounded or floating configuration.[12] Generally, the model is simply described as a floating device which is then subject to stress from the human body model. Conceptually, the device has near infinite impedance with respect to ground. Actually, a floating device has some stray capacitance to ground the value of which depends on many undefined variables in the model. The floating-device model is felt to be a somewhat trivial example that falls within the CCM subset of ESD models, and is worthy of consideration only upon better definition of the stray capacitance.

### 8.8.2 Transient-induced model

Transients such as from a spark discharge or other source can cause temporary upsets of digital systems such as computers. In particular microprocessors are susceptible to upsets. This model deviates from the primary topic of physical damage from ESD but represents a seri-

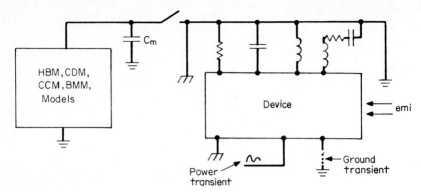

**Figure 8.18** The transient induced model can involve many possible paths.

ous problem. Figure 8.18 is an attempt to depict the many possibilities involved with the transient-induced model (TIM).

## 8.9 Conclusions

The ESD models discussed have significant differences based upon real world encounters in the electronics industry. In spite of these important differences there is a certain amount of overlap and similarity between models. An appreciation of these differences and similarities is important in establishing requirements for analysis, simulation testing, design protection methods, and ESD control.

## 8.10 Susceptibility Classification Testing

Susceptibility or sensitivity classification testing is necessary to quantify the relative degrees of susceptibility of different part types. The stratification of susceptibilities into classes is intended to provide a baseline for instituting progressive ESD control programs where more extensive controls are required for higher sensitivities. In addition, classification testing becomes an influence in part and vendor selection as a user attempts to minimize applications of highly sensitive parts.

The manner in which classification testing is conducted is a controversial and difficult issue. Several important questions must be addressed if one is to undertake classification testing.

1. Which susceptibility or damage models are to be tested?
2. Which pin combinations to test?
3. Which polarities for each pin combination selected?
4. How many parts are statistically representative?

Up to this time, most susceptibility testing has been based on the human body model with the hope that the results were correlatable to the other models. Testing a statistically significant sample of representative pin combinations and polarities is very costly.

### 8.10.1 Human body model testing

Most human body model testing in the United States is fashioned according to the requirements of DOD Standard 1686 or Mil Standard 883C test method 3015.1, or to some variation of the methods in these two documents. As of this writing the revised Military Standard 1686A has recently been released and is just beginning to be imposed on new military contracts. Because of the numerous existing contracts where the older DOD standard 1686 is imposed, requirements of both versions are discussed herein.

**8.10.1.1 DOD Standard 1686 sensitivity classification testing.** DOD Standard 1686 dated May 2, 1980[3] prescribes ESD sensitivity classification methods and categorization levels. Class 1 is defined as items with failure thresholds between 0 and 1000 V when tested to the human body model shown in Fig. 8.19. Class 2 is defined as items with failure thresholds between 1001 and 4000 V. Classification can be done by utilizing Table 2 of App. A of the document; by data from military part specifications, such as Mil-M-38510; or from testing as described in App. B of the standard. Appendix A, shown as Fig. 8.20, is

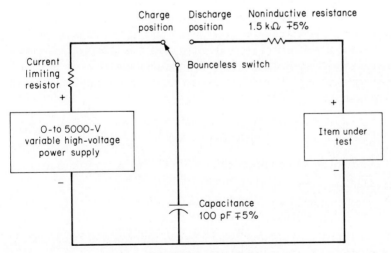

Figure 8.19   Human body model of DOD Standard 1686.

| List of ESDS Parts by Part Type (DOD-STD-1686, 2 May 1980) |
|---|
| Class 1: Sensitivity range 0 to ≤1000 V |
| • Metal-oxide semiconductor (MOS) devices including C, D, N, P, V and other MOS technology without protective circuitry, or protective circuitry having Class 1 sensitivity<br>• Surface acoustic wave (SAW) devices<br>• Operational amplifiers (OP AMP) with unprotected MOS capacitors<br>• Junction field-effect transistors (JFETs) (Ref.: Similarity to: MIL-STD-701: Junction field effect, transistors and junction field-effect transistors, dual unitized)<br>• Silicon-controlled rectifiers (SCRs) with Io < 0.175 amperes at 100°C ambient temperature (Ref.:S Similarity to: MIL-STD-701: Thyristors [silicon-controlled rectifiers])<br>• Precision voltage-regulator microcircuits: Line or load voltage regulation < 0.5 percent<br>• Microwave and ultra-high-frequency semiconductors and microcircuits: Frequency >1 GHz.<br>• Thin-film resistors (type RN) with tolerance of ≤0.1%; power >0.05 W<br>• Thin-film resistors (Type RN) with tolerance of >0.1%; power ≤0.05 W<br>• Large-scale integrated (LSI) microcircuits including microprocessors and memories without protective circuitry, or protective circuitry having Class 1 sensitivity (Note: LSI devices usually have two to three layers of circuitry with metallization crossovers and small geometry active elements)<br>• Hybrids utilizing Class 1 parts |
| Class 2: Sensitivity range >1000 to ≤4000 V |
| • MOS devices or devices containing MOS constituents including C, D, N, P, V, or other MOS technology with protective circuitry having Class 2 sensitivity<br>• Schottky diodes (Ref.: Similarity to: MIL-STD-701: Silicon-switching diodes [listed in order of increasing trr)]<br>• Precision resistor networks (type R2)<br>• High-speed emitter-coupled logic (ECL) microcircuits with propagation delay ≤1 ns<br>• Transistor-transistor logic (TTL) microcircuits (Schottky, low-power, high-speed, and standard)<br>• Operational amplifiers (OP AMP) with MOS capacitors with protective circuitry having Class 2 sensitivity<br>• LSI with input protection having Class 2 sensitivity<br>• Hybrids utilizing Class 2 parts |

**Figure 8.20** Appendix A of DOD Standard 1686.

generally regarded as insufficient to fill most contractor's needs. Most existing Mil-M-38510 data have been based on different classification levels and cannot be correlated very well. The last resort is to test using the method described in App. B. Details of the test procedure and other aspects of DOD Standard 1686 are covered extensively in Chap. 17, and discussed in this chapter only in relation to technical aspects related to tests and models.

**8.10.1.2 Military Standard 1686A[13] classification testing.** Military Standard 1686A contains the following new sensitivity classes:

*Class 1.* Susceptible to damage from ESD voltages greater than 0 to 1999 V as determined in accordance with Par. 5.2.1.1.

*Class 2.* Susceptible to damage from ESD voltages of 2000 to 3999 V as determined in accordance with Par. 5.2.1.1.

*Class 3.* Susceptible to damage from ESD voltages of 4000 to 15,999 V as determined in accordance with Par. 5.2.1.1.

Paragraph 5.2.1.1 of the standard prescribes parts-sensitivity classification as follows:

1. Requirements of applicable part specification
2. Accordance with App. A test data in the Reliability Analysis Center (RAC) ESD Sensitive Items List
3. Classification per App. B of 1686A
4. When specified or at contractor's option by test per App. A of 1686A

Classification by test as called out in App. A of 1686A is to use the test procedure of Military Standard 883, Method 3015. This is a great improvement in that the system level requirements of 1686A and parts level requirements of 883 are now in agreement.

**8.10.1.3 Mil Standard 883C Method 3015.** As of this writing the latest version of this requirement is Military Standard 883C Method 3015.6, notice 7.[14] This method is typically called out in Mil-M-38510 part specifications, used extensively by military contractors in defining integrated-circuit types and requirements. The method uses the same human body model with a few added parametric definitions of the simulator circuit's ESD waveform into a fixed resistor. This test method will be discussed further in Chap. 17.

### 8.10.2 Human body simulators

The standard human body simulation model shown in Fig. 8.19 is a rather simple circuit. However, the circuit is difficult to fabricate in order to achieve the voltage waveform shown in Fig. 8.21 and pre-

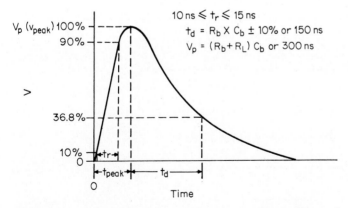

**Figure 8.21** Voltage waveform requirements of earlier revisions of Mil Std 883, Method 3015. (Load resistance = $R_L = 1500 \, \Omega$.)

scribed in earlier revisions of Mil Std 883. In addition, the circuit and waveform requirements have been very controversial.

Historically the commercial industry, for the most part, has not been successful in developing good reliable testers and many users have of necessity fabricated their own custom models. The largest obstacle is finding a fast switching device capable of up to 15 kV with no bounce, parasitic leakage or capacitance, timing, or other problems. In addition, stray capacitance and inductance, as well as leakage associated with the other components can drastically affect the waveform and test results.

**8.10.2.1 Inaccurate test results.** There have been numerous reports of inconsistencies in test results from different simulators. Commercially available testers apparently vary as much if not more than custom made units. Much of this variation had been attributed to differences in the voltage waveforms produced by each individual tester. Critical parameters were believed to be rise time, overshoot, ringing, and other variations of the waveform such as relay contact bounce.

**8.10.2.2 Current waveform.** In 1987 Lin et al.[15] reported results of sensitivity testing using seven commercially available simulators. Their conclusion was that the current waveform was extremely important. As a result of their study the test requirements of Mil Standard 883 Method 3015.1 have undergone revisions to include a properly defined current waveform. Limits on such critical parameters of the current pulse as overshoot and ringing are now specified.

**8.10.2.3 Cumulative damage results—dependency on waveforms.** Later in 1987, Lin et al.[16] reported on a subsequent study of differences in cumulative damage resulting from different simulators. The two testers showing the largest differences in their earlier study were selected for this follow-on effort. Current waveforms were evaluated

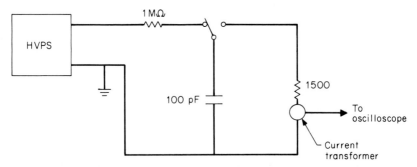

**Figure 8.22** Current waveforms were evaluated from the standard HBM using the test setup as indicated.

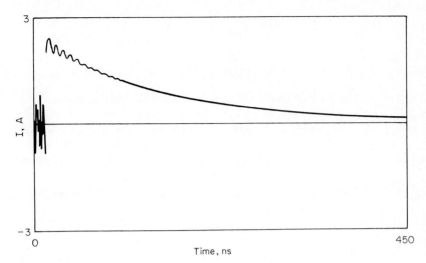

**Figure 8.23** A representation of the current waveform produced by simulator A from a 3750-V pulse.

from the standard human body model as indicated in Fig. 8.22. Current transformer probes used were either the 1-GHz Tektronix CT-1 or the 200-MHz CT-2. Figure 8.23 depicts the current waveform from simulator A from a 3750-V pulse. Figure 8.24 represents the current waveform from simulator B from a 750-V pulse. The theoretical peak currents should be

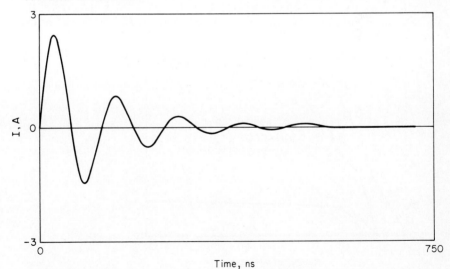

**Figure 8.24** A representation of the current waveform produced by simulator B from a 750-V pulse.

$$I_{PA} = \frac{3750 \text{ V}}{1500 \text{ }\Omega} = 2.5 \text{ A}$$

$$I_{PB} = \frac{750 \text{ V}}{1500 \text{ }\Omega} = 0.5 \text{ A}$$

In actuality the waveforms indicated $I_{PA}$ = 2.5 A, and $I_{PB}$ = 2.2 A due to the overshoot and oscillations on the waveform from simulator B.

Test structures used were from 2 megabit DRAM wafers from the same lot. Chips were selected, burned in, and tested prior to the stressing study to involve eight identical input pins per device. The single exposure thresholds to failure ($\geq 1$-$\mu$A) leakage from input to $V_{DD}$ and $V_{SS}$ (tied to ground) were 4500 and 2000 V for A and B, respectively. Cumulative tests consisted of a single exposure to a fixed level with failure-criteria monitoring after each exposure until failure or 500 exposures. Results are given in Figs. 8.25 and 8.26.

For simulator A, eight of the pins failing at 3750 V were from a device later discounted as being initially defective. Simulator A data were interpreted as little evidence of a cumulative mechanism. That is, devices tended to pass at 3750 V and fail early at 4000 V or above.

Simulator B data clearly indicate cumulative damage, especially when one considers the 500-V level which is only 25 percent of the threshold value. The plots from simulator B having an obvious change of slope at the higher voltages are indicative of two failure mechanisms. This effect is obvious in the 1000-V and 1250-V plots, less obvious in the 750-V plot, but nonexistent in the 500-V plot.

Failure analysis of representative failed pins from selected points on the plots verified the two mechanism hypothesis. Physical scan-

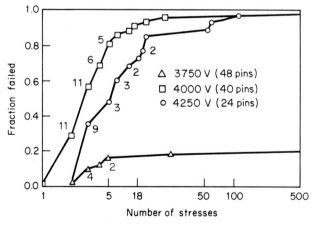

**Figure 8.25** Cumulative failures from simulator A. (*From* Lin,[16] *with permission from the EOS/ESD Association.*)

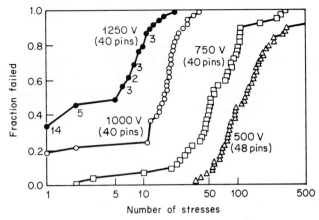

**Figure 8.26** Cumulative failures from simulator B. (*From Lin et al.,[16] with permission from the EOS/ESD Association.*)

ning electron microscope (SEM) examinations of failures from the initial slope region were determined to be due to a combination of electrothermal migration and PN junction thermal overstress in the input protection transistor. Failures from the higher slope region were found to have only junction thermal overstress signatures. The electrothermal migration mechanism was evidenced by damage to the aluminum contact windows. The thermal overstress signature was bird winglike cavities along the junction pattern. Failures from sim-

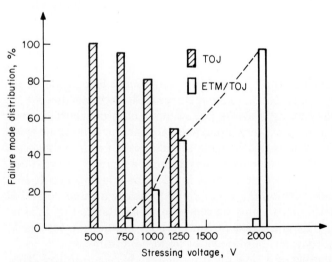

**Figure 8.27** Failure mode ratios for simulator B as a function of voltage. (*From Lin et al.,[16] with permission from the EOS/ESD Association.*)

ulator *A* evidenced both characteristic traits. Figure 8.27 shows the failure mode ratios for simulator *B* as a function of voltage.

The data indicated that the simulator with the most controlled current waveform produced the electrothermal migration and thermal overstress combination failure mechanism that was not particularly cumulative. Therefore a proper test need only include about five exposures to properly determine sensitivity.

### 8.10.3 Factors for consideration in HBM testing

A sensitivity classification method such as Mil Standard 883 Method 3015.3 has certain major areas of concern:

1. Test procedures
2. Pin combinations and polarities
3. Samples required
4. Waveforms whether voltage or current, or both
5. Rise time
6. Simulator calibration method
7. Definition of failure

Method 3015.3 provides detailed procedures defining sample size and pin combinations. The waveform is prescribed in general terms with a rise-time requirement of 15-ns maximum into a matched (1500-$\Omega$) load. An electrometer measurement of the effective stray capacitance is also required to be less than 10 pF. A failure is defined as any out of specification parameter or a change of more than 10 percent of any other parameter.

A test circuit such as the human body model must be considered as a high-frequency circuit. At high frequencies a discrete resistor is better represented as having a series inductance and a parasitic capacitance in parallel to the *RL* series portion. A capacitor will have series parasitic components of resistance and inductance. Of course the relay will add some other parasitic *RLC* constituents as will the device under test. Even if the device under test is taken as a resistance, it is a dynamic resistance with a maximum value typically around 200 $\Omega$. A suggested simulator circuit may be represented similar to Fig. 8.28 at high frequencies. The point is that human body simulation testing is most difficult and parasitic unknowns should be minimized.

The primary objective in classification testing is repeatability. This requires careful characterization of simulation testers. A single-voltage waveform or even a single current waveform definition is in-

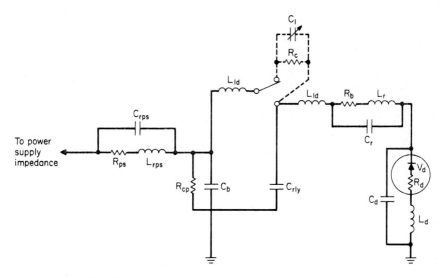

**Figure 8.28** A HBM simulator circuit at high frequencies may be presented as shown. ($C_b$ and $R_b$ = HBM capacitance and resistance; $R_{ps}$ = power supply limiting resistor; $V_d$ and $R_d$ = breakdown voltage and internal junction resistance of the device under test. The remainder of the circuit is made up of possible parasitic elements associated with the HBM simulator. Subscripts are indicative of parasitic elements of varying significance as follows: *rps*, associated with the power supply limiting resistance; *cp*, parallel to $C_b$; *r*, associated with $R_b$; *c*, associated with the switching or contact resistance; *rly*, associated with the relay; *d*, associated with the device under test; *ld*, associated with the leads and connectors.)

sufficient to properly define the source from a complex impedance perspective.

**8.10.3.1 Rise-time definition.** Rise-time variations can affect sensitivity test results. Avery[17] showed that lower thresholds to failure are achieved with faster rise times. This is to some extent a function of the switching speeds of protection devices.

**8.10.3.2 Failure criteria.** The present criteria for failure in classification testing requires examination. If a device goes out of specification limits it is deemed a failure. This is not only true by definition and logic, but seems to provide a defendable legal basis for rejection, changing vendors, or part selection. The 10 percent criteria for any other parameter provides additional margin for discerning and eliminating susceptible product, but seems virtually impracticable. How do you provide instrumentation to monitor unspecified parameters, and which ones do you select?

The possibility of latent damage has been cited in support of more stringent failure criteria. Too little is known about latent failures, but indications are that the damage required is extremely subtle. Classification testing is already quite expensive and prone to problems. The

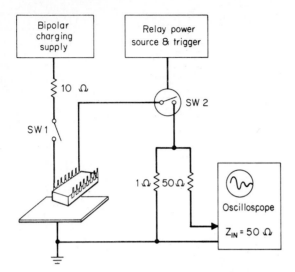

**Figure 8.29** The CDM test circuit proposed by Bossard et al. (*From Bossard,[7] with permission of ITTRI/RAC.*)

magnitude and complexity would increase tremendously if the tests were also expected to identify latency tendencies. The purpose of classification testing must be kept in perspective: to categorize sensitivity; not to screen for subtle damage.

### 8.10.4 Charged-device model testing

With the move toward automation in the industry, it is very likely that charged-device model failures will soon predominate. Thus testing for CDM failure susceptibility may become a necessity. Figure 8.29 shows the CDM test circuit proposed by Bossard et al.[7]. This circuit is pretty straightforward. A power supply is used to apply a fixed voltage to a pin (typically the substrate connection) to simulate the voltage the device might have acquired triboelectrically during processing, such as automatic test. Then the power supply is disconnected by SW1 leaving the device charged. The closing of SW2 discharges the device from another pin to ground through 1 Ω.

**8.10.4.1 Problems with CDM testing.** Any attempt at CDM testing must be done with care to minimize parasitic components, in particular inductance. Inherent inductance from the device is dependent upon the pin discharged and has a large effect on the waveform characteristics. Device inductances depend upon the package and pin count and are typically in the range of 5 to 10 nH for 16-pin DIPs. Device capacitances vary as does the complex proportioning of total capacitance among leads, bond pads, metallization, and substrate. All of these variables affect CDM failures.

### 8.10.4.2 Charged-device model failures in large-scale integrated circuits.

David E. Swenson and Norman P. Lieske[18] discussed results of CDM tests and experiments on a 16-dual in-line 256K DRAM. Their evaluations included effects of cumulative triboelectric generation from sliding on a polyethylene film and then grounding to a metal surface as well as CDM simulation testing. Devices tested were from two distinct sources: group A, manufactured outside the United States; and group B, manufactured in the United States.

**8.10.4.2.1 Triboelectric charging followed by grounding** Initial evaluations showed that devices from group B had a greater propensity for charge generation: ≥2.5 nC versus 1.6 nC for group A. For their approximate equivalent capacitances of 1 pF the voltages are ≥2500 V and 1600 V.

Test results showed no failures of 20 group-B devices after 1, 2, 4, 6, and 10 slides (totaling 23), and four failures out of 20 group-A devices for similar slide and discharge experiments. Interestingly, the four failed devices had 11 failed pins of which only three exhibited damage to pin 9 which was the ground contact pin. This result fortifies theories about internal inductive-capacitive coupling of charged-device waveforms, but negates earlier observations that damage to corner pins could be interpreted as circumstantial evidence of CDM failures. Using $V = Q/C$ to calculate voltage from measured charges indicated a sensitivity level of about 1400 V for group-A devices, whereas group-B devices could withstand greater than 2500 V.

**8.10.4.2.2 CDM simulation test results** Charged-device model testing was conducted on devices from both groups by charging to pin 8 ($V_{CC}$) through 1 MΩ and again grounding through pin 9. Groups A and B showed similar average failure-threshold voltages: approximately 550 V for group A and about 600 V for group B.

### 8.10.5 CDM conclusions

The differences obtained by triboelectric charge testing as compared to CDM simulation testing is significant. Theoretically a charged-device model failure occurs in a manner duplicated by the triboelectric testing so these results are considered most representative. The CDM simulation method of applying a charge from a power supply and suddenly grounding had been believed to give a close approximation to the real-world conditions, but apparently gives pessimistic indications.

### References

1. National Bureau of Mines bulletin 520, 1962.
2. Whitson J. Kirk et al., "Eliminate Static Damage to Circuits," *Electronics Design Magazine,* March 1976.

3. DOD Standard 1686, May 2, 1980.
4. Franklin Institute Research Laboratories, "Electrostatic Hazard to Electroexplosive Devices from Personnel-Borne Charges," February 1965.
5. R. D. Enoch and R. N. Shaw, "An Experimental Validation of the Field Induced ESD Model," *Electrical Overstress/Electrostatic Discharge Symposium Proceedings,* 1986.
6. T. S. Speakman, "A Model for the Failure of Bipolar Silicon Integrated Circuits Subjected to Electrostatic Discharge," *International Reliability Physics Symposium Proceedings,* 1974.
7. P. R. Bossard et al., "ESD Damage from Triboelectric Charged IC Pins," *Electrical Overstress/Electrostatic Discharge Symposium Proceedings,* 1980.
8. D. C. Wunsch and R. R. Bell, "Determination of Threshold Failure Levels of Semiconductor Diodes and Transistors Due to Pulse Voltages," *IEEE Transactions on Electron Devices,* vol. NS-15, no. 6, December 1968, pp. 244–259.
9. B. Unger et al., "Evaluation of Integrated Circuit Shipping Tubes," *Electrical Overstress/Electrostatic Discharge Symposium Proceedings,* 1981.
10. H. Hyatt et al., "A Closer Look at the Human ESD Event," *Electrical Overstress/Electrostatic Discharge Symposium Proceedings,* 1981.
11. O. J. McAteer, "Electrostatic Damage in Hybrid Assemblies," *Annual Reliability and Maintainability Symposium Proceedings,* 1978.
12. W. K. Denson and K. A. Dey, "ESD Susceptibility Testing of Advanced Schottky TTL," *Electrical Overstress/Electrostatic Discharge Symposium Proceedings,* 1982.
13. Military Standard 1686A, August 1987.
14. Military Standard 883C Method 3015.6, Notice 7, 1987.
15. D. L. Lin, M. S. Strauss, and T. L. Welsher, "On the Validity of ESD Threshold Data Obtained Using Commercial Human-Body Simulators," *IRPS 25th Annual Proceedings,* 1987.
16. D. L. Lin, M. S. Strauss, and T. L. Welsher, "Variations in Failure Modes and Cumulative Effects Produced by Commercial Human-Body-Model Simulators," *Electrical Overstress/Electrostatic Discharge Symposium Proceedings,* 1987.
17. L. R. Avery, "IC Technology: Where It Is Going and What It Means for the ESD Industry," *Electrical Overstress/Electrostatic Discharge Symposium Proceedings,* 1985
18. David E. Swenson and Dr. Norman P. Lieske, "Triboelectric Charge-Discharge Damage Susceptibility of Large Scale IC's," *Electrical Overstress/Electrostatic Discharge Symposium Proceedings,* 1987.

# Chapter 9

# Analysis of ESD Failures

## 9.1 Introduction

Interest in the topic of ESD failures has mushroomed in recent years. A motivating factor has been the desire to understand the mechanisms in order to provide proper controls to prevent such failures. ESD controls can be costly, and management must consider implementation tradeoffs. The inevitable question is: "How big is our static problem, anyway?" Unfortunately, many small facilities do not have the failure analysis capabilities to properly identify static failures. Assessment of the extent of static problems is very difficult even with fully equipped laboratory facilities. Therefore ESD control tradeoff decisions are often based upon published data describing the extent of the problem at other companies. A small company or facility can attain a better problem asessment through the use of independent analysis laboratories to determine the cause of failures.

The material contained in this chapter is intended to benefit novices to the field, experienced analysts with extensive laboratory facilities, companies wishing to establish failure-analysis facilities, and those involved with ESD control decisions. The latter group requires a degree of knowledge of ESD failure-analysis procedures and physical traits in order to evaluate the work of others.

The approach to analysis of static failures, and individual techniques and methodologies are discussed. Selected case histories are included in Chap. 10 to illustrate the application of the procedures discussed.

## 9.2 Approach to Analysis of Static Failures

Those unfamiliar with failure-analysis procedures often raise the question, "How do you analyze ESD failures?" Going one step further they often come from two distinct camps. Either they have preconceived notions that it is virtually impossible to positively identify ESD as the cause of failure or that it is simply a matter of removing the device lid and performing a microscopic examination. Those of the first camp have heard about the subtle nature of static failures and in particular the difficulty in distinguishing EOS from ESD failures. Those in the second camp have seen photomicrographs identifying ESD damage and are left with the impression that identification is easy. The assumption that an investigator could, within a few minutes, have the damage photographed under an optical or scanning electron microscope belies the many tedious steps required to reach that point.

The inexperienced cannot be expected to realize the presumption in the very first question as to how ESD failures are analyzed. If the investigator has reached the point of knowing that it is an *ESD failure,* a great deal of the analysis is already complete. To assume a failure is due to ESD could have dire consequences. After all, if the failure were really due to another cause such as excessive moisture content or mobile ions, then the problem would continue uncorrected. The point is that the approach to failure analysis must be general if it is to be objective.

There is no doubt as to the subtle nature of many static failures. Yet the consequences of such failures often mandate that failure-analysis procedures be able to identify the cause. Normally there is little difficulty in distinguishing ESD from EOS. However, extenuating circumstances will occasionally mask the usual distinctive traits. On the other hand the problem of identifying ESD failures is not just a matter of microscopic inspection. The use of scanning electron microscopy must appear to some as a convenient panacea to identify tiny ESD failure traits. Experienced analysts realize that if a scanning electron microscope (SEM) is used at all, it is usually the last of many procedural steps.

Knowledge of failure mechanisms other than static electricity is necessary in order to justify the elimination of other candidate causes during the analysis process. Therefore, techniques outlined in this chapter will provide guidance in the analysis of all failures, but those methods of particular value with ESD problems will be emphasized.

## 9.3 Failure-Analysis Elements

Identification of electrostatic discharge as the unequivocal cause of failure can be difficult. Procedures based upon the same *scientific method* used for failures of any cause can alleviate much of the required effort. Such a comprehensive failure-analysis procedure consists of the following major elements:

- Notification
- Fact gathering
- Part analysis
- Failure-cause identification
- Corrective action

## 9.4 Notification

The designation "notification" is intended to include all the important details associated with observation, recognition, recording, and reporting of the discrepant condition. Some important factors are appropriate test specifications, definition of a discrepancy, and monitoring methodology. The format for reporting discrepancies must facilitate conveyance of the most useful information while remaining brief and avoiding ambiguity.

Certain data should be routinely recorded on the discrepancy reporting form. Identification information such as assembly, subassembly and part drawing number, serial number, manufacturer, and date code are mandatory. Other recorded information would include the test specification and paragraph number, date, person reporting, symptoms, environmental considerations if applicable, and any related failures or parts removed. It is prudent to allow a designated block with sufficient space for additional information at the discretion of the person reporting. Procedures should require inspection of parts prior to removal under good ESD controls as well as careful mechanical-handling conditions. Static-sensitive parts should be placed in static-protective packaging for subsequent handling.

## 9.5 Fact Gathering

The circumstances associated with a discrepant condition are frequently the most pertinent factors in resolving the cause. If practical, the investigator should visit the location of the occurrence to interview personnel involved and observe general conditions in the area. Failure histories of the parts and assembly involved should be reviewed for clues as to the cause of the problem at hand. The fact gathering effort can be thought of as analogous to a detective's sleuthing. Just as one cannot write a general list of items that a detective must check in order to solve a crime, a pat list for ESD failure fact gathering would be self-defeating. However, a few suggested questions applicable to any failure cause might include:

1. What were the symptoms of failure?

2. How was the problem isolated to the subassembly or part level? What diagnostics were used?
3. Did replacement of the isolated part cure the problem?
4. Were special conditions required for observation of the failure or discrepant condition such as specific temperature, switching order, operating time, supply voltage level, or mechanical stress?
5. Are there other pertinent facts that have not been reported? Candid discussions with testers, assemblers, inspectors, storeroom personnel, or others might bring out a crucial bit of information.
6. Have there been similar failures in the past? What were the findings, if any?

Where ESD is a likely causal candidate, questions might include:

1. Were all specified ESD preventive measures being followed?
2. Did personnel attitude and knowledge reflect a keen ESD awareness?
3. Had there been any recent changes in equipment, procedures, or personnel? Subtle changes have a history of contributing to ESD problems.
4. What levels of static potentials can be measured in the area?
5. How sensitive to ESD is the part or assembly?
6. Was there any handling prior to detection of the discrepancy?
7. Is there a history of static problems with the item?

A single answer to a question of the types listed might yield information that could shorten the analysis considerably. For instance, if fact gathering gave strong indications of static damage, then a minimal part analysis to verify the damage may be all that is required.

## 9.6 Part Analysis

Part analysis consists of preliminary analysis, external package-analysis steps as warranted by failure symptoms and importance, and physical dissection. It is important to take great care in physical handling of a failed part in order to avoid misinterpretation of post-failure induced factors as being related to cause of the original malfunction.

Like a physician, the part analyst determines each new procedural step based upon results of the preceding step. Table 9.1 lists typical procedures included in part analysis. In practice, the approach is tai-

**TABLE 9.1 Part Analysis Steps**

A. Preliminary Analysis
- External visual examination (1× to 30×)
- Verification of failure
- Characterization of failure
- Circuit analysis

B. Physical Analysis
- Delidding
- Reverification after delidding
- Internal visual (10× to 1100×)
- Failure site isolation
- Characterization of failure site

lored, and judicious short cuts avoid irrelevant items of this comprehensive list.

### 9.6.1 Preliminary analysis

Disregard for the preliminary steps listed in Table 9.1A can result in loss of important information relating to the cause of failure. For instance, the low-power external-visual examination might reveal physical damage or signs of overheating. Functionally good parts are often erroneously removed, so the failure must be verified. Characterization of the failure by curve tracer and/or other electrical measurements can often provide sufficient information to isolate the failure site by circuit analysis prior to delidding the package. Minimally, pin-to-pin electrical characteristics relating all pins involved in the electrical failure with all other pins should be documented for both bias polarities. The curve tracer (see Fig. 9.1) provides a convenient instrument to accumulate this valuable information. In the absence of such preliminary steps, inadvertent damage during analysis steps, such as delidding, might lead to false conclusions.

### 9.6.2 External package analysis

Most routine failure-analysis procedures require only the preliminary analysis steps prior to physical dissection of a failed part. Sometimes, due to high reliability or other critical requirements, additional analytical steps on the external package are prescribed. In some instances leak tests, particle impact-noise detection (PIND) and/or residual gas

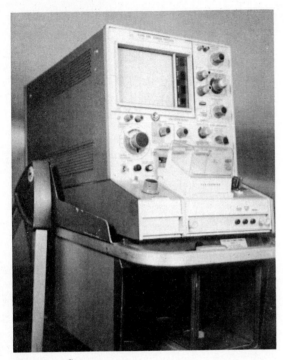

Figure 9.1 Curve tracer.

analysis (RGA), or temperature cycling are advisable based upon symptomatic indications of certain failure causes. These steps relate to ESD only in that they are checks for other failure causes, and could help disprove such suspected causes, as appropriate.

### 9.6.3 Physical dissection

Care must be taken in the physical-dissection steps listed in Table 9.1$B$. Even the delidding operation cannot be taken for granted and considered nondestructive. An example of operational amplifiers altered by ESD during handling associated with the delidding process is discussed in Chap. 10. Electrical characterization following the delidding process can reveal new damage or verify the state of the original anomaly. Careful microscopic examination normally follows the postdelidding electrical characterization, and dissection procedures.

Any further electrical isolation required must be carefully planned and carried out in a stepwise fashion. Each progressive step must be "least destructive" to facilitate recording of pertinent electrical measurement information. In other words electrical measurements are to be made from the bonding pads prior to any intraconnect microprobing or especially any severing of intraconnects for isolation purposes.

**9.6.3.1 Delidding.** The first step of physical dissection is to open or delid the part. The fact that each technique must be applied in a careful manner so that unrelated damage is not induced cannot be overemphasized. Discrete parts such as transistors, diodes, film resistors, and capacitors require various delidding techniques depending on the package type. Integrated-circuit chips in ceramic packages can have Kovar metal lids or be sandwiched between two ceramic lids with a glass seal. Kovar lids can be removed by heating the solder seals or by microsectioning until thin enough to peel back with tweezers. Packages with ceramic lids can be effectively removed using a commercial IC opener as shown in Fig. 9.2. The most difficult package to enter while avoiding damage to the chip is a plastic package with no internal cavity. Chemical etching is usually the best alternative.

**9.6.3.2 Reverification after delidding.** A good practice is to reverify the failure symptoms or electrical characteristics after delidding. This is not just because of the potentially destructive nature of the delidding operation. A relevant example of changes in electrical characteristics indicative of electrostatic damage occurring during the delidding is described in Sec. 10.4.1.1.1. Changes might also indicate other things such as light sensitivity or particle movement.

**9.6.3.3 Internal visual (10X to 1100X).** The next step of physical analysis is internal microscopic inspection. This consists of an overall inspection of the leads, wires, bonds, chip configuration and placement

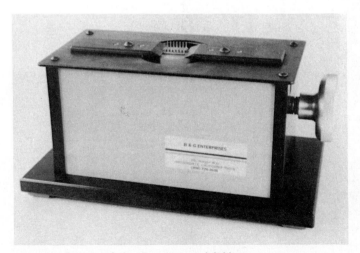

Figure 9.2  Integrated circuit opener, or delidder.

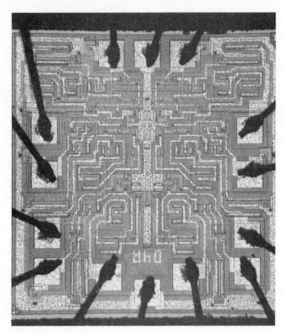

**Figure 9.3** Low-power visual example.

at 1× to 50× magnification. An example is shown in Fig. 9.3. This examination is usually done using a stereomicroscope similar to that shown in Fig. 9.4. Anomalies relating to the failure mode would be photographed.

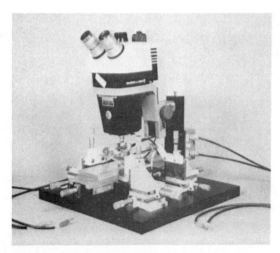

**Figure 9.4** Stereo microscope.

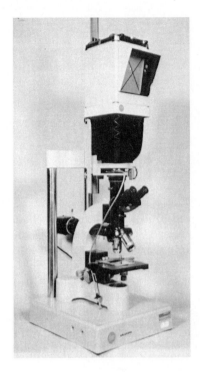

Figure 9.5 Metallurgical microscope.

A higher magnification (50× to 1100×) inspection of the chip surface can be made using a metallurgical microscope such as that shown in Fig. 9.5. Figure 9.6 shows a representative view of an integrated-circuit chip at 100×. The types of defects that can readily be identified include such process, metallization, and dielectric related faults as crystal defects, diffusion faults, masking defects, junction faults, bond defects, scratched metal, metal migration, corrosion, oxide faults, oxide pinholes, particulate, and other contamination. The absence of visible defects related to the electrical problem might be supportive of ESD as a possible cause. This is due to the lower power levels involved resulting in little physical evidence, often below the chip's surface. Obvious signs of electrical stress such as melted gold or aluminum wires, large metallization melts and severe junction damage are typical of higher level electrical overstress. One must be cautious in this regard, however, as ESD damage can often appear to be like EOS under some rarely occcurring conditions. Refer to Fig. 7.21 and the discussion of this example in Sec. 7.6.

Certain enhancements to high-power microscopy can help accentuate surface structural traits. These would include dark-field, oblique, phase-contrast, interference-contrast, and modulation-illumination-contrast attachments.

**Figure 9.6** Representative 100× magnification view of IC.

**9.6.3.4 Failure-site isolation.** Table 9.2 lists common techniques used to isolate the damage site so that it can be related to the cause of failure. Table 9.3 shows that some of these same techniques are included with tools used for characterization of the failure mechanism in order to determine the cause and corrective action.

**9.6.3.4.1 Electrical microprobing.** Electrical pin-to-pin characterization, listed as the first step of electrical-site operation, has already been discussed in Sec. 9.6.3.2. This type of characterization can be extended into the various circuit elements of the chip by microprobing. This technique involves point-to-point probing and isolating on the surface of a chip and with appropriate expertise can be an effective manner of positively locating a fault. ESD problems, which usually result in resistive shorts or leakage paths, are particularly suited for detection by this means. Needle probes connected to a curve tracer and physically positioned by three-dimensional micromanipulators are used to contact device metallization intraconnect lines to make characterization measurements between internal points of the chip. Tungsten is generally chosen for the probe material since it remains sharp longer than steel and can usually break through the overlying glassivation

## Analysis of ESD Failures

**TABLE 9.2  Failure-Site Isolation Methods**

- Electrical pin-to-pin characterization
- Electrical microprobing
- Microsurgery
- Liquid crystals: cholesteric and/or nematic
- Chemical etching
- Chemical isolation
- Plasma etching
- Microsectioning
- Scanning electron microscopy
- SEM voltage contrast
- Electron-beam induced current (EBIC)
- Other special techniques
  Ion etching
  Elemental analysis techniques
  Fine and gross leak tests
  Infrared scanning
  Chemical fault decoration

**TABLE 9.3  Failure-Site Characterization Methods**

- Electrical pin-to-pin measurements
- Electrical microprobing
- Chemical etching
- Scanning electron microscopy
- Other special techniques
  Fine and gross leak tests
  RGA
  PIND
  Biased or nonbiased thermal cycling
  Chemical fault decoration
  Elemental analysis techniques

layer. Tungsten probes can be resharpened by electrolysis in a solution of NaOH.

**9.6.3.4.2 Microsurgery.** The use of the tungsten microprobes to physically isolate portions of the circuitry by severing selected metallization lines is referred to as microsurgery. Perhaps this is because of the degree of skill required to perform this delicate manual task . This destructive step is done only after serious contemplation. Isolation of individual components on an integrated circuit must be thorough as many parallel electrical paths may exist on an integrated circuit, often in addition to the ones indicated by the schematic. A good practice is to verify that metallization cuts are complete by verifying open circuits across the cuts. Typically alternating steps would consist of microprobing before and after microsurgery until the failure site is electrically located. Figure 9.7 shows an integrated circuit after several such steps.

Ultrasonic probes can be effective in cutting conductors through 30,000-Å oxide layers. A common practice that is discouraged is the

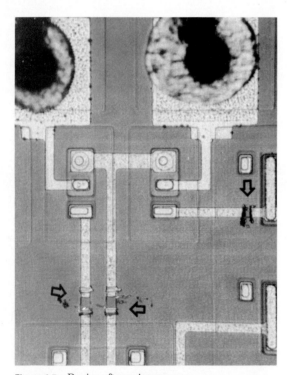

Figure 9.7 Device after microsurgery.

use of capacitive discharge to melt open metal lines for isolation purposes. This could be detrimental to parallel circuitry, especially of ESD-sensitive parts.

### 9.6.4 Special methods and techniques

Visible characteristic traits typical of static failures are discussed in the literature.[1,2] However, microscopic examination often reveals no visible anomaly and combinations of additional special methods such as liquid crystals, chemical etching, and SEM techniques are required to isolate and identify the failure mechanism.

**9.6.4.1 Chemical etching.** Chemical etching is routinely used to access subsurface damage sites. For example, static damage in MOS devices typically occurs in the gate-oxide layer, underneath the gate metallization. A metal, typically aluminum, etch can be used to reveal the gate-oxide layer. Metallization lines can also be chemically etched for electrical isolation purposes. Etching of the silicon dioxide glassivation layer is often used to reveal hidden damage at the silicon–silicon dioxide interface of bipolar devices. Table 9.4 lists formulas for a few commonly used etchants. In practice, experimentation has often resulted in modifications to these formulas for best results in a particular application.

**9.6.4.2 Chemical isolation.** Chemical etching of 3-μm metal lines for electrical isolation has been accomplished by using black wax to mask the adjacent areas from the etchant. This method is atypical and would be used only where microsurgery is felt to be more likely to damage a particular device.

**9.6.4.3 Plasma etching.** The term "plasma" means an ionized and dissociated gas. In plasma etching the reactant gas is excited into an active plasma by an *rf* field. The plasma is preselected to react in such a manner as to remove material from the sample material to be etched. For instance, the standard reaction for removal of silicon is

$$SiO_2(s) + CF_4(g) \rightarrow SiF_4(g) + CO_2(g) \tag{9.1}$$

This plasma-etch method will not attack the aluminum or nichrome, and the reaction temperatures are rather benign at 100°C.

Etch rates usually follow an Arrhenius dependence on substrate temperature:

$$\text{Etch rate} \sim e^{-E_a/kT} \tag{9.2}$$

TABLE 9.4  Formulas for Common Etchants

| Material to be etched | Etchant | Application temperature |
|---|---|---|
| Aluminum | 25 ml H$_3$PO$_4$ (80.6%) | @ Room temp (RT ~ 22°C) |
| | 5 ml acetic acid (16.2%) | |
| | 1 ml HNO$_3$ (3.2%) | |
| Al-Si | | |
| Al-Cu-Si | | Ultrasonic etch sometimes preferred |
| Chromium | 220 g ceric sulfate | RT |
| | 2000 ml H$_2$O | |
| | 200 ml H$_2$SO$_4$ | |
| | or | |
| | 300 g ceric sulfate | 55°C |
| | 3000 ml H$_2$O | Stir 1 h before use |
| | 450 ml HNO$_3$ | |
| Gold | C35 full strength or diluted as desired | RT to 55°C |
| | or | |
| | 150 ml C35 | 40 to 45°C |
| | 150 ml H$_2$O | |
| | 4 g CrO$_3$ | |
| | or | |
| | Aqua Regia: 25 ml HCl | 40°C |
| | 75 ml HNO$_3$ | Can be diluted with H$_2$O if desired |
| | or | |
| | Glyceregia | RT |
| | 50 ml Aqua Regia | Discard after 1 h |
| | 50 ml H$_2$O | |
| | 50 ml glycerine | |
| Palladium | 300 ml H$_2$O (dionized) | 40°C |
| | 144 ml HNO$_3$ | Use within 30 min |
| | 7.5 ml HCl | |
| Polysilicon | 12 ml acetic acid | RT |
| | 5 ml HNO$_3$ | |
| | 1 ml HF | |
| | or | |
| | KOH | Various strengths and temperatures |
| Silicon | Hydrazine (64%) | 80 to 100°C (Anisotropic) |
| | or | |
| | 60 ml Ethylenediamine | 85°C |
| | 40 ml H$_2$O | |
| | 12 g Pyrocathecol | |
| Silicon dioxide | 2400 ml NH$_4$F | RT |
| | 350 ml HF | |
| | 800 ml ethylene glycol | |
| | or | |
| | 1800 ml NH$_4$F | RT |
| | 750 ml H$_2$O | |
| | 600 ml ethylene glycol | |
| | 300 ml HF | |

where the activation energy $E_a$ is a characteristic of the material to be etched, thus preferential etching of different materials also varies exponentially with temperature.

**9.6.4.4 Liquid crystals.** Certain organic compounds exhibit an intermediate "liquid-crystal" state between the solid and liquid phases. This liquid-crystal state can be of the cholesteric, nematic, or smectic types, all of which exhibit unusual optical properties. The liquid-crystal isolation technique has been invaluable in finding ESD failure sites in complex integrated circuits. Both cholesteric and nematic liquid crystals can be useful. The cholesteric type responds to heat by changing color contrast as shown in the black and white photomicrograph of Fig. 9.8. This characteristic is directly applicable to highlighting the sight of resistive shorts by detection of the $I^2R$ heating.

Liquid crystals normally react in a narrow temperature range of just a few degrees Celsius. The appearance changes from clear to fringes of the entire red to violet spectrum as the narrow temperature window is experienced, and then back to clear again outside the reactive range. Several techniques have proven to enhance the effectiveness of isolation of shorts or thermal "hot spots" with liquid crystals.

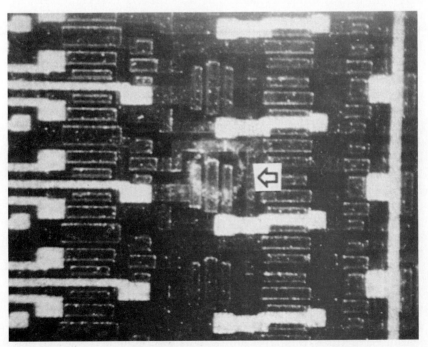

Figure 9.8  Cholesteric liquid crystal.

Obviously, it is necessary to elevate the temperature of the defect site above the temperature of the remainder of the device. Current passing through the defect site will usually accomplish this by $I^2R$ heating. However, if the current is excessive, the defect might be burned "open" destroying the evidence. Thus, the current must be limited (typically to 5 μA or so) depending upon the device and the nature of the short site. Preheating the entire chip to a constant elevated temperature just below the crystal's reactive range will also reduce the defect-current requirement. If a constant current is maintained the temperature will spread rapidly from the defect site to the remainder of the chip producing a *ballooning* effect of color change followed by a clear appearance that prevents resolution of the defect site. This can be overcome by careful determination of a pulsing frequency (typically a few kilohertz) that approximates the proper constant elevated temperature at the short.

Aside from these basic thermal considerations, lighting is of utmost importance. One circular component of incident light is transmitted by the liquid crystal, while the other component is either absorbed or scattered. Thus an oblique (low side angle)-incident light source must be used rather than the direct lighting usually provided by metallurgical microscopes.

A significantly different method utilizes the fact that cholesteric liquid crystals rotate the plane of polarization of the transmitted light. The liquid crystal will appear transparent when viewed through a polarizing microscope having perpendicular polarizers in both the incident light and optical viewing paths. If the short site is heated sufficiently to change from the cholesteric to the liquid phase, the polarized light is no longer rotated in that area and an observed black spot will locate the defect site.

Nematic liquid crystals respond to electric fields by color contrast. Alternating fields such as those produced by chip metallization lines being pulsed cause the nematic liquid crystals to display bright patterns whereas conductors with constant dc voltages or no fields will appear dark. The effect illustrated in Fig. 9.9 is somewhat analogous to the scanning electron microscope voltage-contrast technique. This method is particularly useful in displaying the operating stages of a device to find the stage where the alternating signal path becomes discontinuous.

Liquid crystals have the advantage of being nondestructive as they can be easily rinsed away after use.

**9.6.4.5 Microsectioning.** Microsectioning is analogous to the metallurgist's technique of cross-sectioning such things as weld joints, honeycomb structures, and rivet joints. This technique is frequently used

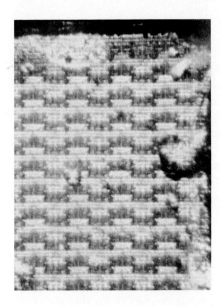

Figure 9.9  Nematic liquid crystal.

for bondability problems, plating thicknesses, seal or die attach evaluations, metallization step coverage, junction depths, oxide thicknesses, but might rarely be used to evaluate subsurface ESD damage.

**9.6.4.6  Scanning electron microscopy.**  High-power optical microscopes are typically limited to a useful magnification of about 1200X. The scanning electron microscope can give useful results up to 100,000× and a resolution at least as good as 200 Å. An Å is $10^{-10}$ m, or one ten-billionth of a meter. To some even more outstanding is the great depth of field providing a near three-dimensional quality to SEM photos.

The scanning electron microscope works on the principle of an electron beam impinging on the surface of the sample and generating secondary electrons that are detected to form an electronic image of the sample surface. The source of primary electrons is typically a thin tungsten wire that is heated by current flow. The electrons are accelerated by high-voltage (up to 25 kV) electron guns and scanned over the sample surface.

There are several problems associated with the use of the SEM due to the manner in which it operates. A highly accelerated electron beam can have a detrimental effect on semiconductor devices due to ionization effects that can degrade electrical properties. The effect is often minimal when low-beam currents and short-exposure times are used. In addition the electron beam tends to negatively charge portions of the sample to varying degrees. The insulative portions are normally the first to become highly charged and then take on a light appearance in the black and white SEM photographic mode. For this reason samples are routinely coated with a conductive material such as vacuum-deposited gold so they can be electrically grounded while under SEM examination. These disadvantages are pointed out as precautions to the failure analyst so that other electrical information will be acquired prior to the potentially destructive SEM operations of sample coating and/or electron-beam exposure.

This instrument is often necessary as the final step in analysis of electrostatic-discharge failures. After the other steps have been utilized to isolate the failure site the high magnification of the SEM may be required to show the tiny physical damage feature that remains.

**9.6.4.7 SEM voltage contrast.** The voltage contrast mode of the SEM can be considered as an extension and practical utilization of the charging effect discussed in Sec. 9.6.4.6. When the electron beam negatively charges a portion of a sample, the degree of lightness indicates the magnitude of charge. This is due to the propensity of a negatively charged area to emit secondary electrons. A positively charged area would have a low propensity for secondary electron emission and would appear dark in comparison. Thus if two electrically isolated metallization tracks on an integrated circuit were biased to voltages of opposite polarities, the positive track would appear dark and the negative track would appear light. Figure 9.10 shows how this technique can be used to find an open circuit in a thin-film resistor. No defect was visible in this resistor at high-power optical or normal SEM examination. Figure 9.11 is an example of an open-metallization oxide step that is invisible to optical microscopy. In the case of complex integrated circuits, the voltage-contrast image of an operating good device can be compared to that of the defective device. Inoperative stages, especially shorts or open circuits, can be identified easily. Detection of degraded parameters are greatly dependent upon the operator's skill, but normally a differential of at least 0.5 V is required for perception. Figure 9.12 shows a defective integrated circuit in the voltage-contrast mode.

**9.6.4.8 Electron-beam induced current (EBIC).** The focused electron beam penetrates the sample to a depth determined by the acceleration voltage. The impingement of this beam will create hole-electron pairs in semiconductor specimens. A current is thereby induced in the specimen. The current magnitude is dependent upon the beam current and the geometries and dopant levels of the sample. The induced current is detected and supplied to a high-gain amplifier for greater differential detection. The amplifier and control circuitry are designed to synchronize the electron-beam induced current with the SEM's cathode-ray

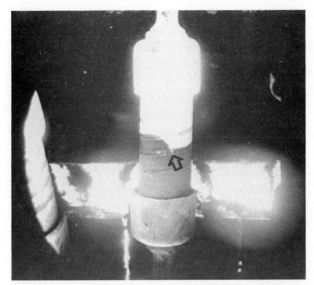

**Figure 9.10** Open circuited thin-film resistor under voltage contrast.

display. The resultant image is varying degrees of bright and dark dependent upon the beam location on the sample. EBIC is virtually unaffected by surface charging so the sample need not be coated. Most practitioners utilize EBIC by comparing response of good versus defective parts. The method can be very useful in detecting subtle subsurface electrostatic-discharge damage in bipolar devices.

### 9.6.5 Other special techniques

Certain other invaluable failure-analysis techniques that have limited or no applicability to ESD analysis are fine and gross leak test, infrared microscopy, infrared thermography, chemical-fault decoration, and elemental analysis techniques.

## 9.7 Failure Cause Identification

The failure-analysis steps discussed thus far, including a gradually acquired repertoire of "characteristic traits," are tools to be used to best accomplish the end purpose of failure analysis. That purpose is to positively identify the cause of failure so that effective corrective action can be defined and implemented. In order to identify the cause of

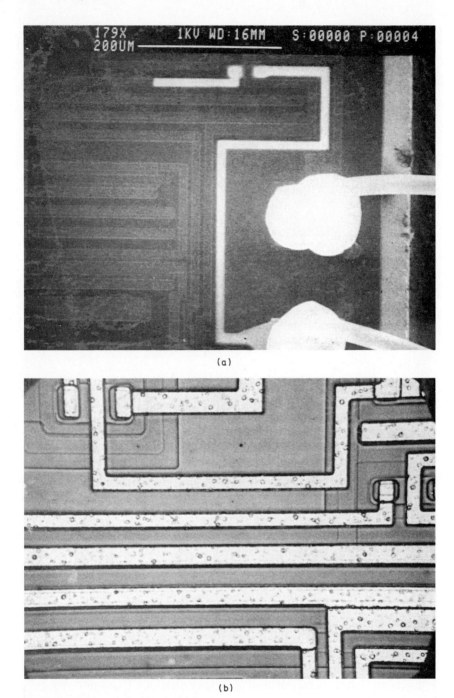

**Figure 9.11** Open metallization step. (*a*) By voltage contrast; (*b*) open circuit not discernable by optical microscopy (400×).

Analysis of ESD Failures   223

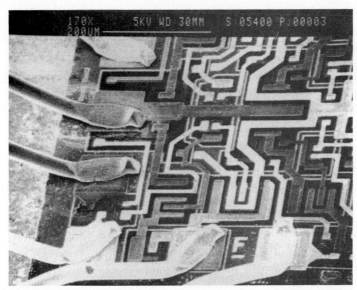

**Figure 9.12** Defective integrated circuit under voltage-contrast mode.

failure the information gained in fact gathering and part analysis is reviewed to deduce the most likely cause of failure.

For ESD failures the cause identification is sometimes straightforward. Suppose fact gathering revealed a blatant case of improper handling, packaging, grounding, or presence of highly generative material and the device had physical traits indicative of ESD. The cause can be assigned to the improper procedure and corrective action is to initiate appropriate handling, packaging, or grounding measures, and/or to remove the generative material.

Frequently, the cause is not so readily determined, and additional steps are necessary. Table 9.5 lists common options to select for additional information to better pinpoint the cause of failure. The steps

**TABLE 9.5  Failure Cause Identification Steps**

A. Analysis of fact gathering and part-analysis data
B. Reinspection of scene of failure occurrence
C. Duplication of failure
D. Study of design
E. Consultation with vendor
F. Outside consultations
G. Control-measure evaluations

listed are used in a selective and iterative process based upon new information gained.

It is important to point out that merely proving that the failure was due to static discharge is not sufficient. Implementation of a reasonable level of ESD control is common, and the analyst must identify the weakness or infraction in these controls in order to recommend corrective measures. For example, suppose part analysis indicated ESD, but there was no obvious infraction or problem source. A more careful reinspection of the scene of failure might disclose that the item had been handled as an ungrounded or improperly grounded work station.

Any doubt about the proper identification of failure cause can best be dispelled by duplication of the failure. That is, identical parts are stressed at the designated pin combinations and polarities indicated by results of all failure-analysis steps taken. When the same symptoms, leakage paths, and physical-damage traits are obtained, one can be confident that the actual failure conditions have been defined.

Additional measures selected from Table 9.5 may be necessary to positively identify the source and contributors to the accumulation of the static level involved. At higher assembly levels additional studies of the design may be necessary to determine the path of the static transient to the failed part. Sometimes the effort will necessitate further evaluation of ineffective ESD control measures such as protective packaging.

## 9.8 Corrective Action

Initial corrective action for first encounters with ESD usually result in implementation of some level of static control where known susceptible items are processed. As additional failures occur there is a gradual improvement and extended usage of control measures until a level that could be defined as "good ESD practices" is attained. Details of such practices are covered in great detail in later chapters.

After a certain degree of control is in place, corrective action resulting from failure analysis must become more specific. This can only come about through precise failure analysis and proper failure cause identification to the localized problem source. Corrective action for ESD can involve a number of areas. Improvement or elimination can be brought about by changes in design, control equipment, or procedures. Table 9.6 lists some common changes affected.

## 9.9 Key Elements

As far as ESD failure analysis is concerned, I have suggested earlier that a generalized approach must be maintained. After all, a failure

TABLE 9.6  Common Corrective Action for ESD Failures

- Design modifications
  Mechanical package (chassis)
  Protective circuitry
  Improved grounding
  Different physical layout
- Improved process
  Better handling procedures
  Streamlined and/or segregated material flow
- Improved preventive tools
  Better materials
  Different or additional preventive measures
- Improved training programs
- Improved monitoring
  Self-checks
  Audits

TABLE 9.7  Key Elements of ESD Failure Analysis

- Circumstantial evidence
- Failure duplication tests
- Chemical etching
- Liquid crystals
- Scanning electron microscopy
- EBIC

analyst must be objective and the cause might be some mechanism other than ESD, even when suspect. At the same time there are certain key elements of failure analysis that are particularly valuable for ESD-caused failures. These elements, listed in Table 9.7, may receive more emphasis as two or more indicators of static as a cause develop.

## References

1. E. R. Freeman and J. R. Beall, "Control of Electrostatic Discharge Damage in Semiconductors," *International Reliability Physics Symposium,* 1978.
2. O. J. McAteer, "Electrostatic Damage in Hybrid Assemblies," *Annual Reliability Physics Symposium,* 1978.
3. Owen J. McAteer and Ronald E. Twist, "Analysis of Electrostatic Discharge Failures," Electrical Overstress/Electrostatic Discharge Symposium, 1981.

Chapter

# 10

# Failure Analysis Case Histories

## 10.1 Introduction

This chapter contains selected case histories to illustrate application of the failure-analysis procedures discussed in Chap. 9, and to fortify discussions of damage traits contained in Chap. 7. The primary purpose is to help guide new entrants into the endeavor of ESD failure analysis in appropriate analytical steps. A secondary purpose is to familiarize individuals not directly involved with failure analysis in the nature of static failures and the analysis process so that realistic expectations from analysis laboratories can be established.

## 10.2 Selection of Case Histories

Case histories were chosen with an attempt to include examples from different technological families and different factory areas, and to illustrate a range of failure-analysis elements appropriate for ESD. The importance of extra-laboratory information is stressed. These efforts are equally as important and complementary to laboratory procedures. Case histories are primarily from first-hand experience, although cited examples from the literature are included to broaden coverage of areas of interest. Those selected are:

1. CMOS devices contained in a metal chassis assembly
2. Catastrophic and subtle damage to bipolar operational amplifier inputs
3. Emitter-coupled logic (ECL) outputs

4. Shorted-hybrid substrates
5. Hybrid breakdown and vaporized metal
6. Charge-coupled device (CCD) failures
7. Wafer level processing
8. Avoidance of possible latent failures on IRAS spacecraft

## 10.3   CMOS Devices Contained in a Metal Chassis Assembly

### 10.3.1   Notification and fact gathering

A custom CMOS integrated-circuit device known to have a static sensitivity of about 200 V was used in a small modular assembly at the Westinghouse Advanced Technology Laboratories (ATL). The assembly was rather complex and contained numerous additional integrated circuits, all of which were enclosed in a metal chassis about 9 by 4 by 1 in in size. After a history of several successful tests on previously processed final assemblies three failures were reported. In the process of fact gathering three pertinent bits of information were disclosed:

1. Each of the three units had functioned properly for some time prior to failure.
2. The tests had been conducted on a newly installed test set.
3. The engineers and technicians involved did not use any static precautions when handling this assembly.

The problem was isolated to damaged CMOS devices at the U1 location in all three units. Although these devices were known to be susceptible to ESD, intuitively it seemed that the metal chassis would have provided protection from any fields or direct contact to the unit. The results of physical analysis were critical to identification of the cause of failure.

### 10.3.2   Part analysis

**10.3.2.1   Microprobe damage site isolation.** By carefully and progressively microprobing further and further into the chip's interior, the problem was isolated to a resistive short between two circuit nodes of the device: the clock signal input and ground. There were 50 CMOS gates in parallel across the two nodes, and no obvious physical damage. Normally gate damage is revealed only after removal of the gate metallization. The damage is often of submicron dimensions requiring a scanning electron microscope (SEM) for observation. The task of

**Figure 10.1** Liquid crystals highlighting CMOS defect site.

examining 50 gates with an SEM to find such a damage site would be too time-consuming to be considered. To attempt to electrically isolate by microprobe measurements and line cutting would also have been impractical. Fortunately, there are other means of problem site isolation that can be used.

**10.3.2.2 Cholesteric liquid crystal isolation.** As described in Chap. 9, cholesteric liquid crystals respond to heat by showing a change in color when observed by optical microscope. Figure 10.1 shows (in black and white) the changed contrast at the defect site due to liquid crystal detection of the $i^2R$ heating at the shorted gate. With the shorted gate identified the analysis could now be carried further. Figure 10.2 shows a photomicrograph of the transistor in question. In Fig. 10.3 the same transistor is shown after chemical removal of the $SiO_2$ glassivation and the aluminum gate. This figure gives an indication of the difficulty faced if the defect site had to have been found by optical examination of the 50 gates in parallel. The short site shown in Fig. 10.4 at 13,000× SEM magnification appears to have the characteristic traits of an ESD failure. Several points are to be emphasized at this juncture of the analysis:

1. The analysis is not over!
2. ESD has not even been fully proved yet.

230  Chapter Ten

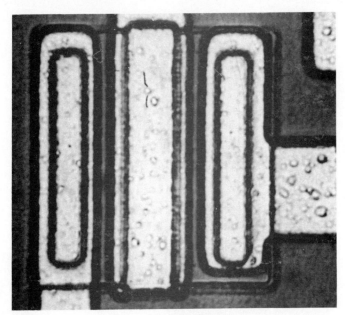

**Figure 10.2**  The shorted transistor at 1000×.

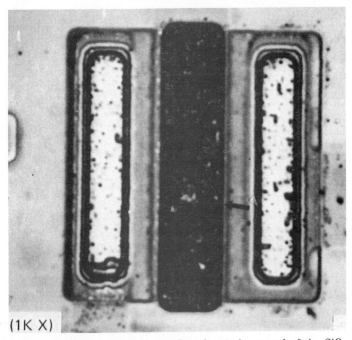

**Figure 10.3**  The same transistor after chemical removal of the $SiO_2$ glassivation and the aluminum gate.

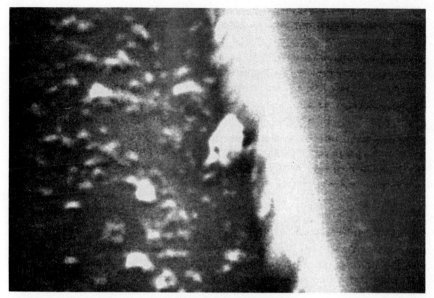

**Figure 10.4** ESD Defect site at 13,000× SEM magnification.

3. Even if ESD had been fully substantiated as the mechanism, the source and conditions of assembly vulnerability have not been determined.
4. Without the fulfillment of 2 and 3, appropriate corrective action cannot be implemented with confidence.

### 10.3.3  Failure cause identification

**10.3.3.1  Failure duplication.**  The difficulties in ascertaining ESD from other types of electrical overstress (EOS) can often be overcome by duplication of the failure under prognosticated conditions. This technique was used at the CMOS-device level by applying a human body model discharge of 200 V between the clock input and ground. Figures 10.5 through 10.7 show the damaged transistor before and after metallization removal and finally under SEM. The damage trait obtained was virtually identical, substantiating ESD as the most likely mechanism.

Next, the assembly design and physical layout were studied for the existence of any unknown points of vulnerability to ESD. The fact that the 9-in long clock signal wire was positioned very near the inner chassis wall was of particular interest. Additionally upon revisiting the test location investigation showed that the new test tool did not

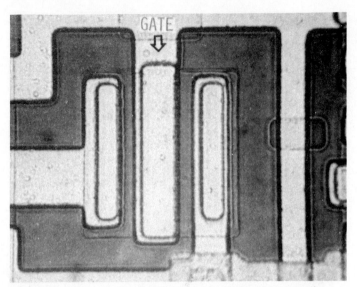

Figure 10.5 Purposely damaged transistor with no visible damage.

ground the assembly chassis during test as had the previous tester. This new information led to static duplication attempts at the assembly level. An experimental model of the assembly was configured to replicate conditions at the test stand. The CMOS failure could be duplicated from human body static discharges to the chassis as depicted in Fig. 10.8. Typical voltages required for failure were between 700 and 800 V. The path was through *capacitive-coupling* from the chassis to the insulated signal wire.

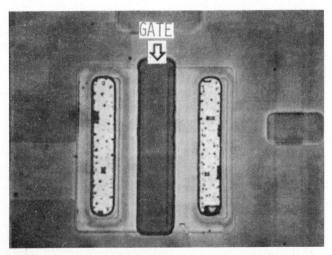

Figure 10.6 Purposely damaged transistor after metallization removal.

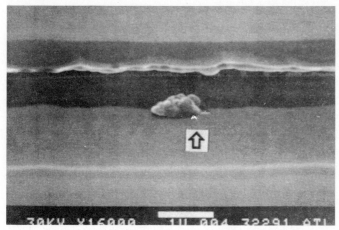

**Figure 10.7** SEM of defect site on failure duplication device.

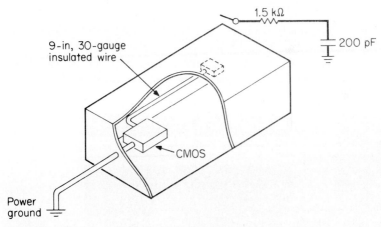

**Figure 10.8** Chassis configuration permitting capacitive coupling between chassis and internal insulated wire.

### 10.3.4 Corrective action

The primary corrective action measure taken was to internally connect chassis ground and signal ground. This measure prevents the possibility of a potential difference between these points, thus shunting the failure path. Additional somewhat redundant measures taken were to route the clock wire at a greater distance from the chassis and to add the chassis ground provision at the test tool.

## 10.4 Catastrophic and Subtle Damage to Bipolar Operational Amplifier Inputs

This example was chosen to give some generalized insight into the nature of typical static damage to bipolar devices. The case at hand is an epitomized version of numerous encounters during failure analysis of custom operational amplifier chips used in hybrid assemblies. The failures were experienced in the winter months of 1972 when static controls were not in place for bipolar devices at the Westinghouse Advanced Technology Laboratories, and institution of such controls began as a consequence.

### 10.4.1 Lot acceptance test failures

Samples were pulled from chip lots and packaged for a series of electrical tests and burn in in order to determine lot integrity. Damaged or degraded input transistors on custom operational amplifier devices was the most common failure of samples processed through these tests, accounting for about 75 percent of the rejects. Proper failure cause determination was extremely important as acceptance or rejection of an entire lot would depend upon relevancy of the failures.

**10.4.1.1 Catastrophically damaged devices.** Approximately 80 percent of the rejected operational amplifiers showed visible damage at one of the input transistors such as that shown in Fig. 10.9. Static electricity was not suspected in 1972 because of prior conceptions of virtual im-

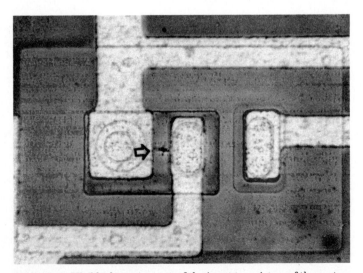

**Figure 10.9** Visible damage at one of the input transistors of the custom bipolar operational amplifier circuits.

munity of bipolar technologies to ESD. For the same reason static controls were practiced only during handling of MOS devices.

**10.4.1.1.1 Part analysis.** Adherence to the part analysis procedures outlined in Chap. 9 led to a proper conclusion in spite of the misconceptions about ESD immunity of the parts. Curve-tracer I-V characteristics were occasionally observed to shift between initial and post-delidding measurements. Experimentation showed that these changes could be brought about by personnel touching a device's metal lids after a slight amount of activity. A person rising from the seated position and then touching the lid would generally bring about a softening of the characteristic knee of the input transistor's I-V curve as was shown in Figs. 2.2 and 2.3. Microscopic examination revealed the same physical trait observed in the original lot acceptance failures. The discharge path can be ascertained from the schematic diagram shown in Fig. 10.10. Because of the metal package, a person touching the lid is effectively contacting the chip substrate. The parasitic diode provides a path with only one forward diode drop ($\approx 0.6$ V) from the P substrate to the N emitter. From the emitter the destructive route is to whichever input base connection is at or near electrical ground.

**10.4.1.2 Subtle beta degradation failures.** The remaining 20 percent of the lot acceptance failures were of a much more subtle nature. In

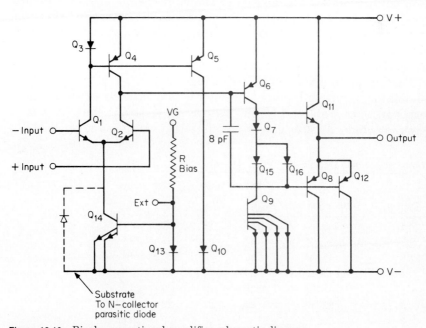

**Figure 10.10** Bipolar operational-amplifier schematic diagram.

these failures one of the input pair transistors would have experienced a "beta degradation" detected at one of the intermediate burn in electrical performance checks. The degree of beta (transistor-current gain) degradation varied between about 15 to 50 percent.

**10.4.1.2.1 Part analysis.** Electrical performance tests showed input-offset current and other parameters out of specification, consistent with degraded beta of an input transistor. Pin-to-pin curve tracer measurements did not reveal any significant shifts in the input-transistor characteristic curves or knee softening. Microscopic examination showed no visible damage.

A review of the analysis results of the catastrophically damaged devices disclosed two facts possibly related to the beta degradation problem:

1. The same input transistors could be damaged by ESD.
2. The resultant catastrophic damage was a form of beta degradation; that is, beta was severely degraded to unity or less.

These considerations led to the premise that lower level static discharges might possibly bring about degradation of beta with no visible physical trait.

Attempts to cause slight damage were made by actual human body movement of less than 90-degrees rotation in a swivel chair and subsequent touching of the device lid. The most common result was catastrophic damage identical to those previously discussed. However, with reduced movement, device failures with electrical parameters similar to the original failures were generated. No visible damage was observed on microscopic inspection.

Further testing was conducted using a 200 pF, 1 k$\Omega$, human body simulator circuit with input transistor beta measured before and after exposure (see Table 10.1) The base of input transistor $Q2$ was at ground potential as the substrate connection was exposed to HBM dis-

TABLE 10.1  Beta Degradation Via ESD

| Device | Voltage (200 pF, 1 k$\Omega$) | Initial betas | | Final betas | |
|---|---|---|---|---|---|
| | | $Q_1$ | $Q_2$ | $Q_1$ | $Q_2$ |
| 1 | 151 V | 12 | 30 | 11 | 2 |
| 2 | 200 V | 32 | 32 | 32 | 25 |
| 3 | 470 V | 30 | 30 | 30 | 19 |
| 4 | Not measured | 23 | 23 | 23 | 16 |
| 5 | 450 V | 35 | 32 | 29 | 0.3 |
| 6 | 320 V | 22 | 22 | 23 | 4.3 |
| 7 | 225 V | 30 | 30 | 30 | 0.68 |

charges as indicated. Note that in each case the beta of $Q2$ was degraded. Most importantly, only devices 5 and 7 showed visible damage. A secondary item to be noted is that the gain of $Q1$ became degraded in devices 1 and 5 even though the base connection was floating. This is because the device was plugged into a test box and the existent lead length at the $Q1$ base socket provided considerable stray capacitance to ground.

Physical analysis was done first on the purposely degraded devices as is the usual practice. With this approach, experience can be gained to help prevent loss of important information in the actual failures from inadvertent damage. On these samples it was discovered that after a partial chemical removal of the $SiO_2$, layer a filamentary trait began to become visible. Figure 10.11 shows such a trait just beginning to emerge on an actual lot acceptance failure after a partial removal of the glassivation layer.

**10.4.1.3 Failure cause identification.** Static-voltage measurements showed that about 600 V on the human body would bring about catastrophic damage, and lesser amounts would sometimes cause beta degradation. The relative humidity in the heated facility ranged between 10 and 40 percent during the winter season. Static measurements on personnel revealed that damaging potentials were easily generated in the lot-acceptance test area as well as the failure-analysis laboratory. The logical conclusion was that the failures were due to static discharges from the human body.

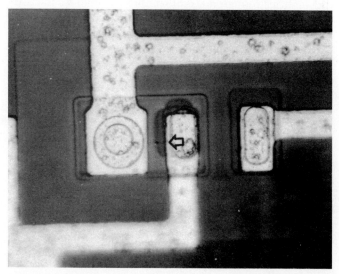

**Figure 10.11** Defect trait just beginning to emerge on lot-acceptance failure after partial removal of $SiO_2$.

**10.4.1.4 Corrective action.** Comprehensive ESD control measures were introduced as a result of this analysis. The measures were applied throughout the ATL facilities and included personnel wrist straps grounded through 1 M$\Omega$, selective conductive flooring, grounded work surfaces and protective packaging for transportation and storage.

## 10.5 Emitter-Coupled Logic (ECL) Failures

### 10.5.1 Failure with no visible damage

Several failures were reported as occurring at printed-circuit-board electrical test. Detailed failure analysis was requested because the ECL dual-quad gate devices in question were no longer available and the small number of spare parts in stores could not tolerate the replacement rate. The reported problem was that the ECL output of the board had shifted by several tenths of a volt. The facts that the output interface failed, it was during winter, and ESD controls were not in strict compliance, led to a suspicion that static might be involved.

**10.5.1.1 Analysis of failed parts with no visible damage.** Curve-tracer measurements showed a resistive short of from 100 to 500 $\Omega$ on failed devices. There was no visible damage trait. The defects were duplicated by an ESD of 500 V from a 200 pF, 1 k$\Omega$ human body model. Table 10.2 lists results along with experiments conducted at higher potentials in hopes of generating visible damage to highlight the failure site. Chemical etching revealed slight visible damage at the silicon and silicon-dioxide interface as shown in Fig. 10.12.

### 10.5.2 Failed parts with visible damage

A subsequent failure occurrence of this same part type showed more severe symptoms:

1. An 80-$\Omega$ short between output pin 1 and $V_{cc}$
2. Visible damage at the pin 1 output
3. Additional secondarily damaged output pins 2 and 3 measured 7.5 k$\Omega$ and 4.5 k$\Omega$ to $V_{cc}$, and output pin 4 was slightly leaky.

**TABLE 10.2 ECL Failure Duplication Attempts**

| Applied voltage | Output pin shorted to $V_{cc}$ | Resistance of short to $V_{cc}$ |
|---|---|---|
| +5000 V | Pin 6 | 114 $\Omega$ |
| +2000 V | Pin 2 | 211 $\Omega$ |
| +500 V | Pin 7 | 300 $\Omega$ |

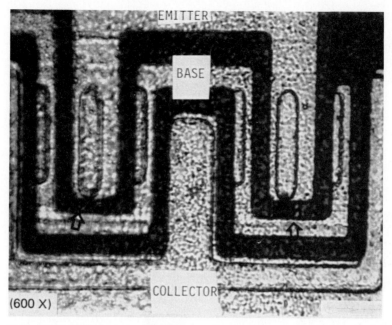

**Figure 10.12** ECL damage visible after etch.

Not only was the damage trait visible at pin 1, but the damage was multiple, occurring at both emitter fingers as shown in Fig. 10.13.

**10.5.2.1 Analysis of parts with visible damage.** Since static experiments had already been conducted up to 5000 V with no visible damage produced, another type of electrical overstress was considered. A device output was exposed to 400 V with pulse widths gradually stepped from 60 ns to 26 μs, at which time an open circuit to $V_{cc}$ was detected. The resultant damage with considerable current melting effects as shown in Fig. 10.14 was quite different from the original failure. Therefore, further static experiments were conducted, including exposure to 10,000 V from the 200 pF, 1 kΩ HBM. The results listed in Table 10.3 show that upon more discerning measurements some multiple output degradation was detected at the 5000-V level. More importantly, rather severe damage occurred to all outputs at the 10,000-V exposure, including visible damage to output pin 2. This experiment not only duplicated the original failure but indicated that we had experienced an actual in-house failure at the rather high level of 10,000 V.

### 10.5.3 Failure cause identification

The assembly and test area was revisited to witness actual handling of the printed-circuit boards. In the final operation an ungrounded

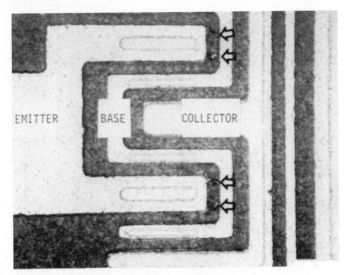

Figure 10.13  ECL with multiple visible damage sites.

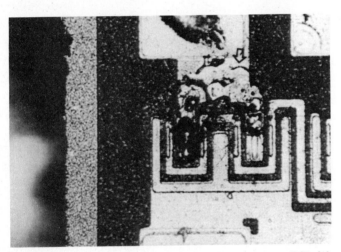

Figure 10.14  Result of EOS failure duplication attempt on ECL differed greatly from the original failure.

technician disconnected a coaxial cable to the ECL output. The other end of the cable had been previously disconnected from the associated test equipment. Failure duplication tests showed that a static discharge of 2500 V from a human body to the shield of the coaxial cable could bring about failure of an ECL output tied to the center conductor. Thus capacitive coupling via the coaxial cable was very likely involved in these failures in a situation that appears benign. In general,

TABLE 10.3  Higher Voltage-Level Duplication Attempt Results

| Device no. | ESD from 200 pF (through 1 kΩ) | Output pin damaged | Result of damage (to $V_{cc}$) |
|---|---|---|---|
| 1 | +600 V to Pin 2 | Pin 2 | 370-Ω short |
| 2 | +5000 V to Pin 2 | Pin 2 | 115-Ω short |
|   |   | Pin 3 | 2-Ω short |
|   |   | Pin 6 | Increase in reverse leakage |
|   |   | Pin 7 | Increase in reverse leakage |
| 3 | +10,000 V to Pin 2 | Pin 2 | 130-Ω short |
|   |   | Pin 3 | 290-Ω short |
|   |   | Pin 6 | 450-Ω short |
|   |   | Pin 7 | 260-Ω short |

static controls in place were fair but incomplete, allowing opportunities for ESD.

### 10.5.4  Corrective action

Corrective action consisted of a general tightening of ESD-control procedures as well as measures to eliminate the capacitive-coupling problem. Specifics included

1. Grounded operators during all board handling
2. Selectively located grounded floor mats and footwear
3. A prescribed order of cable connect and disconnect
4. Cabling bleed off resistors on associated test equipment
5. Protective covers over output ports when not in use

## 10.6  Shorted Hybrid Substrates

Within a 2-day period four failure occurrences of a certain hybrid assembly were reported at a higher assembly test level. The hybrid contained numerous CMOS chips mounted on a sapphire thin-film substrate having two metallization intraconnect layers. These incidents were alarming since numerous higher assembly-level tests had been previously conducted on these and similar units with no failures of the hybrid in question.

### 10.6.1  Part analysis and failure cause identification

Part analysis and failure cause identification are discussed jointly in this example because in actuality the efforts were concurrent and complementary. In each failed hybrid circuit, microprobing revealed a short between signal lines as indicated in Table 10.4. Microscopic examination disclosed scratches at metallization-layer crossover points near the edge of a mounted CMOS chip. One of the more severe scratch examples is shown in Fig. 10.15.

**10.6.1.1 Initial conclusions.** A first natural conclusion might be simply that the scratches caused the shorts. However, the physical configuration at the higher assembly level precluded the scratch occurring at that level without disturbing the many delicate bond wires over the scratch locations. Thus the scratches must have been inflicted at an earlier stage, perhaps during chip mounting, and the shorts developed later. If this were the case, then another mechanism was involved.

**10.6.1.2 Dissection of shorted substrates.** After chip removal, the substrates were given an aluminum etch. The result in Fig. 10.16 indicates that the first (underlying) metal layer was possibly affected as well as the exposed second (top) layer. The etch pattern was indicative of cracks in the insulative $SiO_2$ layer. A second aluminum etch verified that the etchant was penetrating to the bottom layer as shown in Fig. 10.17. Scanning electron microscopy verified the existence of cracks in the silicon-dioxide layer as shown in Fig. 10.18. Of particular interest was the *pseudosquare* anomaly shown in Fig. 10.19. This trait was sometimes characteristic of ESD-type oxide breakdowns in certain crystalline structures.

**10.6.1.3 Fact gathering at the location of failure.** Upon visit to the assembly test-failure location it was discovered that the four assemblies failed in a 2-day period about 1 week after the introduction of stringent clean room garment requirements. The new procedures included complete coveralls and *booty*-type shoe covers. Static measurement showed that voltage on either the coveralls or the shoe covers was typically about 20 kV with a maximum over 30 kV.

**TABLE 10.4 Shorted Signal Nodes on Hybrid Circuit Failures**

| Failed substrate | Shorted signals |
|---|---|
| A | $C_S$ to $B_2$ <br> $C_S$ to $V_S$ |
| B | $C_L$ to $V_B$ |
| C | $V_R$ to $B_1$ |
| D | $V_C$ to $V_B$ |

Failure Analysis Case Histories 243

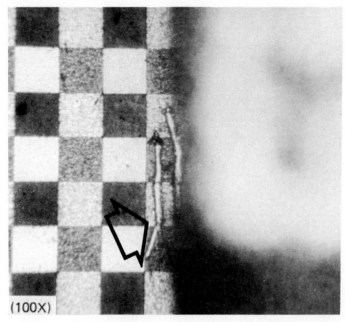

Figure 10.15 Hybrid substrate with scratches at shorted signal crossover site.

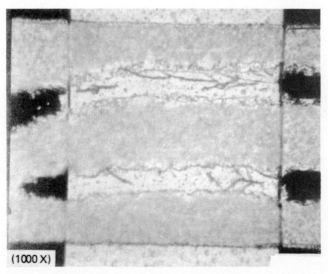

Figure 10.16 Aluminum etch of shorted substrate.

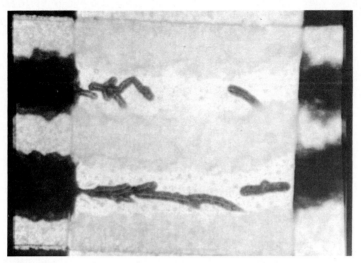

**Figure 10.17** Second aluminum etch showing progressive removal of lower metal.

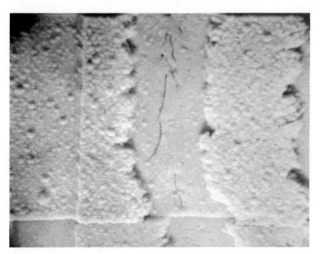

**Figure 10.18** SEM view showing cracks in $SiO_2$ layer.

**10.6.1.4 Failure duplication attempts.** The results of physical dissection and fact gathering both led to suspicion of possible static involvement in the failures. Experiments were conducted to duplicate the failures from nonshorted scratched substrates subsequently exposed to ESD, and to verify the physical effects of scratches of varying degrees of severity. The results follow:

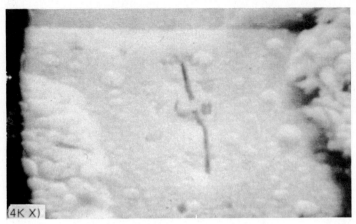

Figure 10.19  Pseudosquare anomaly at damage location.

1. Unscratched crossovers would withstand over 1000 V as anticipated for the 12,000-Å oxide thickness.
2. Purposely induced scratches produced cracks in the $SiO_2$ layer similar to those in actual failures.
3. Shorts could be produced in scratched crossovers at unpredictable ESD levels, sometimes as low as 150 V.
4. Sometimes the static experiments produced a pseudosquare feature in the top metal, such as shown in Figs. 10.20 and 10.21.
5. Some shorts were temporary in nature and were found to have recovered to open circuits on later checks.

**CMOS chip experiments.** A natural question to arise is, "Why didn't the static sensitive CMOS devices fail, if ESD were involved?" The CMOS devices were susceptible to HBM levels of about 200 V, so the question warranted further investigation. CMOS devices were exposed to static discharges across the same signal nodes as were shorted. The discharge levels were stepped in 20-V increments in both polarities up to 200 V, then in 50-V excursions up to 600 V. Curve-tracer characteristics were compared before and after each static exposure. The results listed in Table 10.5 show that each node combination could withstand 600 V or more in one polarity, and at least 250 V in the other. Thus, a substrate at as low as 150-V susceptibility, could easily have been the weaker path.

**Figure 10.20** Experimentally induced pseudosquare anomaly at a crossover with a very light vertical scratch.

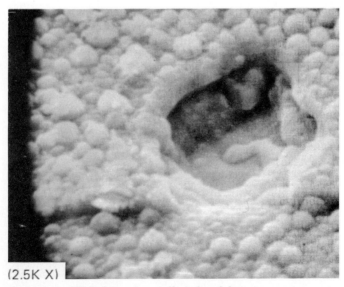

**Figure 10.21** SEM of experimentally induced damage.

TABLE 10.5  CMOS Signal Node Failure Duplication Results

| CMOS signal pads parallel to short | Highest step-stress potential and polarity across pads (V) | Result |
|---|---|---|
| $C_S$ to $B_2$ | +600<br>-300 | No damage<br>Failure |
| $C_S$ to $V_S$ | +600<br>-200 | No damage<br>Failure |
| $C_L$ to $V_B$ | +600<br>-200 | Failure<br>Failure |
| $V_R$ to $B_1$ | +600<br>-200 | No damage<br>Failure |
| $V_C$ to $V_B$ | +600<br>-500 | No damage<br>Failure |

Note: Tested to failure at 40 V, 100 V, 200 V, 300 V, 400 V, 500 V, and 600 V.

### 10.6.2 Corrective action

Primary corrective action was to find a proper laundry treatment for the coveralls. A test of static buildup propensity was made on garments upon receipt. The shoe covers were eliminated. Personnel were grounded via wrist strap and selected conductive flooring with conductive footwear. A stringent inspection criteria was imposed on substrate-metal crossover points. Additionally a 300-V electrical isolation test was imposed between all substrate crossover lines. These corrective actions were apparently effective as there were no reoccurrences of this problem.

## 10.7  Hybrid Breakdown/Vaporized Metal

This example is included to illustrate the occasional difficulties that arise in determining the correct failure path and in differentiating between ESD and EOS. The failed item was a 1- by 2-in hybrid with digital circuitry rejected from board-level test. Three failures were reported of the same hybrid circuit type from two board assemblies. In each case troubleshooting revealed at least one quantizer chip output shorted to pad 7 (3.5 V ps).

### 10.7.1  Part analysis

Microscopic examination revealed oxide breakthroughs near the output pads from the output crossunders to the pad 7 metallization line. The defect sites resembled ESD damage as shown in Figs. 10.22 and 10.23. The diagnosis appeared rather straightforward: an electrostatic discharge from the output to pin 7.

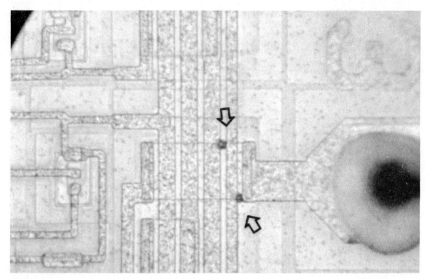

Figure 10.22  Defect sites on quantizer chip.

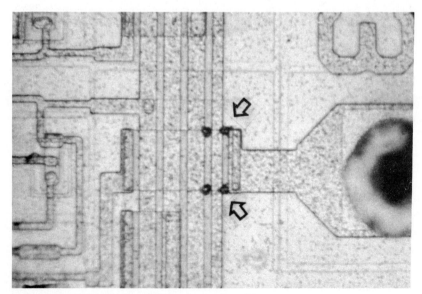

Figure 10.23  Defect sites on different quantizer chip.

## 10.7.2 Failure cause identification

Failure duplication tests using the standard (100 pF, 1.5 k$\Omega$) HBM were conducted between the chip output pad and pad 7. When shorts were produced at 6 kV or greater, the visual traits were considerably different from the original failures. Figure 10.24 shows that damage was produced in the output crossunder area to the pad 7 metallization as desired, but the trail was singular and of a more severe nature. More importantly, as shown in Figs. 10.25 and 10.26 an extensive length of the pad 7 metal was completely vaporized.

The discharge-model series resistance was then modified in hopes of achieving better failure-duplication results. With the series resistance increased to 11.5 k$\Omega$, a short was produced at 15 kV. With the series resistance reduced to 430 $\Omega$, 184 $\Omega$, and finally 0 $\Omega$, shorts were produced at 4 kV, 4 kV, and 2 kV, respectively. In all cases the vaporized-metal line was produced, even though visible crossunder damage resulted only in the 184-$\Omega$ example. The failure was finally duplicated by applying standard HBM discharges of 8 kV to the lid of the device with pad 7 grounded. The same trait was subsequently duplicated at 6 kV. The residual trait produced was similar to the original failures with no associated metal melting.

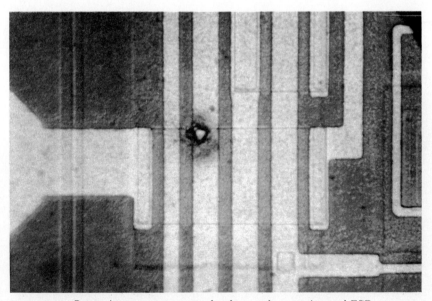

Figure 10.24 Severe intraconnect crossunder damage by experimental ESD.

Figure 10.25  Overview of chip showing vaporized metal line caused by ESD.

Figure 10.26  Vaporized line at higher magnification.

#### 10.7.2.1 Conclusions.
This analysis led to several conclusions. The cause of failure was determined to be human body electrostatic discharge to the package lid at a level of greater than 5 kV. The hybrid package lid was found to be floating rather than tied to ground as required by the design. The vaporized metallization would normally have been attributed to EOS upon first glance. This was one of those rare examples when, after the crossunder-oxide breakdown, the metal had no energy-absorbing elements in series.

### 10.7.3 Corrective action

Corrective action included connecting the lid to the internal ground according to the design as well as improving personnel grounding procedures.

## 10.8 Charge-Coupled Device (CCD) Failures in Hybrid Assemblies

An extremely serious problem was encountered when 90 percent of the CCD shift registers were failing at hybrid-assembly level test. The shift registers had functioned properly at wafer level test.

### 10.8.1 Part analysis

The rejected chips were mounted in dual in-line packages for analysis and testing. The problem was isolated to an output transistor with a low threshold voltage. There was no visible damage and no leakage path detected by curve tracer so a more subtle mechanism such as charge trapping was suspected. The source contact, which was tied to an external lead, was the most likely site of charge injection into the silicon nitride/oxide region. Several failed devices recovered fully after a ½-h bake at 400°C. Such recovery is also characteristic of the charge-injection mechanism.

The source of charge injection is high-voltage stress, so ESD was suspect. Failure duplication attempts were made by the standard HBM at levels of 50 to 150 V. Similar failures were produced over the entire voltage range. Again, a ½-h bake at 400°C brought about recovery.

### 10.8.2 Failure cause identification and corrective action

A review of the hybrid assembly and test area found that thorough ESD controls were in place. Such a high degree of control was not in agreement with a 90 percent shift-register failure rate due to ESD. A careful check was then made on all associated assembly and test equipment for the existence of overvoltages or transients. A wire-bonding machine was found to have a 70-V transient as a result of relay inductive kickback during switching by foot control. Corrective action was to install transient suppression circuits at the bonding machine. The problem was completely eliminated as a result of this measure.

## 10.9 Wafer Level Processing

Very often the wafer level processing area is believed to be immune to ESD problems. The rationale is that packaged devices are much more vulnerable because of exposure of the attached leads and because of the little documented evidence of wafer ESD problems. This example is one of several wafer level static problem case histories reported in the literature.

In a certain fabrication cycle, yields of CMOS wafer test patterns were found to be much lower than expected at the IBM General Technology Division in Essex Junction, Vermont.[1] Investigation revealed that failure density increased toward the center of the wafer. Figure 10.27 shows the "bulls-eye syndrome" with the number of good devices out of six test patterns for each chip. Ruptured-gate oxides as shown in Fig. 10.28 were found during part analysis. Failure duplication by ESD simulation tests produced similar damage sites.

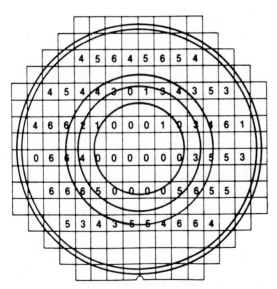

Figure 10.27 Bulls-eye syndrome numbers are total good out of six test patterns. (*From Renaud,[1] with permission from EOS/ESD Association.*)

**Figure 10.28** Ruptured-gate oxide typical of wafer level failures. (*From Renaud,[1] with permission from EOS/ESD Association.*)

Failure cause identification efforts consisted of studying the process variables and conducting appropriate experiments. The addition of a high-pressure jet-deionized water-spray clean operation was found to have been instituted just prior to the decreased yields. Static voltages as high as 2 kV were measured on experimental wafers as they were rotated at the nominal 1000 r/min during the spray operation. Elimination of the clean cycle resulted in excellent yields but unacceptable particulate remained on the wafer surfaces. Finally, experiments were conducted with variations in the cleaning cycle rotation rate, pressure, and application times. Corrective action was then determined by establishing optimum conditions of 1000 r/min, 2500 lb/in$^2$, and 10-s application.

## 10.10  Avoidance of Possible Latent Failures on IRAS Spacecraft

The impact of possible latent failures is of extreme importance in space applications. On such programs failure analysis takes on a special significance because of the potential impact of corrective actions on operational success and reliability. This example,[2] investigated at the Jet Propulsion Laboratory, is another piece of evidence supporting the existence of latent ESD failures and of the criticality of good failure analysis.

Several 3N164 devices had been rejected during functional tests of electronics equipment used on the Infrared Astronautical Satellite

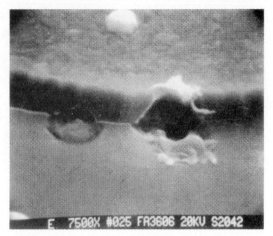

**Figure 10.29** Defect sites of 3N164 devices along metallization edge. *(From Trigonis,[2] with permission of the EOS/ESD Association.)*

(IRAS) program. Part analysis determined resistive shorts in the range of 330 to 2500 Ω either from gate to source or gate to drain. Microscopic examination revealed multiple oxide breakdown sites which varied from 2 to 14 and were found along the metallization edge above the source or drain region. Scanning electron microscope/electron-beam induced current (SEM/EBIC) imaging analysis determined that just a single site accounted for each short. Typical rupture sites are shown in Figs. 10.29 and 10.30 before and after metal removal. The

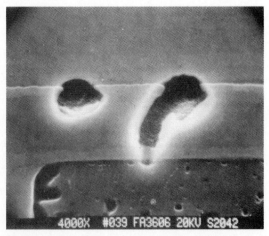

**Figure 10.30** Defect sites after metallization removal. *(From Trigonis,[2] with permission of the EOS/ESD Association.)*

conclusion was that the damage most likely had resulted from multiple ESD exposures and that some of the ruptures had experienced apparent healing. A major concern was that such healing is not regarded as permanent or reliable and could result in latent catastrophic failure if undetected prior to launch.

Failure cause identification efforts via failure-duplication attempts verified that HBM exposures produced similar damage sites. Tests conducted on alternate-device types showed that the G118AL and J230 MOSFETs had susceptibility thresholds 32 and 704 times higher than the 3N165, respectively. Corrective action included a design change to use the J230 device rather than the 3N165.

## 10.11 Importance of Failure Analysis to an ESD Control Program

The examples herein were intended to illustrate the application of the analytical methodology described in Chap. 9. The results of failure analysis are the roots of justification for an ESD control program; after all, the purpose is to prevent static-induced failures. Therefore this subject must not be neglected by those involved with ESD control.

## References

1. D. Renaud and H. Hill, "ESD in Semiconductor Wafer Processing," *Electrical Overstress/Electrostatic Discharge Symposium Proceedings,* 1985.
2. A. Trigonis, "ESD Experience with the IRAS Spacecraft," *Electrical Overstress/Electrostatic Discharge Symposium Proceedings,* 1985.
3. M. H. Woods and G. Gear, "A New Electrostatic Discharge Failure Mode," *IEEE Transactions on Electron Devices,* vol. Ed-26, no. 1, January 1979, pp. 16–21.

# Chapter 11

# Latent ESD Failures

## 11.1 Introduction

The subject of latent ESD failures has been fraught with controversy and misunderstanding. Perhaps the problem begins with the two distinct meanings of the word latent. In one sense a latent failure mechanism implies a time-dependent degradation of an electronic device culminating in a system failure. A second meaning is simply a hidden or subtle defect, with no inference of time-dependent degradation. There is no question that ESD often results in very subtle and hidden defects; however, temporal degradation of electronics devices after ESD exposure is still regarded as highly controversial. Much of the existing confusion stems from authors and readers intermixing the two meanings of latency. Time-dependent latent failures are the primary topic of this chapter, and this meaning is intended unless stated otherwise.

A second and perhaps greater source of controversy has been the proliferation of unfounded claims of high probabilities of latent time-dependent static failures. Unfortunately, little research has been done on the subject and the available data base is much too small to present convincing arguments one way or the other. This chapter includes highlights of published research and the differing conclusions and recommendations concerning the latent ESD question.

## 11.2 Definition

A latent ESD failure is defined as a malfunction that occurs in use conditions because of earlier exposure to ESD that did not result in an immediately detectable discrepancy.[1]

The foregoing definition is intended for all references to latent ESD failures in this book, unless stated otherwise.

## 11.3 Background of the Latency Question

Prior to 1981, the question of latent ESD failures was a high-interest item to those in the electronics industry. At that time a great deal of speculation about static latency existed, particularly on the part of vendors of static-control products. Although this might be attributed to the vendors' vested interest in creating a higher degree of concern for static elimination, such speculation was not completely unfounded. Since ESD failures of varying degrees of severity and subtlety had been well-documented in the literature, a reasonable extension was to conclude that devices with slight damage might degrade with time, in a manner similar to the behavior of certain other failure mechanisms. Some physical theories as to how latent ESD failures might occur were presented in 1980[2] and are discussed in Sec. 11.3.2.

On the other hand, controlled laboratory experiments concerning the latency aspects of ESD have been rare, and for the most part inconclusive. An attestation to the state of research results on this subject was documented in January 1980, by T. Hasegawa[3] of the National Semiconductor Corporation. He surveyed the available literature on this topic and justifiably concluded that suspicions of a latent ESD-failure mechanism were unfounded in the light of the amount of published literature to the contrary.

As a result of the lack of substantive published data, the electronics industry in 1981 was taking a wait-and-see attitude regarding latent ESD failures. This meant that ESD-control expenditures were based upon the perceived detraction on production yields due to ESD. In most instances the subtle nature of ESD problems masked the root causes, and ESD control was minimal at best. At the same time many were apprehensive that ESD exposure would be proven to have detrimental effects on predicted mean time between failures in end use. Therefore the question as to whether or not latent ESD failures were real remained a large interest topic in informal discussions at the annual EOS/ESD Symposium.

### 11.3.1 Early investigations

Published research results leading to the conclusions drawn by Hasegawa and others are discussed in this section.

Gallace and Pujol[4] conducted accelerated life tests on CMOS devices to determine if ESD exposure could result in latent gate-oxide failures. A-series (1 kV sensitivity) and B-series (4 kV sensitivity) CMOS devices were exposed to static discharges of 100, 300, and 500 V; then placed on life test. Control devices with no ESD exposure were included in the life-test sample. The purpose of including control devices is that statistical analysis of any time dependent degradation of parts not exposed to static might indicate contributions of other failure mechanisms rather than ESD. Both plastic and frit type packages were included in the sample. Parametric behavior was then monitored after varying periods of time at storage, operating life, and bias life conditions at several temperatures. No significant difference in failure rates was observed between stressed and unstressed devices. The conclusion was "gate-oxide shorts are not a latent life-related failure mechanism for devices which have previously been subjected to electrostatic discharge."

Schreir[5] tested an extensive variety of semiconductor devices for degradation effects after ESD exposure. Parts included Schottky diodes, high-frequency transistors, dual 3-input NOR gates, TTL quad 2-input NAND gates along with the Schottky and low-power versions, operational amplifiers, ECL OR/NOR gates, N-channel MOS FETs, N-channel JFETs, 1 k CMOS RAMs, and 4 k EAROMS. The human body model was used to deliver 30 pulses at 10-s intervals. The discharge potential was 75 percent of the device-failure threshold. The accelerated life-test sample consisted of 10 stressed and 5 unstresssed devices of each part type. The failure criteria was 10 percent above the specified limit of an electrical parameter. The conclusion was "The data gathered to date on burn-in devices stressed at 75 percent of their ESD degradation threshold suggests that there is no long-term degradation mechanism if ESD levels are kept below the ESD degradation threshold."

Branberg[6] claimed that no evidence of latent ESD failures was evident as a result of his investigations into static effects on CMOS hex inverters. His position was based upon life tests of ESD stressed devices from several manufacturers. First, the catastrophic ESD failure thresholds were determined. Then parts stressed at 2, 10, and 50 percent of the threshold level were put on accelerated (125°C) 3000-h life test, along with unstressed control devices. Life-test failure criteria included degradation of functional integrity or leakage current greater than 1 μA. The highest failure rate was in the 10 percent stressed group, followed by the control devices, then the 2 percent group. Parts stressed at 50 percent of the ESD threshold to failure had the lowest time-dependent failure mechanism of all.

McCullough et al.[7] were given the task by the Rome Air Development Command to develop an ESD characterization and screen for gold-doped dual 4-input CMOS devices. The human body simulation circuit was used for initial familiarization and characterization of the ESD failure threshold of the 4002 NOR-gate devices. The lower limit of two standard deviations from this failure threshold was selected as the ESD screening voltage. Accelerated life tests were then conducted on 14-part samples at 125, 160, and 195°C. Results showed that screened parts behaved as well as nonstressed control parts, indicating that there was no degradation mechanism peculiar to the screening procedure.

Syrjanen[8] found a hint of possible latency during investigations of ESD effects on CMOS devices installed into hybrid circuits. Samples of 78 hybrid circuits were stressed at 100 and 1000 V, then placed on 2000-h 100°C life test. An upward shift of $I_{SS}$ from 2 to 63 µA on a single device was noted after 2000 h. One conclusion: "We did not verify that ESD can cause latent failures. However, the drastic increase of $I_{SS}$ in one device during the test suggests that further study of latent failures is needed."

### 11.3.2 Theories of latency

Theoretical analysis of certain common failure mechanisms might easily lead to the conclusion that ESD exposure could result in time-dependent degradation resulting in eventual failure. Some ideas considered in establishing latent ESD experiments for the Naval Sea Systems Command (NAVSEA) and the National Aeronautical Space Administration (NASA) funded studies are discussed in this section.

**11.3.2.1 Gate-oxide degradation.** A commonly held premise is the concern that MOS-gate oxide exposed to static discharge might suffer very subtle damage resulting in little or no increase in gate leakage. This condition might result from an aluminum-silicon alloy filament that does not completely bridge the gate-oxide thickness. One possible outcome is failure at low-level EOS or ESD in use conditions due to the reduced breakdown potential of the damaged oxide. Gradual spreading or further propagation of the filament under operational electrothermal forces could conceivably result in increased leakage to the point of failure.

**11.3.2.2 Gate-oxide healing/shorting.** The filament formed during EOS/ESD stress might improve from burnout either during the original or subsequent exposure. A "healed" oxide should be regarded as unreliable and would be vulnerable to further propagation or shorting under operational stresses or subsequent low-level static exposure.

**11.3.2.3 Protective circuitry damage.** Many instances of ESD-induced damage to on-chip protection networks have been documented in the literature. If such a device were to survive until final end use, the immunity margin designed into EOS/ESD would be compromised and failure might result from low-level stresses. A more subtle failure could result from waveform timing shifts such as propagation delays due to the *softened* I-V characteristic of a damaged protective diode.

**11.3.2.4 Charge trapping.** A high-voltage transient such as occurs in a static discharge might cause a disruption of the charge equilibrium in a device resulting in "charge trapping." Accumulated charges on the device surface can cause inversion layers that provide leakage paths. Inversion-layer leakage is an Arrhenius-type of time and temperature dependent degradation mechanism that can lead to catastrophic failure.

**11.3.2.5 PN junction degradation.** Silicon PN junction damage from ESD characteristically is evidenced by a *softened* I-V curve (see Figs. 2.2 and 2.3). Localized heating and electrical forces during the transient condition result in a filamentation of an aluminum-silicon alloy. This filament provides a parallel shunt path that rounds out the normally sharp *knee* of the reverse biased I-V curve across the junction. Aluminum-silicon transport mechanisms resulting in alloys that eventually short are usually thought of as high-temperature phenomena. In the case of ESD-caused damage, the cross-sectional area of the filament across a junction might be sufficiently small that significant localized heating could occur during normal operation, thus leading to failure. Burnout of such a small filament during ESD exposure is a possibility that could lead to an operating but unreliable device. Prefilament softened I-V curves and improvement with subsequent ESD are shown in Fig. 7.28 and discussed in Sec. 7.10.2.1.

**11.3.2.6 Metallization failures.** Although the incidence is rare, static discharge has been known to open sites of constriction in integrated-circuit metallization lines. ESD could further reduce the cross-sectional area of these sites to a near-failure condition. During usage conditions an open circuit could result from electromigration, thermal shock, or subsequent electrical transient.

**11.3.2.7 Intermittent failures.** Intermittency or instability is possible in all the aforementioned failure mechanisms. In such an event, detection of the latent condition is dependent upon operational modes or the time of parameter monitoring.

## 11.4 Research Strategy and Rationale for NAVSEA and NASA Studies

In 1980 and 1981, Westinghouse obtained funding from Naval Sea Systems Command and NASA, respectively, to study the issue of latent ESD failures. The primary objective was to settle the question of whether or not latent ESD failures existed. The overall plan was to produce time-dependent static failures by closely controlled experiments on a representative group of susceptible parts. The data showed with statistical certainty that latent ESD failures exist.

Many aspects of the issue have grown in controversy since the time of these studies. A widespread belief has arisen that latent ESD failures are extremely likely, that is, 75 to 90 percent of static failures will be of the time-dependent variety. If the likelihood were that high, then catastrophic failures detected during manufacture and test would be a small part of the problem, merely indicative of the unreliability to be expected.

The facts that many electronics manufacturers have not experienced such poor reliability, and that latent failures from studies remain scant, have convinced many that ESD latency is nonexistent.

In retrospect, the NAVSEA and NASA studies have produced the largest available data base of controlled experiments on the subject. Detailed examination of the results in Secs. 11.5 and 11.6 should not only convince the reader of the existence of latent failures, but provide insight into other facets of the problem.

### 11.4.1 Commonality in earlier investigations

A review of earlier investigations showed the following threads of commonality:

1. Latent failures were defined only at the part level, typically as a part that remained in specification limits after ESD exposure and degraded in life to an out-of-specification condition.
2. Life tests were conducted at an accelerated temperature, typically 125°C.
3. Unbiased life was not considered.
4. Prelife ESD exposure was at some predetermined percentage of the catastrophic failure threshold level.

### 11.4.2 Differences in plan

A different approach in these regards was chosen for the NAVSEA and NASA studies for various reasons.

1. A macroscopic perspective was taken expanding the definition of a latent ESD failure in several ways that gave due consideration to factors affecting assemblies containing ESD-sensitive parts. Macroscopic latent ESD failures can occur in three ways:

*Mac-1:* A device that remains in specification limits after ESD exposure but degrades in life to an out-of-specification condition.

*Mac-2:* A device that remains in specification after ESD exposure, and although it remains in specification during life, degrades sufficiently to cause a discrepancy at a higher assembly level.

*Mac-3:* A device that goes out of specification after ESD exposure, but is not detected at the higher assembly level until further degradation occurs with time.

Of the three types, we can consider Mac-1 failures as the classic latent ESD failure sought by earlier investigators.

Mac-2 failures could occur hypothetically only if the part-specification parametric range had not been properly considered or tested in the assembly design margins. Therefore this type of occurrence is rather unlikely. For instance, consider a part having an input current specification of 100 $\mu$A. Suppose that in several years experience with that part the input current had always been less than 1 $\mu$A. If the input current of a part slightly damaged by ESD increased to 10 $\mu$A, the part would be in specification and the assembly would most likely function properly. If the part degraded further during usage to 80 or 90 $\mu$A, it is conceivable that the assembly might react unfavorably to this unforeseen increased current requirement.

Mac-3 failures on the other hand are somewhat dependent upon good design practice and are regarded as the most likely of the three types. For illustration, suppose a system designer used the same part with the 100-$\mu$A maximum input current limit. Even though past experience had indicated that the input usually drew only a few microamperes, wide design margins were applied to circumvent early system failures. In this case suppose the design could function properly even if the input current were as high as 200 $\mu$A. If input current on a good installed device increased to 110 $\mu$A after ESD, the system would function even though the part were out of specification. Suppose also that this rather severely damaged part degraded even further during usage until the input current required was 220 $\mu$A. The system would then fail because the wide design margin had been exceeded.

2. Theoretical models of latent ESD failure mechanisms consisted of fragile physical entities. Conceivably, accelerated temperatures might tend to cure rather than degrade. For this reason life tests were started at 25°C and raised in temperature for subsequent life tests. Another consideration is that 25°C is representative of many usage conditions.

3. Parametric behavior was monitored after unbiased periods for indications of degradation.

4. The degree of damage necessary to precipitate latent degradation was completely unknown at the outset of these studies. Therefore, limiting the static exposure to a predetermined percentage of the cata-

strophic threshold might lead to inconclusive results if the latency producing level were excluded. A scheme was devised to use parts that had been exposed to a range of single or multiple static discharges. The life-test sample thus formed had a wide range of inflicted damage from very subtle to severe.

### 11.4.3 ESD controls

Strict ESD controls were adhered to by experienced and knowledgeable engineers and technicians during the studies. This included all handling by grounded personnel at static-controlled work stations and static-shielding packaging for transport and storage.

Notable exceptions were in the receipt of some parts from distributors, cited in the parts discussion of Sec. 11.5.1 of the NAVSEA study.

### 11.4.4 ESD circuit

The ESD circuit used was the standard (100 pF, 1.5 k$\Omega$) human body model described in DOD Std 1686. A mercury-wetted contact SPDT relay was used as the switching element. In order to avoid undesirable stray pulses when charging the capacitor, the part was disconnected prior to deenergizing the relay. The resultant ESD pulse was verified to be repeatable over the entire voltage range used in the study. The waveform at 500 V into 1500 $\Omega$ was verified daily and photographed weekly (see Fig. 11.1).

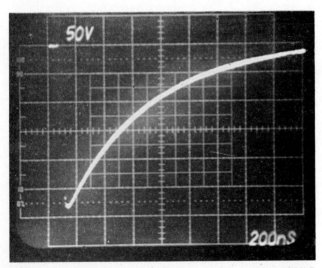

**Figure 11.1** HBM simulator voltage waveform across 1500-$\Omega$ load.

### 11.4.5 Parts selection

The following criteria were used to select parts for this study:

1. Part sensitivities were categories 1 and 2 of DOD Std 1686.
2. Commercial rather than military parts were used for the sake of cost and availability (but from military-qualified suppliers).
3. A variety of generic part types in common use were selected.
4. An attempt was made to purchase all parts of a given type from the same manufacturer and date code.
5. Parts selected were compatible with in-house test capabilities.

### 11.4.6 Pin combinations

Economic considerations required that testing be limited to a single pin combination per part type. DOD Std 1686 describes a procedure to find the most sensitive pin-polarity combination. The method is so rigorous that a large number of devices would be required. Additionally, there is no assurance that the most sensitive pin combination is most likely to produce a latent failure. Therefore, the investigators' extensive background in failure analysis was called upon to select pin combinations and polarities expected to affect designated device structures. For instance, PN junctions are usually more vulnerable to reverse-biased voltage overstress. In addition, test measurements in the reverse direction are often the best indicator of damage.

### 11.4.7 ESD stress plan

The literature contains examples of parts susceptible to multiple ESD exposures below the catastrophic threshold level. Parts having such a propensity were believed to have a tendency for latent degradation as well. The failure mechanisms associated with multiple low-level exposures might very well differ from higher level single ESD pulses. Both types were included to increase the chance of producing latent failures. A comprehensive method of applying ESD stresses was planned in order that the life-test sample would consist of parts with severe, moderate, slight, and no measurable damage.

### 11.4.8 Familiarization tests

A small sample of each part type was used to determine the catastrophic threshold level. For those parts exhibiting cumulative low-level multiple pulse damage, this initial familiarization sample data was extended to determine the voltage level at which the part seemed likely to survive 200 pulses without experiencing a "significant delta."

Significant delta is intended to define those parts having a parametric shift that has been caused by ESD. This delta is taken as any perceptible change after excluding such anticipated variables as measurement error or tolerance, instrument drift, external noise, and temperature variation.

### 11.4.9 Multiple pulse tests

The voltage range between the levels of catastrophic failure and ability to withstand 200 pulses without parametric change was divided by four to determine the voltage increments to be used in multiple pulsing. This resulted in five levels: catastrophic threshold, 200-pulse level, and three intermediate voltage levels.

Multiple pulse tests were performed on groups of parts at each of the five levels until a part experienced a significant delta. Past researchers' data had indicated that parts with no perceptible change did not tend to degrade with time. If an individual part did not show parametric change, stressing was continued to 200 pulses. Current limiting was used on all power supplies to prevent changes due to excessive current following a low-impedance path formed by ESD.

A minimum period of 10 s was maintained during multiple pulsing to allow sufficient cooling to avoid any additional thermal effects. Test parameters were recorded prior to ESD exposure and after each of the first 10 pulses. From 10 to 50 pulses, data were recorded at every 5 pulses. Test parameters were recorded every 10 pulses from 50 to 100 pulses, and every 25 pulses thereafter, up to 200.

### 11.4.10 Single pulse tests

Voltage levels for single ESD pulse tests were determined by taking 90 and 80 percent of the single pulse catastrophic failure threshold. Past data by Schreir[5] indicated that >75 percent was necessary to induce incipient damage.

### 11.4.11 Biased life tests

Standard static life-test bias conditions were chosen for discrete devices, whereas integrated circuits were exercised dynamically, typically at 100 kHz. Devices were connected in parallel in groups of 16 to 20. These groups were coupled to power-supply voltages through series current-limiting resistors to avoid severe voltage drops to the group in the event of device failure. Likewise, certain life-test samples that experienced severe initial ESD damage had individual limiting resistors. A novel feature of the life test was to include all pulsed samples regardless of the degree of damage. In particular, even severely damaged parts are valid candidates for latency, when intermittent conditions occur. Life-test samples included *control* devices that had not been exposed to static in order to discount any failure mechanisms unrelated to the prelife test static exposure.

Initial life testing was for 480 h minimum at 25°C because of commonality with many usage conditions and concern over possible annealing at 125°C, as reported by Schwank et al.[9] Critical test parameters were read at approximately 120, 240, and 480 h. All bias voltages and currents were monitored daily to detect any changes indicative of device degradation. The temperature was then raised to 75°C, and the tests repeated. When at 75°C, the devices were allowed to cool to room temperature under bias in order to detect any extraneous mobile ion phenomena. Whenever devices were without bias for significant periods, performance parameters were remeasured for any changes during the dormant condition.

### 11.4.12 Nonbiased bake

All devices were at room temperature without bias approximately 2 weeks during the period between ESD pulsing and start of life test. In the NAVSEA study, parameters were periodically monitored during a 1700-h 25°C nonbiased bake following biased life tests.

## 11.5 NAVSEA Study Results

The NAVSEA-funded investigation was conducted over the period from September 1980 to June 1981.

### 11.5.1 Parts used

Table 11.1 lists the parts used as well as details relating to the selection criteria. Parts were initially purchased in quantities of approximately 125 of each type. A later decision to purchase additional samples of each contributed to the different date codes and/or manufacturers for some of the parts. These variations were spread throughout the experimental lots so that representative samples were included.

The 3N170 MOSFET was selected because of its unprotected gate. The 1N5711 Schottky diode was chosen because Schreir[5] reported damage from low-level static discharges. Shorted-gate oxide improvement by subsequent ESD exposure in the CD4001A was reported by Branberg.[6] Schwank et al.[9] surmised that the CMOS transmission

TABLE 11.1  Parts Used in NAVSEA Study

| Part | ESDS category | Vendor | Date code | Quantity |
|---|---|---|---|---|
| 3N170 MOSFET | 1 | Intersil | 8012 | 130 |
| 2N441 JFET | 2 | Intersil | 8027 | 25 |
|  |  | Intersil | 8033 | 51 |
|  |  | TI | 8006 | 59 |
| 1N5711 Schottky diode | 1 | HP | 8016A | 178 |
| CD4001A CMOS quad | 2 | RCA | 8018 | 125 |
| NOR gate |  | RCA | 8032 | 50 |
| 5404 TTL hex inverter | 2 | Motorola | 7846 | 123 |
|  |  | TI | 8027 | 75 |
| 54S04 Schottky TTL | 2 | TI | 8022 | 100 |
| Hex inverter |  | TI | 8018 | 75 |

gate structure was likely to function after ESD. The 54L04 low-power TTL hex-inverter selection was based upon Freeman and Beall's[10] experience with subtle ESD failures detectable at −55°C. However, this part was not readily available, so the 5404 TTL version, of which Rutherford and Perkins[11] reported ESD healing softened curves, was substituted. For comparison the 54S04 Schottky TTL hex inverter was used.

Static-control infractions of note follow:

- The MOSFET and JFET devices were received in ordinary plastic bags; however, the MOSFET devices all had shorting clips attached to all leads.
- The Schottky diodes were in antistatic bags.
- The three integrated-circuit types (CD4001A, 5404, and 54S04) were received in antistatic tubes.

### 11.5.2  Pin combinations used and delta criteria

The MOSFET and JFET gates were pulsed in the reverse-biased polarity. Reverse polarity was also selected for the Schottky diode and an input transistor emitter-base junction on the 5404 and 54S04 devices. The gate of an input transistor was targeted on the CD4001A CMOS device. Table 11.2 shows the test parameters, test conditions, specification limits, significant delta criteria, and pin combinations.

### 11.5.3  ESD stress results

Table 11.3 shows the multiple and single pulse levels as applied to all part types. The cumulative damage effect as a result of multiple

**TABLE 11.2 Summary of "ESD Affected" Parts Criteria**

| Part | Test parameter | Test conditions | Specification limits Min. | Specification limits Max. | Significant delta criteria | Pin combination (+) to (−) |
|---|---|---|---|---|---|---|
| 3N170 | $I_{GSSR}$ | $V_{GS} = -35$ V, $V_{DS} = 0$ | | $-10$ pA | $\pm 5$ pA | S to G |
| | $V_{GS(th)}$ | $V_{DS} = 10$ V, $I_D = 10$ μA | 1 V | 2 V | $\pm 0.25$ V | |
| | $I_{DSS}$ | $V_{DS} = 10$ V, $V_{GS} = 0$ | | 10 nA | $\pm 210$ pA | |
| 2N4416 | $I_{GSS}$ | $V_{GS} = -20$ V, $I_D = 10$ μA | | $-0.1$ nA | $\pm 10$ pA | S to G |
| | $V_{(BR)GSS}$ | $I_G = -1$ μA, $V_{DS} = 0$ | $-30$ V | | $\pm 1$ V | |
| 1N5711 | $I_R$ | $V_R = 50$ V | | 200 nA | $\pm 15$ nA | C to A |
| | $V_R$ | $I_R = 10$ μA | 70 V | | $\pm 1$ V | |
| CD4001A | $I_{IL1}$ | $V_{DD} = 15$ V, $V_{IN} = 0$ V | | 45 nA | $\pm 25$ pA | 14 to 1 |
| 5404* | $I_{IH}$ | $V_{CC} = 5.5$ V, $V_N = 2.4$ V | | 40 μA | $\pm 2$ μA | 1 to 14 |
| | $V_R$ | $I_R = 5$ μA | | | $\pm 1.5$ V | |
| | $I_{IL}$ | $V_{CC} = 5.5$ V, $V_{IN} = 0.4$ | 0.7 mA | 1.6 mA | $\pm 2$ μA | |
| 54S04 | $I_{IH}$ | $V_{CC} = 5.5$ V, $I_{IN} = 2.4$ V | | 50 μA | $\pm 2$ μA | 1 to 14 |
| | $V_F$ | $I_F = 5$ μA | | | $\pm 0.2$ V | |
| | $V_R$ | $I_R = 5$ μA | | | $\pm 1.5$ V | |

*$V_{IC}$ and $V_{OH}$ were measured also, but had no significant changes.

TABLE 11.3 Multiple and Single Pulse Levels

| Part | $V_{C(th)}$ | Multiple pulse levels (volts) | | | | Single pulse levels (volts) | |
|---|---|---|---|---|---|---|---|
| | | Intermediate | | | | | |
| | | 1 | 2 | 3 | 200 pulses | 90% $V_{C(th)}$ | 80% $V_{C(th)}$ |
| 3N170 | 190 | 165 | 140 | 115 | 90 | 171 | 152 |
| 2N4416 | 240 | 225 | 210 | 195 | 180 | 216 | 192 |
| 1N5711 | 700 | 600 | 500 | 400 | 300 | 630 | 560 |
| CD4001A | 930 | 898 | 865 | 833 | 800 | 898 | 865 |
| 5404 | 1850 | 1800 | 1750 | 1700 | 1650 | 1800 | 1750 |
| 54S04 | 1500 | 1375 | 1250 | 1125 | 1000 | 1350 | 1200 |

pulsing was observed on all part types, although the effect was almost negligible on the CD4001A and the 5404 devices. Even though the selected multiple pulsing range was narrow for those two parts, they tended to fail on the first pulse or survive 200 pulses with no significant parametric delta. The exceptions were a CD4001A that failed after 25 pulses, and a 54S04 failing after two.

Because of the narrow multiple pulse range derived for the CD4001A and 54S04 parts, the single pulse levels were taken at the first and second multiple pulse level rather than at 80 and 90 percent of the catastrophic level.

The stressing scheme was designed to provide every opportunity for producing barely discernible damage. In fact damage occurrences normally were well beyond the designated delta limits. One exception was $V_{GS}$ threshold changes in the 3N170 MOSFET. For this part type the threshold typically changed by slight amounts until the delta limit was exceeded.

### 11.5.4 Life-test results

The ESD stressing prior to life test resulted in a considerable number of parts that showed no signs of electrical parametric change in addition to those with varying degrees of measurable damage. This section discusses results of life tests and nonbiased bakes on both the degraded and nondegraded parts. For reference, Table 11.4 lists the cat-

**TABLE 11.4 Summary of Categorization Criteria**

| Part | Parameter | Category based upon parameter change | |
|---|---|---|---|
| | | Degraded | Improved |
| 3N170 | $I_{GSSR}$ | > 10 pA increase | > 10 pA decrease |
| | $V_{GS(th)}$ | > 0.2 V from pre-ESD value | > 0.2 V toward pre-ESD value |
| | $I_{DSS}$ | > 10 nA increase | > 10-nA decrease |
| 2N4416 | $I_{GSS}$ | 20% increase with 10-pA min | 20% decrease with 10-pA min |
| 1N5711 | $V_{(BR)GSS}$ | 10% increase with 2-V min | 10% increase with 2-V min |
| | $I_R$ | 20% increase with 200-nA min | 20% decrease with 200-nA min |
| CD4001A | $V_R$ | 2 V change | 2 V toward pre-ESD value |
| | $I_{IL1}$ | > 5% increase with 5-pA min | > 5% decrease with 5-pA min |
| 5404 | $I_{IH}$ | > 20% increase | > 20% decrease |
| | $V_R$ | > 0.5-V decrease | > 0.5-V increase |
| | $I_{IL}$ | > 20% increase | > 20% decrease |
| 54S04 | $I_{IH}$ | > 20% increase | > 20% decrease |
| | $V_F$ | > 0.2-V decrease | > 0.2-V increase |
| | $V_R$ | > 0.5-V decrease | > 0.5-V increase |

TABLE 11.5  Initially Leaky (Unaltered) 3N170 MOSFETs with Degradation

| No. | Initial value (pA) | Begin life (pA) | 25°C, highest value | | 75°C, highest value | | 25°C bake, time in hours | | |
|---|---|---|---|---|---|---|---|---|---|
| | | | (pA) | Hours | (pA) | Hours | 1008 | 2280 | 2664 |
| 6   | $I_{DSS} = 115$  | 120  | 170  | 1508 | 170  | 240 | 150  | 200   | 140   |
| 31  | $I_{DSS} = 800$  | 820  | 180  | 1079 | 1200 | 504 | 1200 | 2500  | 1000  |
| 78  | $I_{GSS} = 25$   | 18   | 3    | 1079 | 3    | 504 | 12   | 13    | 11    |
| 80  | $I_{DSS} = 5000$ | 5800 | 6000 | 1079 | 6000 | 504 | 6000 | 23000 | 5600  |
| 87  | $I_{DSS} = 41$   | 60   | 70   | 1079 | 74   | 504 | 60   | 98    | 62    |
| 97  | $I_{DSS} = 64$   | 65   | 75   | 480  | 78   | 942 | 72   | 100   | 68    |
| 99  | $I_{DSS} = 17$   | 19   | 22   | 1079 | 22   | 504 | 12   | 34    | 15    |
| 101 | $I_{DSS} = 60$   | 56   | 60   | 480  | 900  | 593 | 380  | 720   | 380   |
| 103 | $I_{DSS} = 1600$ | 1600 | 1700 | 1079 | 1800 | 504 | 2000 | 17000 | 17000 |
| | $I_{DSS} = 200$  | 180  | 200  | 248  | 220  | 706 | 190  | 120   | 180   |

egorization limits used. The criteria for degradation and improvement is in addition to specification-limit transitions. If a part did not change sufficiently during life tests to meet the "degraded" or "improved" criteria it is classified as stable. If a part improves and degrades in either order, it is classified as erratic.

**11.5.4.1 Unaltered parts.** Parts with no discernible change usually remained stable during life and nonbiased bake tests. The one notable exception was the 3N170 MOSFET devices. All 23 of the 3N170 devices that were unchanged after ESD exposure degraded to failure ($V_{GS} > 2.0$ V) during subsequent life tests and/or bake. This degradation was discounted as having been due to some undetermined failure mechanism not related to static for two reasons. Twelve of the stressed devices and sixteen controls exhibited parameters that were either out of specification limits or abnormal prior to the ESD exposure and all control devices exhibited similar degradation of $V_{GS}$. Tables 11.5 and 11.6 list the unaltered and control 3N170 devices, respectively, that had other parameters degrade with time. Stressed part number 78 was out of specification limits as received, as were control devices 72 and 107. All others listed had initial leakage parameters that were abnormal but within limits, except control device 122, which was relatively normal at the start of test.

One 1N5711 that was initially marginally acceptable at 200 nA $I_R$ degraded to 230 nA and then improved. This was discounted due to similar behavior in 11 control devices.

**11.5.4.2 Altered parameters exhibiting improvement.** Parts that had shown discernible parametric change after ESD stress either remained stable, degraded further, or showed improvement during life tests and nonbiased bake tests. Parameters that showed improvement tended to approach their initial condition rather than improve beyond the prestress value. The fact that some degraded parameters improved

TABLE 11.6  Initially Leaky (in pA) 3N170 Controls with Degradation

| No. | Start of life test, initial value | 25°C, final value at 1079 h | 75°C, final value at 504 h | 25°C bake, time in hours | | |
|---|---|---|---|---|---|---|
| | | | | 1008 | 2280 | 2664 |
| 72 | $I_{DSS} = 14$ | | | | | |
| | $I_{DSS} = 15000$ | 86 | 80 | 58 | 290 | 54 |
| 107 | $I_{GSS} = 7$ | 32 | 32 | 200 | 160 | 76 |
| 110 | $I_{DSS} = 65$ | 62 | 72 | 54 | 82 | 62 |
| 113 | $I_{DSS} = 1200$ | 1300 | 1400 | 1400 | 36,000 | 1800 |
| 119 | $I_{DSS} = 45$ | 52 | 52 | 45 | 220 | 42 |
| 120 | $I_{DSS} = 80$ | 80 | 92 | 74 | 150 | 82 |
| 122 | $I_{GSS} = 5$ | 150 | 110 | 250 | 160 | 180 |

under normal conditions is a physical reality, perhaps to be expected with such delicate failure structures. Although this improvement does not help to prove latent degradation, neither does it in any way discount those parts that degraded with time.

As a matter of fact, the majority of ESD-altered parts improved during life tests. Perhaps to the consternation of those who consider ESD to be a "CMOS problem," all 66 of the altered CD4001A devices improved during life tests. One of the 66 showed some slightly erratic behavior, while remaining well within the 45-nA specification limit. Generally, the largest improvement occurred during the 25°C bake prior to start of biased life test. The second greatest period of improvement was during the 75°C life test.

All 5404 and 54S04 devices with decreased reverse breakdown voltage tended to improve over the life test.

### 11.5.4.3 Changed parameters exhibiting degradation

*3N170 MOSFET.* 46 out of 75 ESD-altered 3N170 devices degraded to the failed condition of $V_{GS(th)} > 2.0$ V. Eight of these had some out-of-specification condition when received. Another 17 exhibited abnormally high drain to source ($I_{DSS}$) leakage prior to ESD stress. These 46 latent failures must be discounted due to similar gate-to-source threshold failures on all 32 control devices.

Table 11.7 gives $I_{GSS}$ data on two failures not to be discounted. Device 56 was within the 10-pA limit at the start of life test and then degraded to as much as 400 pA. Device 99 was out of specification at 4 nA after ESD stress, then later degraded to 1.2 mA at which point the transistor would no longer function.

*2N4416 JFET.* Of the 62 2N4416 JFETs that were altered by ESD, 48 were immediate failures, usually for excessive gate-to-source leakage. None of these failed devices showed the type of intermittent parametric behavior pertinent to latency. None of the remaining 14 devices degraded.

*1N5711 Schottky Diode.* Of the 84 altered devices, 80 were immediate failures. Most improved with time; however 10 with erratic behavior are listed in Table 11.8.

**TABLE 11.7 Degraded 3N170 Devices ($I_{GSS}$ in pA or As stated)**

| No. | Pre-ESD | Post-ESD | 25°C | | | | 75°C | | | Bake, 25°C | | |
|---|---|---|---|---|---|---|---|---|---|---|---|---|
| | | | 0 | 113 | 268 | 480 | 593 | 708 | 943 | 1008 | 2280 | 2664 |
| 56 | 16 | 15 | 7 | 250 | 340 | 210 | 400 | 240 | 240 | — | 320 | 200 |
| 99 | 2 | 4000 | 3400 | 4000 | 4200 | 4200 | —1.2 mA— | | | —1.4 mA— | | 1.2 mA |

TABLE 11.8  Erratic and Degrading 1N5711 Diodes ($I_R$ in nA; $B_V$ in V)

| | | Parameter | | Test time in hours | | | | | | | Bake | | |
|---|---|---|---|---|---|---|---|---|---|---|---|---|---|
| | | | | | 25°C | | | | 75°C | | 25°C | | 75°C |
| No. | | Pre-ESD | Post-ESD | 0 | 143 | 423 | 621 | 761 | 1090 | 1954 | 2554 | 2675 |
| 19 | $I_R$ | 103 | 5300 | 600 | 600 | 800 | 1200 | 300 | 700 | 680 | 740 | 520 |
|    | $B_V$ | 79.4 | 52.0 | 58.2 | 57.76 | 57.4 | 56.7 | 60.14 | 58.4 | | | |
| 21 | $I_R$ | 89 | 130 | 5000 | 6000 | 21,000 | 17,000 | 17,000 | 16,000 | 16,000 | 17,000 | 17,000 |
|    | $B_V$ | 83.8 | 23.0 | 22.78 | 36.95 | 30.64 | 32.5 | 32.28 | 32.54 | | | |
| 22 | $I_R$ | 154 | 155 | 190 | 250 | 200 | 130 | 130 | 130 | 130 | 150 | 120 |
| 25 | $I_R$ | 120 | 22,000 | 18,000 | 50,600 | 56,000 | 56,000 | 56,000 | 55,000 | 56,000 | 60,000 | 54,000 |
|    | $B_V$ | 83.95 | 28.13 | 36.2 | 0.007 | 0.007 | 0.006 | 0.004 | 0.006 | | | |
| 28 | $I_R$ | 180 | 27,000 | 22,000 | 50 | 400 | 260 | 215 | 180 | 190 | 200 | 170 |
| 32 | $I_R$ | 200 | 14,000 | 4600 | 11,000 | 11,000 | 9600 | 9000 | 9000 | 9200 | 9200 | 9200 |
|    | $B_V$ | 84.38 | 34.77 | 48 | 38.84 | 38.42 | 42.08 | 43.31 | 42.96 | | | |
| 41 | $I_R$ | 160 | 7200 | 1600 | 3600 | 3000 | 1800 | 1600 | 1000 | 1300 | 1300 | 1200 |
|    | $B_V$ | 80.66 | 47.63 | 57.46 | 60.63 | 62.64 | 67.87 | 68.24 | 71.85 | | | |
| 78 | $I_R$ | 125 | 10,000 | 2400 | 1000 | 4200 | 2300 | 2000 | 1200 | 1350 | 1400 | 1200 |
|    | $B_V$ | 75.79 | 40.58 | 48.36 | 54.88 | 57.3 | 65.35 | 66.34 | 68.78 | | | |
| 80 | $I_R$ | 190 | 4800 | 1200 | 2400 | 2400 | 1200 | 1200 | 780 | 780 | 820 | 600 |
|    | $B_V$ | 82.73 | 60.04 | 65.26 | 70.99 | 71.82 | 79.58 | 79.83 | 82.23 | | | |
| 103 | $I_R$ | 190 | 620 | 620 | 600 | 130 | 290 | 250 | 250 | 225 | 250 | 200 |

*5404 TTL and 54S04 Schottky TTL Hex Inverter.* None of the hex-inverter circuits exhibited degradation with time.

**11.5.4.4 Altered parameters remaining stable.** Of the ESD-altered devices, 16 2N4416s, 6 1N5711s, 7 5404s, and 33 54S04s remained stable during life tests.

## 11.6 NASA Study

The NASA funded efforts were conducted between January and November, 1981.

### 11.6.1 Parts used

Larger samples of just two part types were used rather than smaller samples with a greater variety. There were indications that MOSFET devices with gate leakage might have a propensity for latency. Therefore a larger sample of MOSFET devices having no gate protection were used. Specifically, the 3N128 device was selected because of ESD susceptibility, cost, and availability. The second part type included in this study was the 54L04 low-power TTL hex inverter.

The 54L04, of particular interest because of subtle ESD failures reported by Freeman and Beall,[10] had not been available for the NAVSEA study. Failures of this part were experienced in the Viking '75 Lander camera system. The failed devices had an output stuck in the high state. Some of the failures were evident during −55°C testing, but functioned properly at room ambient. Failure analysis disclosed that the beta of $Q2$ had been degraded. The schematic of an inverter in Fig. 11.2 shows that $Q2$ drives the output stage. When $Q2$ beta was sufficiently low, the output transistor $Q3$ remained on, $Q4$ remained off, and the output was stuck at the high level.

The fact that this damage was sometimes detectable only at low temperature gave assurance that such a failure could be expected to operate at room temperature. The damage was different from most

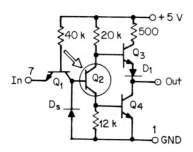

**Figure 11.2** Schematic of a single inverter of the 54L04 device.

static failures in that an internal transistor stage was damaged rather than one at the input or output. The subtle nature of the failure and the critical dependence on the value of $Q2$ beta made this a good candidate for latent ESD damage. Once slight damage was inflicted catastrophic latent failure depended only upon an appropriate degree of degradation of $Q2$.

### 11.6.2 Pin combinations and delta criteria

The 3N128 MOSFET pin combination used was gate (+) to source (−). The objective was to inflict gate-oxide damage in accordance with indications from the NAVSEA study that this might result in latency. Parameters monitored were $I_{GSSR}$ and $V_{GS(off)}$. The 54L04 was stressed from input (pin 7) to ground (pin 1) with the input negative. This was the combination reported by Freeman and Beall to cause subtle damage to the base emitter junction, and thus beta degradation of $Q2$. Parameters used to monitor degradation were $I_{IH}$ and $V_{OL}$. Test conditions, specification limits, and delta criteria are given in Table 11.9.

### 11.6.3 ESD stress results

Initial familiarization data of 3N128 devices indicated a cumulative effect from multiple exposures below the catastrophic threshold, but over a rather narrow range, somewhat dependent upon individual device endurance. Multiple pulse results for this part are given in Table 11.10.

Familiarization data of 54L04 devices indicated a cumulative effect over a wider voltage range. Multiple pulse results from the first 50 devices are given in Table 11.11. Since all 10 devices survived 200 exposures at 2000 V, and only 5 out of 10 devices failed the 5000-V level, the level of the range was raised by 750 V. Results of the second 50 devices are given in Table 11.12. The fact that all 10 failed at the 5750-V level and only 4 survived 200 pulses at the 2750-V level indicated that the two ranges were about equally valid.

Single pulses of 90 percent and 80 percent of the catastrophic failure threshold were applied to 20 parts at each level. This resulted in significant changes in all 20 of the 3N128 devices at 103 V. However, 10 were substantial changes in $V_{GS(off)}$ that did not meet our significant delta criteria. The other 10 met the $I_{GSS}$ delta criteria. At 92 V, there were 5 $I_{GSS}$ significant deltas, and 12 with >0.2 V change in $V_{GS(off)}$ out of 20 devices. The 54L04 results were within expectations: 17 out of 20 at 4500 V and 11 out of 20 at 4000 V.

**TABLE 11.9 NASA Study Test Parameters and Delta Criteria**

| Part | Test parameters | Test conditions | Specification limits Min. | Specification limits Max. | Significant delta criteria | Pin combination (+) to (−) |
|---|---|---|---|---|---|---|
| 3N128 | $I_{GSS}$ | $V_{GS} = 8$ V<br>$V_{DS} = 0$ V | | −50 pA | ≥ −25 pA | G to S |
| | $V_{GS(off)}$ | $V_{DS} = +15$ V<br>$I_D = 50$ μA | −0.5 V | −8.0 V | ≥ −0.5 V<br>≤ −8.0 V | |
| 54L04 | $I_{IH}$ | $V_{CC} = 5.5$ V<br>$V_{IN} = 2.4$ V<br>All unused inputs at ground | | 10 μA | ±20 nA | IN(7) to G(1) |
| | $V_{OL}$ | $V_{CC} = 4.5$ V<br>$I_{OL} = 2$ mA<br>$V_{IN} = 2.0$ V<br>All unused inputs at 5.5 V | | 0.3 V | ±50 mV | |

TABLE 11.10 Multiple Pulse Results with 3N128

| Voltage level | Number of devices where delta was exceeded in 1 pulse | Mean number of pulses to exceed delta | Highest number of pulses to exceed delta |
|---|---|---|---|
| 115 | 12 | 3.6 | 1 at 25 |
| 110 | 11 | >16 | 1 at 200 |
| 105 | 6 | >22.8 | 1 at 200 |
| 100 | 7 | 30.35 | 2 at 175 |
| 95 | 6 | >64.4 | 4 at 200 |

TABLE 11.11 Multiple Pulse on First Sample of Fifty 54L04 Devices

| Voltage level | Number of devices where delta was exceeded in 1 pulse | Mean number of pulses to exceed delta | Highest number of pulses to exceed delta |
|---|---|---|---|
| 5000 | 5 | >61.79 | 3 at 200 |
| 4250 | 7 | >41.5 | 2 at 200 |
| 3500 | 5 | >100.5 | 5 at 200 |
| 2750 | 1 | >152.6 | 7 at 200 |
| 2000 | None | >200 | All 10 at 200 |

TABLE 11.12 Multiple Pulse on Second Sample of Fifty 54L04 Devices

| Voltage level | Number of devices where delta was exceeded in 1 pulse | Mean number of pulses to exceed delta | Highest number of pulses to exceed delta |
|---|---|---|---|
| 5750 | 10 | 1 | 1 |
| 5000 | 7 | >38.4 | 1 at 200 |
| 4250 | 9 | >20.9 | 1 at 200 |
| 3500 | 4 | >62 | 3 at 200 |
| 2750 | 1 | >129.2 | 4 at 200 |

### 11.6.4 Life-test results

Biased life-test conditions at 25°C and 75°C were as shown in Table 11.13. Results were categorized in accordance with Table 11.14 in addition to specification-limit transitions.

**11.6.4.1 Unaltered parts.** Both device types having no initial significant parametric delta according to Table 11.4 tended to remain stable throughout life tests. This accounted for 4 stressed and 68 control 3N128 devices and 53 stressed and 50 control 54L04 devices.

However, in the case of the 3N128 devices, considerable change in $V_{GS(off)}$ was permitted during ESD stressing for the following reasons:

**TABLE 11.13 Life-Test Conditions**

| Part | Life-test conditions |
|---|---|
| 3N128 | $V_{GS} = -8$ V, $V_{DS} = 0$ V, 1 MΩ series resistance |
| 54L04 | $V_{CC} = 5$ V, $V_{SS} = 0$ V, $V_{IN} = 0$ to 3-V peak |
| | $F_{IN} = 100$ kHz, duty cycle = 50% |

**TABLE 11.14 NASA Study Results Categorization Criteria**

| Part | Parameter | Category based upon parameter change | |
|---|---|---|---|
| | | Degraded | Improved |
| 3N128 | $I_{GSSR}$ | >20% increase with 10-pA min | >20% decrease with 10-pA min |
| 54L04 | $V_{GS(off)}$ | >0.2 V from pre-ESD value | >0.2 V toward pre-ESD value |
| | $I_{IH}$ | >20% increase | >20% decrease |
| | $V_{OL}$ | >50-mV increase | >50-mV decrease |

1. The primary objective was to stress until gate-oxide leakage.
2. The specification limits on $V_{GS(off)}$ are quite wide (−0.5 V to −8.0 V).

Those with upward shifts of $V_{GS(off)}$ included seven parts that had survived 200 pulses, and 22 parts that had received single pulses. The multiple-pulsed devices in Table 11.15 showed a tendency toward a decrease in $V_{GS(off)}$ as the life test proceeded. There are two points of particular interest to be made from this data:

1. Device number 105 continued to decrease in $V_{GS}$ value until it was out of specification at 840 h at 75°C.
2. The specification limit on the 3N128 is sufficiently tolerant to allow such drastic changes in $V_{GS}$ as those caused by 200 exposures to ESD.

Single-pulsed 3N128 devices that improved more than 0.2 V after marginal changes in $V_{GSS}$ are listed in Table 11.16.

Four single-pulsed 3N128 devices that showed degradation greater than 0.2 V during life test are listed in Table 11.17.

**11.6.4.2 Altered parameters showing improvement.** There were four 3N128 MOSFETs having $I_{GSS}$ > 180 μA after ESD stress that showed some improvement during life test. All of the 42 $V_{GSS(off)}$ failures that began life test at greater than −8.0 V showed a trend toward improvement, especially after exposure to 75°C life. There were inter-

TABLE 11.15 Marginally Changed $V_{GS(off)}$ (in −volts) Life-Test Data

| 3N128 device | Initial $V_{GS(off)}$ | Post-200 pulses | Biased life-test hours | | | | | | | |
|---|---|---|---|---|---|---|---|---|---|---|
| | | | At 25°C | | | | At 75°C | | | |
| | | | 0 | 120 | 312 | 648 | 304 | 465 | 840 | |
| 6   | 1.81 | 5.75 | 3.82 | 4.78 | 4.93 | 4.84 | 2.63  | 2.56  | 2.67  |
| 54  | 1.45 | 4.26 | 2.25 | 3.83 | 3.97 | 4.3  | 0.95  | 0.85  | 0.91  |
| 57  | 1.72 | 5.94 | 3.56 | 4.6  | 4.7  | 4.7  | 2.9   | 2.4   | 2.6   |
| 58  | 2.12 | 7.5  | 5.6  | 6.3  | 6.3  | 6.5  | 3.6   | 3.8   | 3.7   |
| 75  | 1.53 | 6.2  | 3.7  | 6.1  | 5.2  | 5.4  | 2.8   | 2.9   | 2.9   |
| 87  | 1.7  | 6.8  | 6.8  | 5.8  | 6.0  | 6.1  | 2.8   | 2.9   | 2.9   |
| 105 | 1.4  | 3.8  | 3.8  | 3.3  | 3.5  | 3.8  | 0.752 | 0.503 | 0.294 |

TABLE 11.16 Single Pulsed 3N128s with Marginally Changed $V_{GS(off)}$ (in −volts)

| 3N128 device | Initial $V_{GS(off)}$ | Post-1 pulse | Biased life-test hours | | | | | | |
|---|---|---|---|---|---|---|---|---|---|
| | | | At 25°C | | | | At 75°C | | |
| | | | 0 | 120 | 312 | 648 | 304 | 465 | 840 |
| 21 | 1.37 | 1.80 | 1.53 | 1.52 | 1.52 | 1.50 | 1.45 | 1.41 | 1.40 |
| 113 | 2.10 | 3.36 | 3.36 | 3.47 | 3.41 | 3.48 | 3.21 | 3.13 | 3.12 |
| 118 | 1.74 | 2.51 | 2.51 | 2.37 | 2.38 | 2.41 | 2.09 | 2.07 | 2.07 |
| 119 | 1.47 | 2.15 | 2.15 | 2.06 | 2.08 | 2.14 | 1.57 | 1.67 | 1.52 |
| 132 | 1.71 | 2.29 | 2.29 | 2.29 | 2.27 | 2.28 | 1.98 | 1.96 | 1.95 |
| 135 | 2.06 | 3.20 | 3.20 | 3.31 | 3.31 | 3.39 | 2.87 | 2.71 | 2.72 |
| 138 | 1.74 | 2.60 | 2.60 | 2.48 | 2.50 | 2.50 | 2.01 | 1.98 | 1.98 |
| 139 | 1.72 | 2.92 | 2.92 | 2.86 | 2.87 | 2.91 | 2.20 | 2.16 | 2.15 |

TABLE 11.17 $V_{GS(off)}$ (in −volts) in Marginally Altered Single-Pulsed 3N128s

| 3N128 device | Initial $V_{GS(off)}$ | Post-1 pulse | Biased life-test hours | | | | | | |
|---|---|---|---|---|---|---|---|---|---|
| | | | At 25°C | | | | At 75°C | | |
| | | | 0 | 120 | 312 | 648 | 304 | 465 | 840 |
| 117 | 2.72 | 4.15 | 4.18 | 4.63 | 4.64 | 4.67 | 4.17V | 4.11 | 4.13 |
| 120 | 2.26 | 3.84 | 3.84 | 4.14 | 4.21 | 4.25 | 3.81 | 3.73 | 3.68 |
| 128 | 2.76 | 3.47 | 3.47 | 3.63 | 3.64 | 3.69 | 3.55 | 3.57 | 3.57 |
| 129 | 3.26 | 3.93 | 3.93 | 4.09 | 4.11 | 4.18 | 4.09 | 4.05 | 4.07 |

mittencies in 12 of those *improving,* exhibited by occasions of acceptable readings followed by subsequent values which exceeded specification limits (see Table 11.18). Device 12 shifted $V_{GS}$ from $-1.6$ V to the minimum acceptable level of $-0.5$ V from the ESD exposure. This part degraded to 0.4 V at the start of life test then gradually improved to 0.45 V during life test. There also were improvements in 13 of the 29 3N128s with marginal $V_{GS}$ shifts.

Of the 117 54L04 devices that were altered, 33 improved during life tests. The 21 $V_{OL}$ parameters that changed are listed in Table 11.19. The other four improved devices were $I_{IH}$, typically shifting to about 200 µA, then improving about 30 µA over the life test.

**11.6.4.3 Changed parameters remaining stable.** There were 76 3N128 devices with increased $I_{GSS}$ after ESD exposure. Of these, 49 remained stable, showing no significant degradation or improvement throughout the life test. Of the 29 3N128s with marginal shifts of $V_{GS(off)}$, 13 remained stable. Seventy of the 117 altered 54L04 devices remained stable during life test.

**11.6.4.4 Changed parameters that degraded during life test.** Table 11.20 shows 3N128 devices with increased $I_{GSS}$ that degraded during life test. Changed 54L04 hex inverters that degraded during life test are given in Table 11.21.

## 11.7 Analysis of Data from NAVSEA and NASA Studies

Although the analysis of this valuable data base reveals much about latent failures, the need for further work in this area is overwhelming.

### 11.7.1 Available data base

The results of the NASA project are of primary interest since the study plan was based upon prior knowledge gained during the NAVSEA project. Therefore, most meaningful results were anticipated from the 3N128 and 54L04 data. Table 11.22 gives a summary total of the data provided by the sample population of the two studies.

### 11.7.2 Control device behavior

Behavior of *control* devices in a study such as this is of utmost importance. The controls were not exposed to ESD stressing and are intended to represent the population of *good* parts from which the entire study sample was taken. If stressed parts showed degradation during

TABLE 11.18  Improving $V_{GS(off)}$ (in −volts) on 3N128s with Intermittencies

| 3N128 device | Initial $V_{GS(off)}$ | Post-200 pulses | Biased life-test hours | | | | | | |
|---|---|---|---|---|---|---|---|---|---|
| | | | At 25°C | | | | At 75°C | | |
| | | | 0 | 120 | 312 | 648 | 304 | 465 | 840 |
| 6   | 2.2  | 8.48 | 7.3  | 8.24 | 8.01 | 8.05 | 5.85 | 5.73 | 5.51 |
| 41  | 2.3  | 8.13 | 7.07 | 8.16 | 7.92 | 7.83 | 6.02 | 5.8  | 5.62 |
| 51  | 2.14 | 9.23 | 7.38 | 8.45 | 8.15 | 8.15 | 5.43 | 5.7  | 4.49 |
| 60  | 2.15 | 8.56 | 6.76 | 7.64 | 8.04 | 8.1  | 5.69 | 5.11 | 5.17 |
| 64  | 2.20 | 8.23 | 7.02 | 7.64 | 8.04 | 8.2  | 5.9  | 5.86 | 5.52 |
| 67  | 2.70 | 8.37 | 7.39 | 8.26 | 8.40 | 8.48 | 6.75 | 6.42 | 6.33 |
| 68  | 2.21 | 8.19 | 7.59 | 8.15 | 8.29 | 8.74 | 6.15 | 5.90 | 5.35 |
| 74  | 2.46 | 8.44 | 7.49 | 8.11 | 8.16 | 8.23 | 6.48 | 6.24 | 6.16 |
| 88  | 2.62 | 8.04 | 8.04 | 7.79 | 8.12 | 8.27 | 6.48 | 6.37 | 6.20 |
| 95  | 2.48 | 8.17 | 8.17 | 7.66 | 8.05 | 8.28 | 5.45 | 5.84 | 5.74 |
| 96  | 3.38 | 8.11 | 8.11 | 7.95 | 8.04 | 8.14 | 5.62 | 5.18 | 5.08 |
| 110 | 2.69 | 8.43 | 8.43 | 7.98 | 8.51 | 8.54 | 6.73 | 6.56 | 6.51 |

TABLE 11.19  54L04 Devices with Degraded $V_{OL}$ (volts) That Improved

| 54L04 device | ESD stress results | | | | Biased life-test hours | | | | | | |
|---|---|---|---|---|---|---|---|---|---|---|---|
| | $V_{OL}$ | | Pulses | | At 25°C | | | | | At 75°C | |
| | Pre | Post | K-volts | (N) | 0 | 120 | 456 | 696 | 134 | 294 | 648 |
| 4 | 0.18 | 5.0 | 7.5 | (1) | 4.9 | 0.26 | 0.20 | 0.26 | 0.22 | 0.20 | 0.15 |
| 22 | 0.18 | 1.18 | 3.0 | (5) | 1.15 | 0.83 | 0.83 | 0.83 | 0.72 | 0.57 | 0.55 |
| 24 | 0.18 | 0.67 | 2.75 | (5) | 0.59 | 0.42 | 0.49 | 0.47 | 0.60 | 0.28 | 0.27 |
| 33 | 0.18 | 0.26 | 5.0 | (6) | 0.23 | 0.23 | 0.23 | 0.19 | 0.23 | 0.22 | 0.17 |
| 34 | 0.18 | 1.10 | 5.0 | (7) | 0.84 | 0.64 | 0.68 | 0.63 | 0.54 | 0.36 | 0.32 |
| 38 | 0.18 | 0.24 | 5.0 | (1) | 0.20 | 0.20 | 0.20 | 0.17 | 0.20 | 0.19 | 0.16 |
| 41 | 0.18 | 0.25 | 4.25 | (8) | 0.26 | 0.25 | 0.22 | 0.22 | 0.23 | 0.22 | 0.20 |
| 68 | 0.17 | 0.26 | 2.75 | (50) | 0.27 | 0.16 | 0.16 | 0.17 | 0.17 | 0.17 | 0.17 |
| 83 | 0.18 | 2.73 | 5.75 | (1) | 0.21 | 0.21 | 0.21 | 0.20 | 0.21 | 0.21 | 0.21 |
| 91 | 0.17 | 5.0 | 5.0 | (1) | 0.20 | 0.20 | 0.20 | 0.20 | 0.20 | 0.20 | 0.20 |
| 92 | 0.18 | 5.0 | 5.0 | (1) | 5.0 | 2.0 | 1.09 | 1.23 | 0.77 | 0.49 | 0.56 |
| 93 | 0.19 | 2.73 | 5.0 | (175) | 2.06 | 2.04 | 1.92 | 1.86 | 1.78 | 1.45 | 1.33 |
| 98 | 0.17 | 5.0 | 5.0 | (1) | 5.0 | 0.96 | 0.95 | 0.74 | 0.59 | 0.37 | 0.39 |
| 104 | 0.17 | 5.0 | 4.25 | (1) | 1.68 | 0.68 | 0.65 | 0.49 | 0.37 | 0.23 | 0.28 |
| 111 | 0.18 | 3.59 | 3.5 | (1) | 0.19 | 0.18 | 0.19 | 0.18 | 0.19 | 0.19 | 0.19 |
| 118 | 0.16 | 5.0 | 3.5 | (8) | 5.0 | 0.21 | 0.21 | 0.21 | 0.20 | 0.20 | 0.20 |
| 122 | 0.17 | 1.30 | 2.75 | (130) | 1.25 | 1.07 | 1.12 | 1.07 | 0.95 | 0.82 | 0.79 |
| 136 | 0.17 | 0.62 | 4.0 | (1) | 0.32 | 0.20 | 0.21 | 0.21 | 0.20 | 0.20 | 0.19 |
| 138 | 0.17 | 5.0 | 4.0 | (1) | 1.89 | 1.75 | 1.77 | 1.80 | 1.56 | 1.40 | 1.36 |
| 158 | 0.17 | 5.0 | 4.5 | (1) | 5.0 | 5.0 | 5.0 | 5.0 | 5.0 | 4.59 | 0.93 |
| 162 | 0.17 | 5.0 | 4.5 | (1) | 2.49 | 2.49 | 2.50 | 2.50 | 2.50 | 2.50 | 2.50 |

TABLE 11.20  3N128 Devices with Altered $I_{GSS}$ That Degraded Further

| 3N128 device | ESD stress results | | | | Biased life-test hours | | | | | | |
|---|---|---|---|---|---|---|---|---|---|---|---|
| | $I_{GSS}$* | | Pulses | | At 25°C | | | | | At 75°C | |
| | Pre | Post | Volts | (N) | 0 | 120 | 312 | 648 | 304 | 465 | 840 |
| 2  | <0.5pA | 126 | 150 | 1   | 150 | 150 | 175 | 180 | 175 | 175 | 175 |
| 36 | 1.6pA  | 38  | 105 | 1   | 58  | 59  | 59  | 62  | 62  | 62  | 60  |
| 49 | <0.5pA | 9.5 | 100 | 175 | 10  | 10  | 10  | 11  | 14  | 14  | 14  |
| 52 | 2.2pA  | 76  | 100 | 2   | 220 | 190 | 195 | 190 | 190 | 190 | 185 |
| 63 | 1.0pA  | 38  | 115 | 10  | 50  | 46  | 48  | 48  | 56  | 54  | 55  |

*All $I_{GSS}$ data is in microamperes except the pre-ESD measurements.

TABLE 11.21  Altered 54L04 $V_{OL}$ (volts) That Degraded during Life Test

| 54L04 device | ESD stress results | | | | Biased life-test hours | | | | | | |
| --- | --- | --- | --- | --- | --- | --- | --- | --- | --- | --- | --- |
| | $V_{OL}$ | | Pulses | | At 25°C | | | | | At 75°C | |
| | Pre | Post | K volts | (N) | 0 | 120 | 456 | 696 | 134 | 294 | 648 |
| 5   | 0.18 | 0.33   | 7.0  | (2) | 0.38 | 0.22 | 5.0  | 5.0  | 5.0 | 5.0  | 5.0 |
| 16  | 0.17 | 0.92   | 4.0  | (1) | 0.96 | 0.96 | 1.41 | 0.85 | 5.0 | 5.0  | 5.0 |
| 27  | 0.17 | 0.12   | 4.25 | (1) | 0.15 | 5.0  | 5.0  | 5.0  | 5.0 | 5.0  | 5.0 |
| 43  | 0.18 | 2.68   | 4.25 | (1) | 2.78 | 5.0  | 5.0  | 5.0  | 5.0 | 5.0  | 5.0 |
| 81  | 0.17 | 0.17*  | 5.75 | (1) | 1.4  | 2.12 | 5.0  | 5.0  | 5.0 | 5.0  | 5.0 |
| 84  | 0.17 | 0.27   | 5.75 | (1) | 0.22 | 0.22 | 3.08 | 0.28 | 5.0 | 5.0  | 5.0 |
| 113 | 0.18 | 3.56   | 3.5  | (1) | 5.0  | 5.0  | 5.0  | 5.0  | 5.0 | 5.0  | 5.0 |
| 133 | 0.17 | 5.0    | 4.0  | (1) | 5.0  | 4.8  | 2.48 | 0.35 | 5.0 | 4.73 | 3.0 |
| 134 | 0.17 | 0.21*  | 4.0  | (1) | 0.21 | 0.20 | 0.23 | 0.22 | 5.0 | 5.0  | 5.0 |
| 165 | 0.17 | 0.26   | 4.5  | (1) | 0.33 | 4.97 | 4.69 | 5.0  | 5.0 | 5.0  | 5.0 |

*Devices 81 and 134 had $I_{IH}$ changes that exceeded delta criterion.

TABLE 11.22 Summary of Data Base from NAVSEA and NASA Studies

| Parts | NAVSEA study ||||||| NASA study ||
|---|---|---|---|---|---|---|---|---|
| | 3N170 | 2N4416 | 1N5711 | CD4001A | 5404 | 54S04 | 3N128 | 54L04 |
| Total parts | 129 | 134 | 140 | 160 | 136 | 161 | 152 | 170 |
| Controls | 32 | 36 | 37 | 42 | 55 | 60 | 68 | 50 |
| Initially ESD altered life test | 75 | 62 | 84 | 66 | 49 | 92 | 148* | 117 |
| Stable | 0 | 16 | 3 | 0 | 7 | 33 | 62* | 70 |
| Improved | 28 | 39 | 65 | 65 | 42 | 59 | 67* | 33 |
| Degraded | 47 | 1 | 1 | 0 | 0 | 0 | 2* | 10 |
| Erratic | 0 | 6 | 15 | 1 | 0 | 0 | 16* | 4 |
| Initially unaltered life test | 22 | 36 | 19 | 52 | 32 | 9 | 4 | 53 |
| Stable | 0 | 36 | 17 | 52 | 31 | 9 | 4 | 53 |
| Degraded | 22 | | | | | | | |
| Erratic | | | 2 | | 1 | | | |
| Improved | | | | | | | | |

*Includes 30 marginally altered $V_{GS}$ devices: 13 stable, 13 improved, 1 degraded, and 3 erratic.

TABLE 11.23  Control Device Behavior

| Control devices | Behavior |
|---|---|
| 3N170 | All 32 devices degraded to $V_{GS(th)} > 2.0$ V. Six of these were initially out of specification, another nine had abnormally high leakage as received. |
| 2N4416 | All 36 remained stable. |
| 1N5711 | 11 out of 37 degraded in reverse leakage, typically to slightly over the 200-nA maximum limit. |
| CD4001A | All 42 remained stable. |
| 5404 | All 55 remained stable. |
| 54S04 | All 60 remained stable. |
| 3N128 | All 68 remained stable. |
| 54L04 | All 50 remained stable. |

life tests, the relationship to ESD would not be proven if control devices showed similar degradation. On the contrary, this would be indicative of the presence of another failure mechanism. If, on the other hand, control devices remained stable while stressed parts degraded, a statistical analysis could be made to quantify the causative relationship to ESD. Control-device behavior is given in Table 11.23.

### 11.7.3  Life test results of devices unaltered by ESD stress

In general, devices that were not altered by the ESD stress phase of the study remained stable showing no tendencies for degradation, improvement, or erratic behavior. The only exceptions were the 1N5711 devices and some abnormal 3N170 devices. Note that control devices for these two part types were also unstable. The overall study data suggest that normal devices, with parameters that were not affected by ESD stress, will remain stable throughout life.

### 11.7.4  Life-test behavior of ESD altered devices

Those devices having some initial effect due to the ESD stress provide the data source for possible latent failures.

### 11.7.5  The case for latent failure existence

Candidate devices to be considered as possible latent failures are listed in Table 11.24.

**11.7.5.1  Mac-1 or classical latent failures.** The "classical" latent failure sought by earlier investigators was the type befitting the Mac-1 definition: a part that is within acceptable limits after ESD exposure but

TABLE 11.24  Latent Failure Candidates

| Part | Latent failure candidates | Latent type | Device numbers |
|---|---|---|---|
| 3N170 | 2 devices from Table 11.4 | Mac-1 | 56 |
|  |  | Mac-3 | 99 |
| 1N5711 | 10 devices from Table 11.8 | Mac-1* | 22*, 103* |
|  |  | Mac-3 | 19,21,25,28,32,41,78,80 |
| CD4001A | None |  |  |
| 3N128 | 2 devices from Table 11.15 | Mac-1 | 105 |
|  |  | Mac-2 | 54 |
|  | 12 devices from Table 11.18 | Mac-1 | All 12 |
|  | 5 devices from Table 11.20 | Mac-3 | 2,36,49,52,63 |
|  | 4 devices from Table 11.17 | Mac-2 | 117,120,128,129 |
| 54L04 | 10 devices from Table 11.21 | Mac-1 | (5),27,81,84,134,165 |
|  |  | Mac-3 | 16,43,113,133 |

*Discounted because of similar control device behavior.

degrades to an out-of-specification value during life. Since many consider this the purest definition of a latent failure, perhaps the strongest cases should be based upon any Mac-1 failures evidenced in the data. Pertinent specifics about such failures are listed in Table 11.25.

Data for each part type will be analyzed separately to determine the statistical relationship of the induced ESD stress to the failure occur-

TABLE 11.25  Mac-1 Latent Failures

| Mac-1 failures | Description of failure conditions and behavior | Data location |
|---|---|---|
| 3N170 #56 | $I_{GSS}$ (10 pA max) was 7 pA at the start of life test and degraded to 250 pA after 113 h at 25°C life, continued to degrade to 400 pA at 113 h and 75°C | Table 11.4 |
| 1N5711, #22, 103 | Increased in reverse leakage to beyond maximum limit. Discounted because of similar behavior in control devices. | Not applicable |
| 3N128 #105 | After marginal increase in $V_{GS}$ (−0.5 V to −8 V) from 200 ESD pulses, remained acceptable until 0.294 V at 421 h and 75°C. | Table 11.15 |
| 6,41,51,60, 64,67,68,74 | All were within the 8-V max limit at the start of 25°C life, 60 and 64 exceeded 8 V at 312 h, the others at 120 h. | Table 11.18 |
| 88,95,96,110 | All were slightly over 8-V max at start of life, below 8 V at 120 h, and over at 312 and 648 h. | Table 11.18 |
| 54L04 #5 | Barely over 0.3-V max for $V_{OL}$ after ESD at 0.33 V; catastrophic level of 5 V at 456 h. | Table 11.21 |
| 27,84,134,165 | All were in limit after ESD but 2 were above at 120 h and 1 at 456 h (25°C), and 1 at 134 h at 75°C. | Table 11.21 |
| 81 | Within $V_{OL}$ limit at 0.17 V after ESD, failed at start of life test after 2 weeks unbiased 25°C. |  |

rences during life tests. The most impressive data appear to be from the parts used in the NASA study which resulted in six catastrophic failures of 54L04 hex inverters and 13 3N128 MOSFETs.

Of the 170 54L04 devices exposed to ESD, 99 were immediate catastrophic failures. Since these are not valid candidates for latency, they were excluded from the latent failures statistics. Only 19 of the remainder showed parametric shifts while remaining within specification after ESD exposure. In addition, 40 devices withstood 200 pulses and 12 devices withstood single pulses with no significant delta. A case could be made for excluding the unaffected parts based upon stability of such parts throughout the studies. Regardless, for the purpose of analysis, they shall be considered as valid latency candidates.

Let

$$x_i = 1 \quad \text{(if the } i\text{th device is a failure)} \tag{11.1}$$

$$= 0 \quad \text{(if the } i\text{th device is not a failure)} \tag{11.2}$$

$$S_e^2 = \text{variance} \quad \text{(ESD exposed)} \tag{11.3}$$

$$F = \sum_{i=1}^{n_e} x_i^2 = \sum_{i=1}^{n_e} x_i \quad \text{(since all } x_i = 1) \tag{11.4}$$

$$S_e^2 = \frac{F - F^2/n_e}{n_e - 1} \tag{11.5}$$

where $F$ = total number of latent failures = 6 and $n_e$ = candidate exposed parts = 71.

$$S_e^2 = \frac{6 - 36/7}{70} = 0.078471$$

$$S_c^2 = \text{variance} \quad \text{(non-ESD-stressed controls)} \tag{11.6}$$

$$C = \sum_{i=1}^{n_c} x_i^2 \tag{11.7}$$

$$= \sum_{i=1}^{n_c} x_i = 0 \quad \text{(since all } x_i = 0) \tag{11.8}$$

$$S_c^2 = \frac{C - C^2/n_c}{n_c - 1} \tag{11.9}$$

$$= \frac{0 - 0/n_c}{n_c - 1}$$

$$= 0$$

where $c$ = total number of latent failures in control sample = 0

$S^2$ = pooled variance

$$= \frac{(n_e - 1)S_e^2 + (n_c - 1)S_c^2}{n_e + n_c - 2} \quad (11.10)$$

$$= \frac{70(.078471) + 49(0)}{71 + 50 - 2}$$

$$= 0.046159$$

The "$t$" test is used to check the hypothesis that the ESD exposure had no effect on the parts that degraded with time.

$$t = \frac{F/n_e - C/n_c}{[S^2(1/n_e + 1/n_c)]^{1/2}} \quad (11.11)$$

$$= \frac{6/71 - 0/50}{[0.046159(1/71 + 1/50)]^{1/2}}$$

$$= 2.1305$$

for $n_e + n_c - 2 = 119$ degrees of freedom

$t_{119, 0.02} = 2.359$

$t_{119, 0.05} = 1.980$

By interpolation, $2.1305 = t_{119, 0.038}$. Thus, there is a 96.2 percent probability that the 54L04 latent failures were due to the ESD exposure.

Table 11.26 also contains similar data on the other part types. So a strong case for latency can be made based upon the classical type failures alone, especially in the cases of the 3N128 and 54L04 devices from the later study.

**11.7.5.2 Mac-2 failures.** Mac-2 failures included only devices within the specification limits, but with large parametric shifts that the assembly design would not tolerate. Hypothetical system-level failures can be possible from the parametric data of the Mac-2 examples listed in Table 11.24. However, these are the weakest examples of possible latency and will not be factored into the statistics, but are worthy of consideration in this imperfect world.

**11.7.5.3 Mac-3 failures.** This type of failure may be even more likely to affect system reliability than the classic Mac-1 type latent failure. The Mac-3 type failure occurrence begins with a part installed in a properly functioning assembly. An ESD stress is applied to the assembly and the part is affected. If the part were to be removed and functionally tested, one or more parameters would be found to be out of acceptable limits. Nevertheless, with good design margins it is possible that the assembly or system would continue to function properly. In the event of further parametric degradation with time, the system could eventually fail. Design applications where such a failure might occur can be hypothesized for the Mac-3 parts listed in Table 11.24. If these are considered as valid latent failures the modified statistical analysis as shown in Table 11.27 is even more convincing.

**11.7.6 Conclusions**

The following conclusions can be made based upon the NAVSEA and NASA studies:

1. Latent ESD failures have been proved to exist.
2. Cumulative damage from multiple exposures to ESD at levels below the damage threshold has been observed throughout the studies.
3. Failure in end use could result at low-level ESD exposure because of prior cumulative damage.
4. A universal method to screen for latent ESD damage is not evident.
5. The studies have not indicated a high likelihood of a latent failure from ESD occurring by chance.

The probability of a latent failure occurring after ESD stress has been applied seems to be emerging as the "latency controversy" of the late 1980s. With extremely careful monitoring of change due to a comprehensively planned variable ESD stress application, the number of latent failures was relatively small. Only 21 classical latent failures occurred out of 970 stressed parts. Under ordinary handling conditions the ratio would be expected to be much less.

The statistics from Tables 11.26 and 11.27 support this argument in that experience gained in the first study may have contributed to the greater significance of the follow-on (NASA) study. The probabilities associated with Mac-1 failures from the NAVSEA parts are not convincing. Although the ESD stressing for this initial study was well-planned, it can be considered to be closer to the random conditions of ordinary handling than the second study. The data from Table 11.27 indicates that Mac-3 failures may be much more likely under ordinary manufacturing and usage conditions.

The fact that latent ESD failures can occur is, nevertheless, of utmost importance to the electronics industry. The risk of ESD exposure

**TABLE 11.26  Mac-1 Latent Failure Statistics**

| Part type | No. Mac-1 | Controls $n_1$ | $n_2$ | $S_e^2$ | $S_c^2$ | $S^2$ | $t$ | Deg frdm | Probability due to ESD |
|---|---|---|---|---|---|---|---|---|---|
| 54L04 | 6 | 71 | 50 | 0.078471 | 0 | 0.046159 | 2.1305 | 119 | 96.2% |
| 3N128 | 13 | 67 | 68 | 0.158752 | 0 | 0.078779 | 4.0159 | 133 | 99.9% |
| 3N170 | 1 | 72 | 26 | 0.013889 | 0 | 0.010272 | 0.598936 | 96 | 44.7% |

**TABLE 11.27  Mac-1 and Mac-3 Latent Failure Statistics**

| Part type | Mac-1 + Mac-3 | Controls $n_1$ | $n_2$ | $S_e^2$ | $S_c^2$ | $S^2$ | $t$ | Deg frdm | Probability due to ESD |
|---|---|---|---|---|---|---|---|---|---|
| 54L04 | 10 | 71 | 50 | 0.122736 | 0 | 0.072198 | 2.83932 | 119 | 98.7% |
| 3N128 | 18 | 67 | 68 | 0.199457 | 0 | 0.098979 | 4.96080 | 133 | >99.9% |
| 3N170 | 2 | 72 | 26 | 0.027386 | 0 | 0.020255 | 0.853051 | 96 | 67.0% |
| 2N4416 | 5 | 62 | 36 | 0.075357 | 0 | 0.047883 | 1.75882 | 96 | 92.5% |
| 1N5711 | 8 | 84 | 37 | 0.087206 | 0 | 0.060824 | 1.957 | 119 | 94.65% |

must be carefully assessed where high reliability is required. Conceivably, there may be certain applications or even part types discovered where latency is so likely as to be intolerable. The data base presented here is much too small to warrant an absolute assertion on likelihood of occurrence.

## 11.8  Results of Other Latency Studies

### 11.8.1  Martin Marietta study

In 1983, J. S. Bowers et al.[12] reported on the results of another latent ESD failure study. This study also used the 54L04 low-power TTL hex inverter and the 3N171 MOSFET. The ESD stressed life-test sample consisted of two groups of 40 of each part type. Each group of 40 was from different manufacturers, and was accompanied by three unstressed control devices.

The 3N171 precatastrophic degradation mechanism immediately following ESD stress was determined to be due to charge trapping in the gate oxide. The damage would partially anneal in 5 min at 300°C. A constant voltage stress of $1.5 \times 10^6$ V/cm was used rather than accelerated temperature. This voltage provides an acceleration factor of $2.5 \times 10^6$ at 25°C. In this case the actual voltage used was 15 V applied to the gate, rather than the normal 3 V. Life-test results for this part showed annealing with operating life rather than degradation.

Life test for the 54L04 devices was conducted at 125°C with $V_{CC}$ at 5 V and input signal frequency at 100 kHz. These devices were stressed in a similar manner to that used in the NASA study, resulting in possible degradation of the gain or *beta* of $Q2$. The minimum $V_{CC}$ operating level was shown to be a good gauge of the beta of $Q2$ and was used to measure ESD degradation. All six gates per device were stressed during the exposure phase. Increments of 100 V were used to best achieve a target of 10 percent beta degradation. Repeated stresses at the same level were observed to cause a nearly linear increment of degradation to $Q2$.

There were five catastrophic failures of each gate location during initial stressing. During life testing there were 105 additional gate failures at 20 h. The researchers discounted these failures because "preliminary evaluation indicated they failed due to emitter-to-collector punch-through in $Q1$." This failure mode was apparently not related to the expected increased degradation of $Q2$. The $Q1$ punch-through was attributed to two factors: an unusually narrow $Q1$ base width; or ESD stressing with the $V_{CC}$ terminal open.

## 11.8.2 Studies of cumulative aspects of latency

The following discussion of two studies into latency are included because the results are felt to be of considerable value. These studies do not address the controversial issue of time-dependent failures, but give greater insight into cumulative ESD damage.

**11.8.2.1 University of Southhampton study.** Results of a study on ESD latency effects of the 74 HC series HCMOS devices was reported in 1984 by R. G. M. Crockett, J. G. Smith, and J. F. Hughes.[13] These devices are high-speed CMOS that are pin-compatible equivalents to the 74 series TTL devices. The primary advantage over TTL is low-noise immunity. This device type is being used in many applications so ESD behavior is important.

The study did not address time-dependent degradation in the pure sense, but gave insight into the cumulative ESD degradation of polysilicon resistors. Step-stress tests were conducted starting at 500 V with 250 V increments up to 3 kV or open circuit. This data was compared with repetitive single-level pulse data at 2.5 kV and 3 kV. Surprisingly, the lower level step-stress test caused failure in fewer exposures. It was then shown that initial single-pulse stressing at 2.5 kV and 3.0 kV appeared to harden the devices to better survive step-stress testing. On the other hand initial single-pulse stressing at 500 V and 1 kV weakened the parts which increased the likelihood of failure from repeated 3 kV pulses.

This high-voltage hardening effect is called transient annealing. The effect is analogous to pulsed laser annealing, commonly used to stabilize polysilicon resistors. The 2.5 kV to 3 kV ESD pulse peak-power level as well as total energy dissipated is similar to pulsed laser values. The grain structure is likely to be affected by these energy levels and electromigration due to the high field present.

**11.8.2.2 RIT Research Corporation study.** In 1985, P. S. Neelakantaswamy et al.[14] discussed models of cumulative ESD damage. The researchers extended the models to latent time-dependent failures although there were no experimental data cited for correlation.

## 11.8.3 Sandia National Laboratories study

P. E. Gammill and J. M. Soden[15] conducted a latency study using CD4035A four-stage shift registers. In addition, their investigation disclosed a well-documented field occurrence of a latent ESD failure on a satellite program. Sandia reportedly had an LSI failure-rate goal of no more than 50 FITs. One FIT represents one failure in $10^9$ h. This type of reliability requirement places a high importance on virtually any probability of latent failure.

The CD4035A shift registers were stressed from a human body simulator circuit with $V_{CC}$ (+) and the pin-three input (−). The average voltage to cause damage was determined to be 4020 V ± 263 V as the result of stressing 66 devices. Single pulses at $V_{CT}$, 0.9 $V_{CT}$, and 0.8 $V_{CT}$ were applied to 28, 23, and 20 parts, respectively. This resulted in 31 immediate failures which were discarded from the life-test sample. The remainder consisted of 11 parts with increased $I_{DD}$ and 29 with no apparent damage (<1 μA change). The 11 $I_{DD}$ increases, as high as 756 μA, were actually failures since they exceeded the 10-μA limit.

These devices were then placed on biased life test for 234 h at 25°C followed by 509 h at 125°C. A control sample of nine unstressed devices was also included. Life-test results were one catastrophic $V_{THN}$ failure at 65 h into the 125°C test; 20 stable devices; 14 with $I_{DD}$ that decreased (improved); and 5 with $I_{DD}$ that increased.

Failure analysis showed the catastrophic latent failure to be a gate-to-drain oxide rupture short in a PMOS transistor. The short was evidenced by a silicon nodule at damage location. This short was not located at a stress-preferential site typical of ESD-induced failures. In other words, it was not at an oxide step nor at the gate edge which are high field-intensity areas. Two similar anomalies tend to discount the possibility of ESD as the cause of failure. In the same vicinity there was a relatively large gate-oxide patterning defect which occurred during wafer processing. Additionally, a second rupture-type short with a silicon nodule was found in the same vicinity, again at a nonpreferential site. This second short exhibited diode action rather than a purely resistive characteristic. The diode was reverse biased, thus it did not contribute to the failure. The investigators concluded that ESD caused the latent failure by overstressing an existing oxide defect. The defect did not become a short until exposed to the life-test stresses of time, temperature, and electrical bias.

Sandia also investigated a field failure of a CD4041A quad true-complement CMOS buffer that occurred after more than 2000 h successful operation in a system intended for future satellite usage. The failure was intermittent when first detected: failing, then recovering twice before becoming a *hard* or constant failure with the complementary output stuck at 4 V. Electrical connections to this device were inaccessible at the system level where failure occurred. The unit had operated properly during the 7 months since the chassis had last been opened.

Failure analysis showed a 2.5 kΩ resistive path between the input (pin 3) and output (pin 2), as well as an $I_{DD}$ leakage of 1 mA at $V_{DD}$ = 5 V. Physical dissection revealed a gate-to-drain rupture causing a shorted output stage. The location was a site of high field intensity between the gate-metallization edge and the drain. The conclusion was that the failure occurred latently because of ESD stress sometime early in the life of the system.

### 11.8.4  British Telecom studies

**11.8.4.1  Literature search in 1986.** A 1986 literature search by R. D. Enoch and R. N. Shaw[16] concluded that the evidence on time-related ESD failures was inconclusive. The authors criticized the short life-test durations and relatively low temperatures of 25°C and 75°C used in the NAVSEA and NASA studies. The concern was that these conditions did not produce very much acceleration of any ESD degradation mechanism present.

**11.8.4.2  Failure-analysis investigation of 1988.** Results of a latent failure study were reported in 1988 by J. Woodhouse and K. D. Lowe.[17] The study consisted of exposing parts to approximately 90 percent of their ESD damage thresholds and then conducting life tests on all parts (91 out of 102) that did not fail catastrophically. A low-power FET having a 200-V failure threshold was found to sometimes function properly after higher stresses, up to 2 kV. This was attributed to filament forming and subsequent fusing, a possibility discussed in Chap. 7. Table 11.28 lists the particulars on the parts used, including stresses and life-test conditions.

Reported results were that all electrical failures were found to be unrelated to the ESD stresses. "There was no evidence of drift in the electrical parameters of any of the components in the latency investigation at the end of life testing. There were no functional failures and all of the components operated to specification." Failure analysis revealed that parts selected from the latency study showed damage traits indicative of dielectric breakdown, junction breakdown, or polysilicon resistor damage. The authors concluded that a part may receive an electrostatic discharge sufficiently close to the damage threshold to escape initial failure and affect long-term reliability. They felt however that the probability of this occurrence is small enough to be discounted.

### 11.8.5  TriQuint semiconductor study on MESFET-based GaAs integrated circuits

In 1988, Anthony L. Rubalcava and William J. Roesch[18] reported results of a study on metal semiconductor field-effect transistor

**TABLE 11.28 British Telecom Latent ESD Failure Study**

| Part type | Tech | $V_{th}$ (kV) | Stress (kV) | No. on life test | Life test conditions |
|---|---|---|---|---|---|
| Low-power FET | NMOS | 0.2 | 0.1 | 5/5 | 150°C, 1500 h |
| | | | 0.2 | 5/5 | |
| | | | 0.5 | 5/5 | |
| | | | 1.0 | 5/5 | |
| | | | 2.0 | 5/5 | |
| 4000 series buffer | CMOS | 1.7 | 1.5 | 5/5 | 125°C 3500 h |
| HC hex inverter | CMOS | 3.3 | 3.0 | 5/5 | |
| LS hex inverter | LSTTL | 0.6 | 0.5 | 5/5 | |
| HC decoder (vendor A) | CMOS | 2.1 | 2.0 | 4/5 | |
| HC decoder (vendor B) | | 2.1 | 2.0 | 5/5 | |
| HC decoder (vendor C) | | 2.1 | 2.0 | 5/5 | 150°C dynamic, 2000 h |
| HC decoder (vendor D) | | 1.3 | 1.1 | 5/5 | |
| HC decoder (vendor E) | | 1.1 | 1.0 | 5/5 | |
| HC tranceiver (vendor A) | CMOS | 2.2 | 2.0 | 4/5 | |
| HC tranceiver (vendor B) | | 2.4 | 2.2 | 4/5 | |
| HC tranceiver (vendor C) | | 2.4 | 2.2 | 2/5 | |
| High-power FET (vendor A) | NMOS | 0.4 | 0.3 | 3/5 | 85°C 85% RH, 1000 h |
| SRAM | CMOS | 1.0 | 0.9 | 7/8 | 125°C dynamic, 2500 h |
| SLIC | CMOS | 1.1 | 1.0 | 4/4 | 130°C dynamic, 2000 h |

Note: All life tests static-biased except the ripple counter, which was unbiased.

(MESFET)-based gallium-arsenide integrated circuits and associated elements such as air bridges, metal-insulator-metal (MIM) capacitors, and resistors of nichrome and ion implanted silicon types. The major part of the study consisted of life tests on ICs and MESFETs subjected to HBM stresses just below the failure threshold. Parts used were a 4-bit ripple counter, a 300-gate digital application specific integrated circuit (ASIC), a 1–8 GHz monolithic microwave integrated circuit (MMIC), and a MESFET with a gate of 1 μm length.

**11.8.5.1 Cumulative effects.** Pulsing was routinely done in a step stress fashion with multiples of five pulses in each polarity at each voltage level. In general there were no cumulative damage effects observed. A typical result reported was a MESFET pulsed from gate to drain. With five pulse steps in approximate 25-V increments beginning at 75 V, there was no change in breakdown voltage until the thirty-ninth step (195 pulses) at the 1050-V level. A similar lack of cumulative damage was observed on the ICs as well.

**11.8.5.2 Latent effects.** Latency tests consisted of life tests conducted on unpulsed control samples and samples stressed at 90 percent of their failure thresholds. Tests conducted are summarized in Table 11.29. The results showed virtually no difference between pulsed and unpulsed samples. The authors concluded that:

1. Unlike silicon MOSFETs, MESFETs have no gate oxide that is vulnerable to latent damage.
2. MESFETs and MESFET-based ICs show no cumulative or latent effects from ESD below the instantaneous damage levels.

## 11.9 An Assessment of the Latent ESD Failure Controversy

Apparently, the literature has not convinced industry that latent ESD failures exist. Perhaps the detailed coverage of the NAVSEA and NASA studies has provided sufficient evidence for those in doubt. The statistics from the NASA study results of the 3N128 and 54L04 should be sufficient. If one acknowledges only the classic Mac-1 failures, the data supports 99.9 percent and 96.2 percent probabilities, respectively, that the latent failures were due to the ESD stresses.

### 11.9.1 Studies prior to 1980

The studies of ESD latency prior to 1980 were, taken as a whole, inconclusive. Several possible contributing factors to the lack of success in producing latent failures were

TABLE 11.29  MESFET-Based GaAs Latent Study

| Part type | No. stressed | No. pulses | No. controls | Life-test conditions |
|---|---|---|---|---|
| 1 μm × 300 μm MESFET | 4 | 1 | 1 | 1000 h 260°C channel temp |
| 1 μm × 300 μm MESFET | 7 | 1 | 7 | 100 h 275°C channel temp |
| 4 bit ripple counter | 4 | 1 | 1 | 1100 h 200°C channel temp |
| MMIC amplifier | 2 | 50 | 5 | 500 h 240°C channel temp |
| 300 gate ASIC | 7 | 10 | 7 | 1000 h 250°C channel temp |

Note: All life tests static-biased except the ripple counter, which was unbiased.

1. High temperature (≥125°C) life tests bringing about annealing,
2. Setting fixed percentages of the catastrophic failure threshold that were too low to condition the parts for degradation with time
3. A sample size that was too small for statistical significance
4. An insufficient number of control devices to discount irrelevant failures
5. Selection of the most sensitive pin combination rather than the combination most apt to lead to latent failure
6. A definition of latent failure that included only the Mac-1-type failure

The problem with annealing from higher acceleration temperatures was the very reason lower temperatures were used in the NAVSEA and NASA studies. The Martin Marietta and Sandia investigators shared this same concern. An additional reason for conducting life tests at 25°C is the fact that many real-world end-usage applications are at room temperature. Low temperature may have been a factor in the failure occurrences in the Martin Marietta and Sandia studies. The validity of *latent* failures within several hundred hours at these lower temperatures has been questioned on the basis of insufficient time. By definition any failure that occurs after acceptance and delivery is latent.

Most investigations have shown a variety of behavior during life tests following ESD stress: improvement, stability, degradation, and erratic. The tendency might be somewhat dependent upon the degree of initial damage. The degree of damage required for a given part type is generally not known. Therefore, a wide range of initial stressing seems preferable over using a few predetermined percentages of the catastrophic damage level.

Similarly, the pin combination most likely to bring about latent conditions is not known. It might not be simply a matter of selecting the most sensitive pair. For instance, the 54L04 low-power TTL hex-inverter pin combination used in the NASA study had a catastrophic threshold level greater than 5000 V. Yet this combination was quite successful in producing latent failures.

Establishing criteria for latency that considers only acceptable parts that go out of specification with time is self-limiting. This discounts parts with slightly out-of-specification parameters from the life-test sample which would be valid candidates for a Mac-3-type failure. This also eliminated the possibility of detecting certain failed parts with intermittent behavior. Wide design margins that permit parametric shifts beyond specification bounds are common. In such an application the classic Mac-1 part-level failure might operate properly at the system level, and a true latent failure in end use would not occur. On the other hand the same system might pass acceptance tests with an out-of-specification part, then degrade with time beyond design margins until the system fails catastrophically.

### 11.9.2 Studies since 1980

The studies since 1980, taken as a whole, have been successful in producing only a small number of latent failures. Latent failures produced during the Westinghouse, Martin Marietta, and Sandia studies are supportive of the existence of latent ESD failures. The low incidence, however, may indicate a low probability of occurrence.

The two study results (British Telecom and TriQuint Semiconductor) reported in 1988 tended to discount latent ESD failures. However, precaution is advised against premature disregard for the possibility of ESD latency in silicon or GaAs devices based upon these studies alone. Some possible reasons that the two cited studies were not conducive to producing latent failures may be one or more of the following:

1. A too small sample size of each part stressed under certain conditions
2. High temperature ≥125°C bringing about annealing
3. Common use of fixed percentage of catastrophic damage threshold with no requirement for parametric change

Both studies have helped place some pieces in the latent ESD puzzle. The British Telecom study was particularly informative in showing definite physical ESD damage to pulsed devices that were able to survive life tests equivalent to ≥20 years of operation. The TriQuint study has provided much needed data on GaAs devices, a technology of growing usage. While not justifying complete disregard for ESD latency, the data contained in the two studies should dispel some of the widespread belief that ESD latent failures are more easily produced than instantaneous failures.

### 11.9.3 General conclusions

The small data base available on this topic precludes strong generalizations on many considerations of interest. A few conclusions that can be asserted are

1. Latent ESD failures exist.

2. The presence of slight parametric shifts or softened I-V characteristics does not assure further degradation with time.
3. System-level latent failures can result from parts that might be regarded as out of specification after ESD stress.

### 11.9.4 Interpretation of existing data

The following discussions relate to issues subject to opinions based upon experience and interpretations of existing data. They are apt to be modified through knowledge gained in later studies and newly reported case histories.

1. Experimental data seems to indicate that latent failures are unlikely under the random conditions of actual handling. The lack of substantive latent field failures (where field handling is not possible) supports this contention.

2. The author knows of no universal screen to identify incipient damage that might lead to latency. Low-temperature ($-55°C$) test would show promise in the specific case of the 54L04 device, and might have other applicability. Supply voltage variation is another avenue worth exploring.

3. Damage to internal nodes of a device might be more apt to lead to catastrophic latent failures because the incipient damage is somewhat electrically isolated and difficult to detect.

4. In light of the apparent significance of damage to internal nodes, the charged-device model (CDM) might be more likely to bring about latent failures.

5. Latent failures are likely to be the result of very subtle physical damage and not apt to be properly identified by failure analysis.

### 11.9.5 Recommendations

The following are recommendations to users of static-sensitive electronic parts and assemblies:

1. ESD must be considered as a threat to reliability.
    The fact that latent failures have been proven to exist places a high degree of importance on the overall ESD control program. Although current indications are that probability of latent static failures is low, that probability is unknown. Furthermore, it may be high for certain parts or for certain applications.

2. The reliability engineering community must develop methods to assess the reliability impact of potential latent ESD failures.

   This will have future impact on decisions as to whether or not ESD jeopardized parts, assemblies, or even systems are shipped. The risk impact must be carefully weighed against the consequences of failure.

3. A screen or screens must be developed to detect incipient damage most likely to lead to latent failures.

   Such a screen would be worthwhile on a high-reliability program where ESD control infractions have placed an assembly or system at risk. For most non-high-reliability applications the best approach might be to deliver the hardware and correct the infraction to avoid future occurrences.

4. ESD controls must not be compromised.

   The latency aspect of static mandates strict and thorough ESD control. These controls are of particular interest to the end user who is vulnerable to ESD stress during many tiers of manufacturing. Thus, these controls must not only be self-imposed but extended and monitored at major subcontractors, assembly and subassembly manufacturers, and hybrid and part fabricators.

### 11.9.6 Additional studies needed

Latent ESD failures is one of the most important potential problems facing the electronics industry today and has not received the attention warranted. Some suggested areas of further endeavor are

1. Physics of latent ESD failures.

   Research to date contains virtually nothing in the way of detailed studies into the failure mechanism of physical latent failure specimens. The models and theoretical ideas must be compared with actual part dissection results in order to fully understand the properties or design variations that lead to latency.

2. Additional part types.

   The sample data base on this topic must be expanded to include many other part types and part technologies. Without this work the fear remains that still untried part types might have a greater propensity for latent failure development. This type of effort would also complement work that has been completed as well as help in the understanding of the physics involved.

3. Charged-device model.

   Latency investigations to date have considered only the human body model failure. Charged-device model failures often cause damage to internal nodes. The apparent tendency of latent failures to develop from damage to internal nodes makes the charge-device model an ideal source of initial ESD stress. In addition, industrywide ESD controls have for the most part accomplished grounding of personnel handling sensitive items. This factor coupled with increased automation has brought about a trend that minimizes the threat of static failures from the human body. In the fully automated factory of the future charged device failures may predominate.

## References

1. O. J. McAteer, R. E. Twist, and R. C. Walker, "Latent ESD Failures," *Fourth Electrical Overstress/Electrostatic Discharge Symposium Proceedings*, September 1982.
2. O. J. McAteer, R. E. Twist, and R. C. Walker, "Identification of Latent Failures," *Second Electrical Overstress/Electrostatic Discharge Symposium Proceedings*, September 1980.
3. T. Hasegawa, "Latent Effects of Electrostatic Discharge Stress on Microcircuits: Not Proven," National Semiconductor Reliability Physics Brief no. 07, January 1980.
4. L. J. Gallace and H. L. Pujol, "The Evaluation of CMOS Static Charge Protection Networks and Failure Mechanisms Associated with Overstress Conditions as Related to Device Life," *International Reliability Physics Symposium Proceedings*, 1977.
5. L. A. Schreir, "Electrostatic Damage Susceptibility of Semiconductor Devices," *International Reliability Physics Symposium Proceedings*, 1979.
6. G. A. Branberg, "Electrostatic Discharge Sensitivity Level of Commercial CMOS Logic," *First Electrical Overstress/Electrostatic Discharge Symposium Proceedings*, September 1979.
7. D. T. McCullough, C. H. Land, and R. A. Blore, "Reliability of EOS Screened, Gold-Doped 4002 CMOS Devices," *First Electrical Overstress/Electrostatic Discharge Symposium Proceedings*, September 1979.
8. J. R. Syrjanen, "Electrostatic Discharge Damage to CMOS Chip When Connected into a Functional Hybrid Circuit," Medtronics Incorporated, Minneapolis, Minn., Internal Report,
9. J. R. Schwank, R. P. Baker, and M. G. Armendariz, "Surprising Patterns of CMOS Susceptibility to ESD and Implications on Long Term Reliability," *Second Electrical Overstress/Electrostatic Discharge Symposium Proceedings*, September 1980.
10. E. R. Freeman and J. R. Beall, "Control of Electrostatic Discharge Damage to Semiconductors," *International Reliability Physics Symposium Proceedings*, 1974.
11. D. H. Rutherford and J. F. Perkins, "Effects of Electrical Overstress on Digital Bipolar Microcircuits and Analysis Techniques for Failure Site Location," *First Electrical Overstress/Electrostatic Discharge Symposium Proceedings*, September 1979.
12. J. S. Bowers, M. C. Rossi, and J. R. Beall, "A Study of ESD Latent Effects in Semiconductors," *Fifth Electrical Overstress/Electrostatic Discharge Symposium Proceedings*, September 1983.
13. R. G. M. Crockett, J. G. Smith, and J. F. Hughes, "Latency Effects Observed in HCMOS Input Protection Circuits," *Sixth Electrical Overstress/Electrostatic Discharge Symposium Proceedings*, September 1984.
14. P. S. Neelakantaswamy, T. K. Sarkan, and R. Turkman, "Residual Fatigues in Microelectronic Devices Due to Thermoelastic Strain Caused by Repetitive Electrical Overstressing, A Model for Latent Failures," *Seventh Electrical Overstress/Electrostatic Discharge Symposium Proceedings*, September 1985.
15. P. E. Gammill and J. M. Soden, "Latent Failures Due to Electrostatic Discharge in

CMOS Integrated Circuits," *Eighth Electrical Overstress/Electrostatic Discharge Symposium Proceedings,* September 1986.
16. R. D. Enoch and R. N. Shaw, "Event Dependent ESD Latent Failure Behavior of Bipolar Integrated Circuits," *Electrostatic Discharge in Electronics Seminar,* May 8, 1985.
17. J. Woodhouse and K. D. Lowe, "ESD Latency: A Failure Analysis Investigation," *Electrical Overstress/Electrostatic Discharge Symposium Proceedings,* 1988.
18. Anthony L. Rubalcava and William J. Roesch, "Lack of Latent and Cumulative ESD Effects on MESFET-Based GaAs ICs," *Electrical Overstress/Electrostatic Discharge Symposium Proceedings,* 1988.

# Chapter 12

# Design Techniques

## 12.1 The Design Problem

Designers involved with electronic parts and equipment are faced with many obstacles in combatting ESD:

1. Static electricity presents a hostile environment for electronics.
2. Most commonly used parts are susceptible to ESD.
3. Ignorance and myth prevail among users as well as part and subassembly designers.
4. Design techniques for ESD immunity are not well developed, especially at the assembly level.
5. Most design techniques have an associated negative tradeoff in one or more factors such as space, weight, function (normally speed or frequency), human factors, environmental restraints, added maintenance, and even reliability.

## 12.2 Definition of Design-for-ESD-Immunity Techniques

Design-for-ESD-immunity techniques are defined herein to be those methods used by electrical and mechanical designers to reduce the vulnerability of electronics parts and equipment to the deleterious effects of electrostatic discharge. The foremost vulnerability reduction method normally considered is design hardening achieved by changes in hardware. However, a broader approach from an overall systems and economics perspective is recommended. Much can be accomplished by extending vulnerability reduction methods to include special associated equipment interface requirements, maintenance or operational procedures, fault detection and redundancy, and/or environmental or functional mode conditional restraints.

## 12.3 On-Circuit Protection Devices

On-circuit protection usually means a specially designed protective circuit that is added to the chip of an integrated circuit. In a broader sense, on-circuit protection can be extended to include other factors relating to ESD immunity such as process variables, layout considerations, geometries and spacings, packaging considerations, test considerations, built-in test (BIT), and fault tolerance.

### 12.3.1 Protection circuit development

Protection networks are sometimes fabricated onto an integrated circuit's surface with the remainder of the device. Such a protection network is usually intended to shunt the transient away from sensitive nodes and often has resistive elements to slow the transient to manageable frequencies. The designer of on-chip protective networks must achieve immunity goals while

1. Avoiding slowdown of functional circuitry
2. Using up a minimium of chip space or "real estate," which is at a premium
3. Retaining compatibility with device design rules of the technological family

### 12.3.2. Ideal protection circuit

An ideal ESD protection circuit would provide a low impedance to a high voltage and infinite impedance (open circuit) to low or normal operating voltages. The change from high to low impedance and back would occur with zero time delay. Such circuits, if truly ideal, would take up no space on the chip and would require no extra processing. The circuit in Fig. 12.1 represents an ideal protection circuit. $D_1$ should be high impedance except during a positive transient input, at which time it should be a very low impedance. $D_2$ should also normally be a high impedance, but become a very low impedance in the event of a negative transient. $D_3$, which is in series with the susceptible node being protected, should be low impedance except during a potentially damaging transient of either polarity. If these three elements were ideal, nothing more would be needed and the circuit would function during transients except for the short duration transient period when the high impedance of $D_3$ might cause a signal interruption.

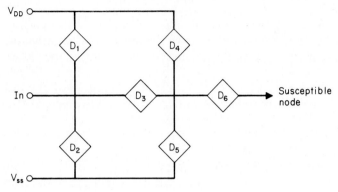

**Figure 12.1** A representative ideal protection network.

To compensate for imperfections additional elements, $D_4$, $D_5$, and $D_6$, are cascaded for redundancy. Similar ideal protection networks have been described[1,2] by others.

Actual protection networks are designed with attempts to approach the ideal conditions. That is, minimal space and processing changes are sought, as are fast switching times to low impedance with minimal effect on normal operation. No one single structure has been found to best accomplish this feat. Most protective networks are some combination of diodes, diffused or polysilicon resistors, thin or thick oxide devices, field plate structures and punch-through devices. Nodes to be protected may involve susceptible metallization, bipolar, resistor, MOS capacitor, and various parasitic structures as well as MOS gates. However, for simplification, illustrations will generally indicate the protected node as the classic *gate*.

### 12.3.3 Diode protection

One of the first protection networks consisted of a diode with reverse breakdown voltage $V_{bd}$ somewhere between the normal gate operating and breakdown voltages. Such a diode, being normally reverse-biased, introduces minimal leakage current and a parasitic capacitance that usually has no deleterious effects. When voltages outside the normal operating range occur, the diode is to protect the gate. If the applied voltage $V_T$ goes above the gate voltage, the diode breaks down and maintains the voltage below the gate oxide threshold. If the voltage goes to the opposite polarity, the diode is forward-biased, again protecting the gate. The source resistance and the dynamic resistance $R_s/R_d$ of the diode during breakdown constitute a divider circuit.

$$V_{\text{gate}} = V_{bd} + (V_T - V_{bd}) \frac{R_d}{R_s + R_d} \quad \text{for } V_T > V_{bd} \quad (12.1)$$

From the resistor ratio in this equation it is obviously desirable to keep the dynamic resistance much smaller than $R_s$. Two important variables affecting dynamic resistance are the diode junction area and the depletion layer width. Smaller depletion layer widths and larger diode junction areas will result in lower dynamic resistances. Efforts to decrease the dynamic resistance are limited to a degree by the fact that there is an associated increase in capacitance. This capacitance must be kept significantly less than the gate capacitance in order to retain high-frequency characteristics. Figure 12.2 shows a single diode network including the diode resistance in series. Suppose a circuit of this type is used to protect a gate that is susceptible to 100 V from the human body. Figure 12.3 is a plot of voltage variation with HBM voltage and the HBM potentials providing protection to $\leq 90$ V, both as a function of $R_D$. This plot verifies the disadvantage that the diode itself is usually too susceptible if used alone.

In many applications where the gate will see both positive and negative voltages in normal operation, such as in CMOS circuitry, paired diodes are used in such a way that one will always be in reverse breakdown if the applied voltage exceeds the breakdown voltage. A paired combination is shown in Fig. 12.4.

### 12.3.4 Resistive protection

A diffused resistor is a distributed resistor in a substrate layer of opposite dopant polarity; thus it has a distributed diode characteristic as well. Figure 12.5 shows the equivalent circuit of a diffused resistor. The diagram shows elements of resistance per unit length $r_l$ as well as the distributed diodes with their associated dynamic resistances $r_d$.

During an overvoltage $V_T$ greater than the distributed diode's breakdown voltage $V_{bd}$, the voltage across the gate is given by[3]

$$V_{\text{gate}} = V_{bd} + (V_T - V_{bd}) \frac{\sqrt{r_l r_d}}{R_s + \sqrt{r_l r_d}} \left[ \cosh \sqrt{\frac{r_l}{r_d}} L_r \right]^{-1} \quad (12.2)$$

where  $r_l$ = resistance per unit length
  $r_d$ = distributed dynamic resistance per unit length
  $\sqrt{r_l r_d}$ = input impedance
  $L_r$ = resistor's length

When $\sqrt{r_l/r_d} \times L_r$ is much greater than 1, then

$$\left[ \cosh \sqrt{\frac{r_l}{r_d}} L_r \right]^{-1} \approx 2 \exp\left(-\sqrt{\frac{r_l}{r_d}} L_r\right) \quad (12.3)$$

If the term from Eq. (12.3) is inserted into Eq. (12.2), the effect is readily seen to be an exponential reduction of the overvoltage

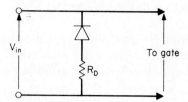

**Figure 12.2** A single diode protective device.

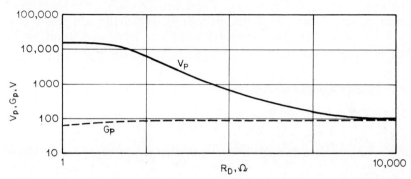

**Figure 12.3** Human body model protection level $V_P$ provided by a single diode and the protected gate's maximum voltage $G_P$, both as a function of diode resistance $R_D$. $V_{bd}$ is 50 V.

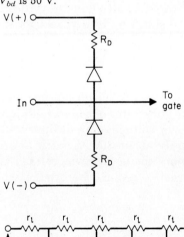

**Figure 12.4** Paired protective diodes to $V(+)$ and $V(-)$.

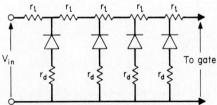

**Figure 12.5** Equivalent circuit of a diffused resistor includes distributed resistor elements ($r_i$'s) and distributed diode elements with their associated resistances $r_d$'s.

($V_T - V_{bd}$). This exponential attenuation is a desirable characteristic of distributed networks.

In spite of the added diode characteristic, for most effective protection the diffused resistor is used in conjunction with other protective structures Resistors can affect speeds due to increased $RC$ time constants.

A nondiffused polysilicon resistor does not have the distributed diode, is not effective in protecting dielectric structures, and is of limited value in protecting PN junctions. Polysilicon resistors are sometimes used in combination with other structures to provide additional current limiting and to slow down transients to facilitate shunting by other devices.

### 12.3.5 Diode resistor combinations

A resistance can be added to a diode circuit to help minimize $V_{\text{gate}}$:

$$V_{\text{gate}} = V_{bd} - (V_T - V_{bd})\frac{R_d}{R_s + R_{\text{add}} + R_d} \qquad (12.4)$$

The rise time required and the gate capacitance can limit $R_{\text{add}}$. With this improvement the protection provided is still limited. In practice the added resistor is usually a diffused resistor, although polysilicon resistors are sometimes used. Figure 12.6 shows a diffused resistor in combination with diodes.

### 12.3.6 Field plate

An ideal protective network would have low breakdown voltage and low dynamic resistance. MOS devices use substrates of somewhat high resistivity, about 5 Ω · cm. The breakdown voltage of an ordinary bipolar junction in this material is typically about 300 V. With 3-μm depth (typical for drain to source diffusion) the breakdown voltage is about 80 V. Several methods can be used to reduce the breakdown

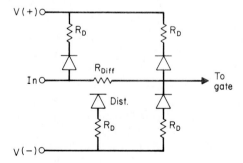

**Figure 12.6** Schematic diagram of diffused resistor in combination with diodes.

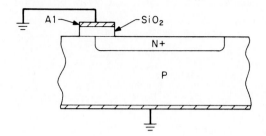

**Figure 12.7** Construction of a grounded field plate.

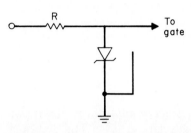

**Figure 12.8** The usual schematic designation for a grounded field plate.

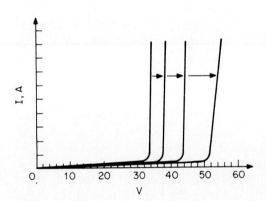

**Figure 12.9** Walkout of field plate breakdown characteristic with repeated transient exposures.

voltages. Early breakdown can be induced at the surface by using a grounded field plate as depicted in Fig. 12.7. Fabrication consists of a thin oxide layer over the diode junction in addition to a metal or polysilicon electrode over the oxide. This conductive electrode is connected to the substrate. The diode's breakdown voltage can be controlled by selecting the thickness of the oxide. The field plate is usually depicted schematically as in Fig. 12.8. An inherent problem is the high series dynamic resistance of the device. Perhaps more serious is the device's tendency toward breakdown "walkout" with repeated stresses. This effect, shown in Fig. 12.9, makes the device less protec-

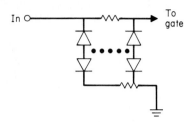

**Figure 12.10** Circuit diagram of punch-through device.

tive with repeated transient exposures. The problems associated with the field plate render the device insufficient protection unless it is used in conjunction with other measures.

### 12.3.7 Punch-through devices

A punch-through device is fabricated by placing adjacent to the input diffusion a grounded diffusion of the same dopant type. The distance between to the two diffusions is sufficiently narrow as to allow punch-through during an overvoltage. Punch-through is used to describe the reverse-bias condition whereby the junction's depletion region extends across the separation (of opposite dopant) between the two adjacent diffusions. A punch-through device circuit is designated as in Fig. 12.10.

### 12.3.8 Thin oxide transistor

A thin oxide transistor is similar to the punch-through device. The basic structure consists of two adjacent diffusions with an added gate, which is usually tied to the substrate. Figure 12.11 shows the circuit for the thin oxide protective circuit to $V_{SS}$. This has been the most common protection network for NMOS technology. It protects by gate-aided avalanche breakdown and punch-through to the source diffusion. High stress protection is by parasitic transistor turn-on as covered in Sec. 12.4.1.

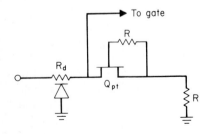

**Figure 12.11** Thin oxide device connected to $V_{SS}$ (ground).

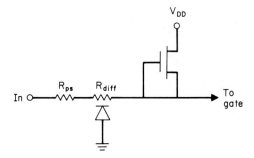

**Figure 12.12** A thin oxide device connected to $V_{DD}$.

The thin oxide device is sometimes connected to $V_{DD}$ as shown in Fig. 12.12. The advantage is that the device turns on when the input exceeds the $V_{DD}$ pin by about 1 V. The structure must be designed for rapid turn on, or the thin oxide gate will break down. The combination with a diffused resistor provides excellent protection from both polarities. Diffused $N^+$ resistances must be kept reasonably small ($\sim 1$ k$\Omega$), or the pad connection will tend to avalanche, resulting in junction alloying. Thin oxide structures protect to a threshold of around 2000 V. For high-speed applications, i.e., where the input drives large capacitances, Hulett[4] suggests omitting the polysilicon resistor and using a carefully designed layout with parallel redundant paths to $V_{SS}$ to better spread the high current.

### 12.3.9 Thick oxide transistor

A cross section of a thick oxide protective device is shown in Fig. 12.13. This is very effective for either NMOS or PMOS technologies. The gate metallization is typically tied to the drain. The source is usually grounded. Under positive ESD stress the parasitic lateral NPN transistor goes into second breakdown providing excellent protection to the sensitive node. The drain acts as collector, the substrate as base,

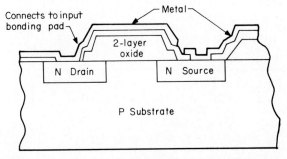

**Figure 12.13** Construction layout of a thick oxide device.

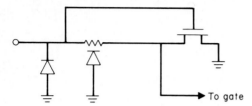

**Figure 12.14** Circuit diagram of a thick oxide enhanced punch-through transistor.

and the source as emitter. During negative ESD the drain substrate diode is forward-biased.

**12.3.9.1 Thick oxide enhanced punch-through device.** Figure 12.14 is the circuit diagram of a thick oxide enhanced punch-through device described by Keller.[5] This device is fabricated in a manner similar to the simple punch-through device described in Sec. 12.3.7. A series diffused input resistor is placed adjacent to another diffusion connected to the substrate or ground. The region is then covered with a thick oxide (~ 10,000 Å) and a metal gate tied to the input. A diffused diode is placed at the input pad connection. For a negative (reverse-biased) transient the input and distributed resistor diode are forward-biased and conduct current from the substrate. The nodal voltage is thereby clamped to one diode drop below ground. Positive transients either turn on the thick oxide transistor at about 25 V or cause punch-through. In either event the node is protected. In the event of high currents a parasitic lateral transistor turns on as shown in Fig. 12.15.

**12.3.9.2 Problems with thick oxide transistors.** The thick oxide device as shown in Fig. 12.13 will be modified with process variables. One such variable is that diffusions are often covered with silicide, such as titanium disilicide ($TiSi_2$). This practice, called *silicide cladding,* has developed as a measure to reduce diffusion sheet resistance and to improve speeds. In device fabrication utilizing this process the titanium disilicide layer would be present above the source and drain of the

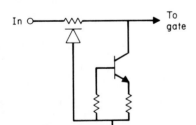

**Figure 12.15** Parasitic lateral transistor that turns on in event of high current.

thick oxide transistor. Silicide cladding has been reported[6-8] to reduce the effectiveness of protection against positive ESD stresses by 70 percent. The recommendation was that either the silicide process be masked out of the protective devices or new devices free of the silicide type of protection degradation be developed. Another reported problem[9-11] with thick oxide transistors is metal-silicon alloy shorts from drain to either the channel, source, or substrate resulting from ESD or electrical overstress (EOS).

### 12.3.10 Spark gaps

The spark gap[12,13] typically consists of two closely spaced (~ 50 μm) metal pads placed close to the bonding pads. Another method is to use the input pads in conjunction with a substrate metallization running the periphery of the chip. Typically it functions in the range of 300 to 400 V, so other means are needed to limit voltages below gate thresholds. Figure 12.16 shows a spark gap used in conjunction with a diffused resistor and diode pair. An additional drawback lies in the fact that the gap region must be left unpassivated to minimize discharge impedance. Metal erosion through repeated transient stresses is also a problem.

## 12.4 Application of Protection Devices

Protection devices are used extensively to protect virtually all susceptible integrated-circuit and discrete device structures. Certain protective devices and schemes are more appropriate for some technologies than others. General comparative results of selected protective networks are given in Table 12.1.

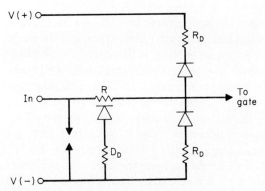

**Figure 12.16** A spark gap device used in conjunction with a diffused resistor and diode pair.

**TABLE 12.1 Comparison of HBM Protection Provided by Selected Structures**

| Protective circuit device | Technology protected | HBM protection level attained (V) | Ref. |
|---|---|---|---|
| Thick oxide | CMOS, NMOS | >6000 | 30 |
| Diode-resistor-diode | CMOS gate array | 4000 | 31 |
| Diffused resistor-gated diodes and spark gap as in Fig. 12.20 | CMOS/SOS type A CMOS/SOS type B | 800 | 15 |
| Diode-resistor circuit of Fig. 12.23 | CMOS/SOS type C | 1800 to 2000 | 15 |
| DIFIDW of Fig. 12-27 | CMOS | 8000 | 26 |
| Thick oxide-diffused resistor-field plate | DRAM (junction depth 0.4 µm; 20-V breakdown; 1-µm field oxide) | 5000 | 10 |
| | EPROM (junction depth 0.8 µm; 26-V breakdown; 1.4-µm field oxide) | 6000 | 10 |
| Diode-diffused resistor-field plate | EPROM (junction depth 0.8 µm; 26-V breakdown; 1.4-µm field oxide | 4500 | 10 |
| Thick oxide-polysilicon resistor-field plate | EPROM (junction depth 0.8 µm; 26 V breakdown; 1.4 µm field oxide) | 3000 | 10 |

Note: CMOS/SOS types: A had 6-µm gate length and 100-nm gate oxide, B had 3-µm gate length and 50-nm gate oxide, C had 1.25-µm gate length and 35-nm gate oxide

### 12.4.1 NMOS protection via thin or thick oxide transistor

NMOS devices frequently have inputs protected by "field-effect" transistor structures of either the thin or thick oxide types discussed in Secs. 12.3.8 and 12.3.9. As mentioned earlier, it turns out that bipolar transistor action of the parasitic elements predominates in ESD protection. Figure 12.17 shows a representative NMOS protection network with parasitic elements indicated by dashed lines. A bipolar NPN transistor is formed by the source, substrate, and drain for either thin or thick oxides. The base resistance is due to substrate bulk resistance. The diode represents the distributed diode formed by the diffused resistor and substrate. This diode protects the active circuit from negative transients, whereas the parasitic transistor turns on and for high transient levels goes into avalanche limiting voltages at the active circuit. The $I$-$V$ characteristic is shown in Fig. 12.18. If current

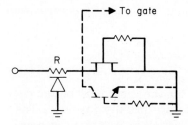

**Figure 12.17** A representative thin or thick film NMOS protection network with parasitic elements indicated by dashed lines.

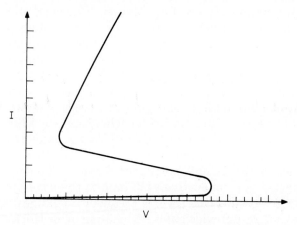

**Figure 12.18** Representative characteristic $I$-$V$ curve of parasitic PNP during overvoltage, exhibiting snapback.

density reaches certain threshold levels, snapback, or a negative resistance region, occurs after which the $I$-$V$ characteristic takes on a different slope. The transient durations of ESD stresses are too short to cause damage in the avalanched device.

### 12.4.2 CMOS

Complementary metal-oxide-semiconductor devices use an assortment of protective networks to guard the highly susceptible gate oxides. Included are junction diodes, distributed diodes, zener diodes, transistors, dual transistors forming bilateral devices, and resistor diode networks, sometimes with spark gap devices added. However, the use of thin and thick oxide devices is increasing because of the high levels of ESD immunity provided. Extreme design precautions must be taken to guard against latchup in CMOS circuitry from the added parasitic elements.[1,14]

### 12.4.3 SOS

Silicon on sapphire (SOS) technology is for the most part better defined as CMOS/SOS since CMOS circuits are usually involved. The primary process involves fabricating transistors on epitaxial, or more appropriately heteroepitaxial, silicon islands on a sapphire ($Al_2O_3$) substrate. The silicon islands are thin, typically 500 nm or so. Advantages are the highly insulating properties of sapphire, reduced substrate capacitances, elimination of need for isolation diffusions, and resultant increased densities and speeds. The small silicon volumes are also more immune to radiation effects. Even bonding pads on sapphire have less capacitance than on silicon substrates. Latchup is avoided because of the lack of parasitic junctions between transistors.

The edges of the silicon islands have been a problem source, however. As a result of anisotropic etching the edges can have a different crystalline structure (111 versus 100) from the top island surface. This crystalline difference results in thicker $SiO_2$ growth and a tendency for charge trapping at the edges as shown in Fig. 12.19. The effect of the edge charges can act in an unstable mobile ion fashion. The degradation mode is formation of a conductive path around the edge increasing leakage and possibly affecting reverse breakdown voltages.

Protective networks for 1.25-μm-wide gates with 35-nm oxide thickness were discussed by Palumbo and Dugan.[15] The same input protection network had been used since 1977 when developed for 6-μm-wide, 100-nm-thick oxides. The circuit had originally achieved an ESD immunity level of 1500 V, but had degraded to about 800 V with the smaller dimensions of newer devices. The protection network outlined in Fig. 12.20 consisted of a silicon diffused epitaxial input resistor, gated diodes to both $V_{DD}$ and $V_{SS}$, and a spark gap from the input bondpad to ground. The separate island CMOS chips were of "edgeless" variety, which provides 30 to 40 percent better ESD immu-

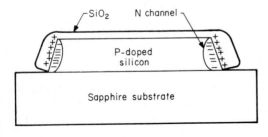

**Figure 12.19** SOS island edge effects resulting in erratic leakage paths.

Design Techniques    323

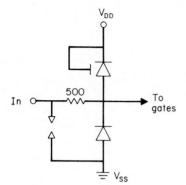

**Figure 12.20** Previously used SOS protective network.

nity. Figures 12.21 and 12.22 show the difference between edged and edgeless transistors. An improved test evaluation network was designed that would withstand at least 1800-V pulses. The circuit is shown in Fig. 12.23. The first epitaxial silicon resistor (350 Ω) is followed by either PIN or gated diodes (determined to be equivalent), a second epitaxial silicon resistor (700 Ω), and a pair of smaller diodes to $V_{DD}$ and $V_{SS}$. Spark gaps were also evaluated with the protective network circuit. Configurations consisted of aluminum or tantalum silicide sawtooth structures spaced from 3 to 7 μm and vertical open bias between two metal layers. All broke down at 320 to 360 V regardless of spacing. The aluminum structures were severely damaged during arcing, whereas the tantalum silicide structures were apparently unaffected. Since the breakdown or "ignition" voltage is so high, spark

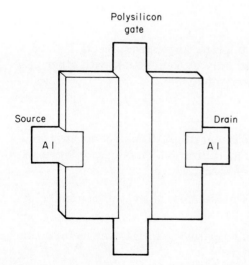

**Figure 12.21** "Edged" transistor island configuration.

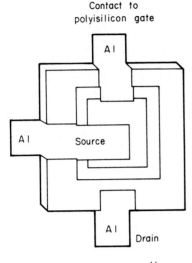

**Figure 12.22** "Edgeless" transistor island configuration.

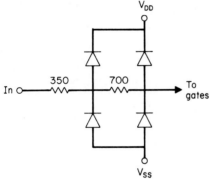

**Figure 12.23** Improved SOS protection network.

gaps alone would be ineffective in protecting gates susceptible to 35 V or so. Their use is recommended only in conjunction with other protective networks for greater effectiveness.

### 12.4.4 VLSI

The problems in very large-scale integration (VLSI) technologies transcend the smaller geometries of metallization, junctions, and gate oxides. Multimetal layers are more likely, which brings about

1. An increase in the number of metal steps of smaller nominal cross sectional area
2. An increase in the likelihood of parasitic metal over metal structures with thin oxides in between

Simply because of the increased complexity, other parasitic bipolar structures are more common. Thus latchup and other breakdown mechanisms are possible. Two major aims of downscaling are faster speeds and lower power dissipation. Protective circuits are therefore required to clamp to lower potentials while providing little if any impact on speed parameters. Other desired features are immunity of protective circuitry to ESD, less spatial or geometry requirements, immunity to latchup, and stable process variables.

## 12.5 Geometric Considerations

Protection circuits should be placed as close to the bonding pad as possible and isolated on the layout as far as possible from unassociated circuitry. Adjacent circuitry should be kept at a distance of at least 50 to 100 µm. Power-supply and output leads should include diffused resistors. Sharp bends are to be avoided in lines. In other words, rounded corners will minimize field concentrations. Crystalline defects such as precipitates, cracks, and irregularities are to be avoided.

### 12.5.1 Improved contact designs

Current crowding conditions can be minimized by proper design considerations.[4] Several small metal to diffusion contacts as shown in Fig. 12.24 are recommended to spread the current and decrease localized current densities.[16] The alternatives shown in Fig. 12.25 suggest layout considerations to further reduce current crowding. Significant improvements can also be made by using rounded contacts instead of square contacts to avoid field concentrations.

Petrizio[17] reported that CD4001A failures were occurring to one side of the devices when tested per Mil M38510/05202. This was later

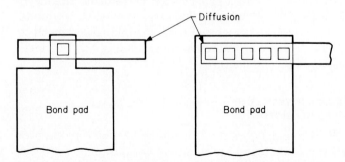

**Figure 12.24** Contact designs showing: (a) single-contact window and (b) preferred multiple-contact windows to spread current.

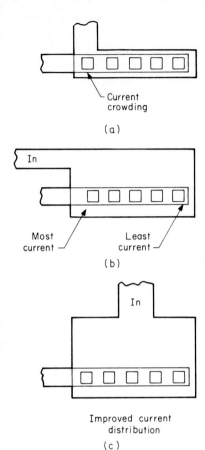

**Figure 12.25** Layout variations: (a) worst, inviting current crowding at first contact window; (b) better, allowing for spreading the current to other windows still with greater concentration at first pad(s); and (c) best, having well-distributed current.

determined to be due to failures of the intrinsic diodes between the $V_{SS}$ P well and the $N^+$ channel of the transistors. The device side where the $V_{SS}$ contact was located experienced no failures: the other side experienced ≈ 40 percent failures. This was proven to be due to the increased diode resistance associated with the increased length to the $V_{SS}$ contact. The device was improved by extending the $V_{SS}$ metallization and adding a second contact.

### 12.5.2 Considerations in protective circuitry layout and geometries

Unfortunately, protective network duplication by blindly copying previously used structures can lead to problems if the design and process parameters are not understood. Process variables affecting protection capabilities include contact resistance, etch times and temperatures, oxide integrity, and irregular geometric shapes. Metal-to-metal or metal-to-polysilicon crossovers separated by insulating oxides are undesirable voltage-sensitive structures. Rounded or curved geometries are generally preferred over sharp angles and squared shapes.

### 12.5.3 Metallization burnout

The submicrometer to 3-µm linewidth range is particularly susceptible to burnout during transient conditions. As discussed in Chaps. 7 and 10, when there is no energy-absorbing element in the circuit except the metal line, even greater widths can vaporize from ESD. With effective shunting protective networks in place, the end result is minimal energy absorption that can result in melting or vaporization of the lines. Assume a 600-V transient that goes across a 3-µm line to a protection network. The network may turn on at about 20 V, leaving 580 V across a low-resistance line. From the standard human body model the current would be approximately

$$i = \frac{580}{1500} = 387 \text{ mA}$$

If the line were 8000 Å thick, the current density would be

$$J = \frac{i}{A} = \frac{387 \times 10^{-3}}{3 \times 8 \times 10^{-9}} = 1.6125 \times 10^7 \text{ A/cm}^2$$

which is potentially damaging. Step coverage[18] is also critical, as discussed in Chap. 7.

### 12.5.4 Channel length

Wilson et al.[6] reported that the failure voltage threshold of NMOS devices increased with channel length in the range of 2 to 3 µm, decreased with channel length in the range of 3 to 8 µm, and showed no correlation or significant change in the range of channel lengths between 8 and 20 µm. These results were in mixed agreements with some earlier studies on this dependency. The important indication is that an optimum channel length for a given NMOS process can be determined from test pattern evaluations.

### 12.5.5 Polysilicon resistors

The same type of burnout problem exists with polysilicon resistors. They are inferior to diffused resistors not only because of the absence of the distributed substrate diode but because of thermal isolation. Polysilicon resistors are surrounded by deposited glass or a layer of insulating glass. Diffused resistors on the other hand have a good thermal path to the substrate. Step coverage is a consideration for polysilicon as well. If used, geometries must be made sufficiently large to handle anticipated currents. This geometric requirement places a corresponding limit on the resistance value, typically about 500 Ω. Transient current will find the path of least resistance. This is often the first avalanched diode of the first diffused resistor in the path to the substrate via the nearest contact at the bonding pad. Sometimes this contact will crowd the current, heat up, and result in an aluminum-silicon alloy through the junction either vertically or laterally.

## 12.6 Processing Variations and Susceptibility

Overalloying of the contacts can result from excessive time and/or temperature extremes during processing. The end result is deeper contact alloys usually characterized by spikes as shown in Fig. 12.26. These conditions not only concentrate the electric field but also make it easier to develop a short through a more shallow junction. The alloying can apparently continue[19,20] from electrothermal conditions of repeated ESD stresses in a cumulative fashion. Thick oxide devices at the input bonding pad are a good solution to overcoming the slow reaction time of diodes to ESD transients. Processing problems can leave defects such as cracks in the thick oxide resulting in greatly degraded ESD immunity of the protection device. Cracks are prone to occur at step coverage sites. Typically the field oxide consists of 10 to 15 kÅ of combined thermal and deposited glass, which theoretically should withstand ~ 1000 to 1500 V. As discussed in the hybrid circuit example of Chap. 10, damaged $SiO_2$ will have unpredictably lower breakdown levels. A recent paper[21] discusses process variables' effects on ESD thresholds for NMOS devices.

### 12.6.1 Corrective measures

When metal crosses metal or polysilicon or visa versa, layouts must ensure sufficient energy-absorbing elements in series in any transient

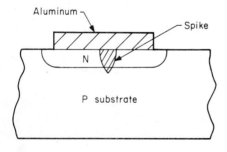

Figure 12.26 Deep alloyed contact spike.

event. Metal-to-diffusion contacts should be avoided or at least minimized. A short polysilicon strap of adequate width and thickness should be used to connect aluminum to diffused resistors. The addition of about 1 percent silicon to aluminum metallization serves as an effective inhibitor of undesirable aluminum-silicon diffusion at contact sites. The entire path of a transient must be considered in designing protective circuitry.

### 12.6.2 Use of phosphosilicate glass as top layer

An observation reported in the investigation of reversible charge induced failures[22] discussed in Chap. 7 related to the selection of the type of glassivation to be used in MOS circuits. The higher conductivity of phosphosilicate glass (PSG) may mean that it is a better choice for the top glassivation layer than ordinary $SiO_2$. Experiments with reversible charge-induced failures in CMOS matrix switches showed that devices cured spontaneously in 48- to 72-h periods if the $SiO_2$ layer had been removed leaving PSG as the top glassivation. Devices with normal $SiO_2$ intact would not cure unless exposed to x-rays.

## 12.7 Special Input Protection Devices

### 12.7.1 Double implant field isolation device in well

The double implant field isolation device in well (DIFIDW) is a thick oxide with a deep well junction. The structure is shown in Fig. 12.27. The device is an MOS transistor utilizing a thick field oxide as the gate oxide. A significant difference from regular thick oxide devices is that the source and drain have $N^+$ areas diffused into deep N wells. The depletion regions are now much larger and farther from the aluminum contacts. Most of the energy dissipation occurs at the drain depletion region. The larger region means less temperature rise, and the distance makes the normally vulnerable contact heat even less, thus reducing the likelihood of aluminum spikes across the junction. In addition the layout is such that source and drain gate overlap is preserved. The thin oxide regions associated with the "bird's beak" and $N^+$ regions are outside the gate coverage and also at a larger distance from the depletion region of maximum temperature. Additionally the structure allows larger separation of the source and gate electrodes which are likely to have maximum ESD potentials. The process can be of reverse dopant polarity as well. The large geometries mean an associated high capacitance, which is a disadvantage in VLSI or VHSIC technologies.

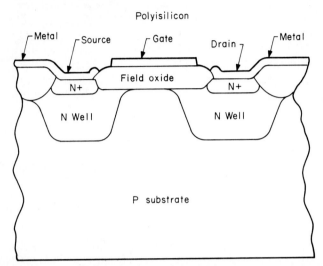

**Figure 12.27** DIFIDW structure cross section.

#### 12.7.1.1 Effectiveness of DIFIDW structures.
Nelson et al.[23] reported on characterization of a DIFIDW structure with a field-oxide thickness of 7000 Å, total N-well depth of 3.8 μm, P-substrate resistivity of 30 Ω · cm, and a threshold of 23 V. The device had a second breakdown type of operational $I$-$V$ characteristic similar to that shown in Fig. 12.18. The negative resistance region began in the neighborhood of 30 V and 300 mA. After the short snapback region the secondary breakdown slope indicated nearly zero resistance.

Since as much as 30 V or higher could be developed across the device during transient rise time, the device would not provide sufficient protection for many fragile gates. The protective device was therefore cascaded with a polysilicon resistor and a vertical bipolar transistor. Characterization testing showed this device to be capable of withstanding greater than 8 kV from two different commercially available human body simulators.

#### 12.7.1.2 High density, 1.2-μm protection circuitry.
Hu[14] of Taiwan, Republic of China, reported excellent results with near-micrometer CMOS technology protective circuitry. Modifications to the DIFIDW resulted in a protective device that did not require any additional clamping device. The major modification was to extend the $N^+$ region of the input drain beyond its deep N well. This device is capable of protecting 250-Å gate oxides from up to 5000 V ESD. The size is only 75 × 130 μm².

During a positive transient a parasitic lateral PNP made up of the drain as collector, substrate as base, and the source as emitter turns on protecting the gate. During a negative transient the P substrate to N input is forward-biased. In the case of both polarities, the deep well tends to reduce localized current densities and hot spots. A second modification was the addition of an extra $N^+$-N well tied to $V_{CC}$, in addition to the one normally present. This "pseudocollector" was added primarily to prevent latchup. An additional feature of this pseudocollector is that it provides a parasitic NPN path to $V_{CC}$. During positive transients the input drain acts as the collector, the substrate as base, and the pseudocollector as emitter. For negative transients the collector and emitter positions are reversed. Figure 12.28 shows the circuit diagram of the DIFIDW with the added pseudocollector path to $V_{CC}$.

### 12.7.2 The phantom emitter

Minear and Dodson[24,25] reported a novel manner in which to improve immunity of bipolar transistors from ESD. Most of the reported problems and the highest susceptibilities are with reverse-biased base-to-emitter junctions (b-e). As it turns out, the b-e junction is damaged whether the base is back-biased with respect to the emitter, collector, or both.

With a simple b-e reverse-biased electrostatic discharge, the least resistant path is directly across the junction sidewall between the base and emitter as was shown in Fig. 7.14. When the collector-base junction (c-b) is reverse-biased, the current initially will enter the base region on a broad front. Some of the current at the extreme emitter end will develop a voltage drop across the base region as indicated in Fig. 12.29. If this voltage drop exceeds the e-b breakdown by one forward drop, typically about 7 V plus about 0.7 V, the underside of the emit-

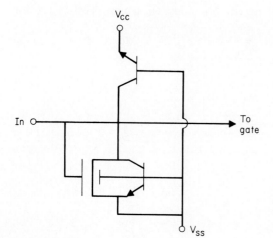

**Figure 12.28** Circuit diagram of DIFIDW device.

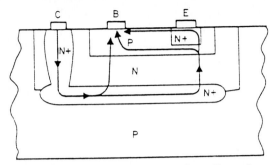

**Figure 12.29** Bipolar transistor under collector-base reverse-bias condition. *(From Minear and Dodson,[24] with permission from ITTRI/RAC.)*

ter junction must be forward-biased. Current entering the emitter must leave through the reverse-biased e-b sidewall. After this set of conditions has been established, the current is sustained by the much lower potential $BV_{CEO}$. If the usual condition of $BV_{CBO} > BV_{CEO} + BV_{EBO}$ exists, the emitter-to-base path carries the predominate current. The end result can be thermal runaway and failure.

Figure 12.30 shows the currents after a phantom emitter added in the vicinity of the base contact. The effective schematic is shown in Fig. 12.31. During reverse-biased e-b ESD some of the current is diverted through the buried collector region. The added phantom emitter means that sustained base-to-collector potential is limited to $BV_{CEO}$ rather than the higher $BV_{CBO}$. With lower resistance collector diffusions the collector path will divert most of the current away from the e-b sidewall resulting in higher thresholds to failure. With c-b reverse-biased, current will flow to both emitters by dropping the sustaining $BV_{CEO}$. The phantom emitter path is the least resistant so will predominate, again protecting the b-e sidewall. Susceptibility testing

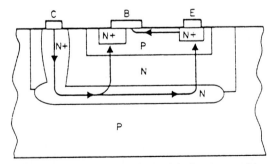

**Figure 12.30** Bipolar transistor with phantom emitter under collector-base reverse-bias condition. *(From Minear and Dodson,[24] with permission from ITTRI/RAC.)*

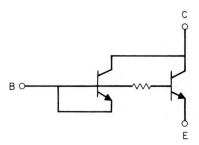

**Figure 12.31** Circuit diagram of transistor with phantom emitter.

showed about a fourfold improvement in the threshold to failure on devices fabricated with the phantom emitter. The geometric location of the base between the emitter and collector is preferred.

### 12.7.3 Static induction transistor

Nishazawa et al.[26] reported on a device called the static induction transistor (SIT) in 1975. The SIT has voltage current characteristics similar to a vacuum tube triode. An ordinary junction field-effect transistor (JFET) has characteristics similar to a vacuum tube pentode. The SIT is a JFET with a much shorter channel length, as depicted in Fig. 12.32. The voltage current characteristics are compared in Fig. 12.33. Of primary importance is the tendency of the SIT to conduct higher gate-to-source currents as a function of drain voltage, whereas the ordinary JFET tends to reach a limit that is not exceeded even when the drain voltage is exceeded. The SIT has low impedance, current capability of several thousand amperes per square centimeter, thermal stability, and response time of approximately 1 ns. These properties are ideal for an ESD protection device.

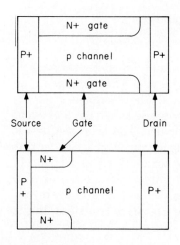

**Figure 12.32** Ordinary JFET and SIT device.

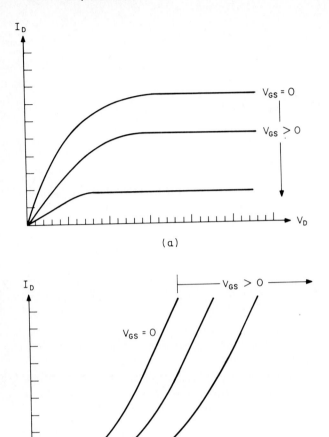

**Figure 12.33** Pentode- and triodelike characteristics. (a) JFET devices; (b) SIT devices, respectively.

A novel scheme devised by Turkman and Neelakantaswamy[27] used the SIT in order to circumvent the problem of ESD damage to the PN junctions involved in protective circuits. The design was an attempt to sink the transient current directly to the substrate and avoid any lateral flow through junctions or otherwise. The proposed layout of the SIT directly under a contact pad is shown in Fig. 12.34. In this manner the normal operating condition is normally off and $V_{GS} = 0$. Note that the depth of the two diffused $N^+$ gate regions is small with respect to the separation between the two $P^+$ (source and drain) regions. Figure 12.35 shows operation during a positive transient. In the case of a negative transient the forward-biased diode formed by the $P^+$

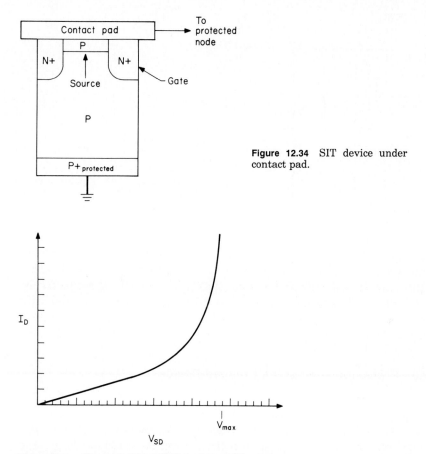

**Figure 12.34** SIT device under contact pad.

**Figure 12.35** Characteristic of SIT during positive transient.

drain and P channel to the $N^+$ gate clamps the pad to less than 1 V below ground.

This new scheme was investigated because most protective networks have several PN junctions along the discharge path that can fail because of localized heating in reverse bias. Thick field oxide transistors have long been recognized as very effective protection for NMOS and CMOS devices. Even in the thick field oxide transistor, localized heating in the reverse-biased drain-to-substrate junction can be a problem.

## 12.8 Merged Input Protection Circuits

An excellent technique to conserve space is to share protective networks between pins if possible. In the case of the circuit of Hu et al.,[14]

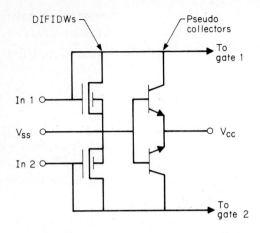

**Figure 12.36** Protective circuit sharing DIFIDW sources and pseudocollectors.

a shared circuit was devised with only slight compromise of performance. Figure 12.36 is a schematic resulting from their model with a shared pseudocollector and source, tied to $V_{CC}$ and $V_{SS}$, respectively. One-half of this shared cell size was 45 × 130 μm as compared with 75 × 130 μm for individual circuits. Effective protection was 4000 V versus 5000 V.

## 12.9 Protection against the Charged-Device Model

The charged-device model (CDM) produces a damped sinusoidal waveform the characteristics of which are dependent on the complex components of $L$, $C$, and $R$ within the device, its leads, package, and

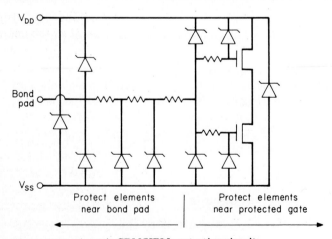

**Figure 12.37** Avery's CDM/HBM protective circuit.

interfacing parameters to the outside world (or ground). The entire transient is usually 10 ns or so. Avery[28] reported excellent results for both CDM and HBM stresses using the circuit shown in Fig. 12.37. Important features of this circuit are

1. Processing included rounded polysilicon and diffusions, known as closed cell devices, to minimize field concentrations.
2. Clamping to $V_{SS}$ and $V_{DD}$ near the active gate to be protected.
3. Clamping diodes near the bonding pads to protect field oxides under the bonding pads, and poly resistors, typically in series with the pad and the remainder of the protective circuitry.

Reported protection levels exceeding 3 kV for the CDM and 8 kV for the HBM were achieved when this circuit was incorporated in 1.25-μm P well, 1.5-μm N well, and 3-μm CMOS devices.

## 12.10 Adverse Effects of Protective Circuits

Although prior discussions of protective networks were made with emphasis on the positive gains made in prevention of static damage, there are certain adverse effects. The most important negative aspect is the degradation of high-frequency limits caused by added capacitive and resistive elements. Additionally, a price is paid in "real estate" or space allocations on the semiconductor chip's surface. The added space can be appreciable when all pins are provided with protection. The end result is diminished functional complexity per chip, so more chips are required per circuit.

## 12.11 Assembly Protection

ESD protection at the assembly level should receive more attention for the sake of economically achieving system immunity. Certain measures that require little if any cost can greatly reduce the likelihood of failures. Assembly protection can involve mechanical design as well as electrical design, associated equipment as well as the primary item, and procedures as well as hardware modifications.

### 12.11.1 Mechanical design

Much can be done to avoid ESD problems through the mechanical design. As a matter of fact these measures are believed to have the highest return on a relatively low investment. Changes would involve such areas as physical placement of parts, chassis design, human engineered features, wiring layout and locations, and shielding. The most effective human engineering effort is to place grounds where they are most likely to be grabbed.

**12.11.1.1 Strategic placement.** Mechanical design includes the location or placement of parts and assembly structures. Thus placement of susceptible items is critical. These items can sometimes be strategically placed where there is existent electrostatic shielding from mechanical structures or at least away from external interfaces. Just as important but often overlooked are the many interfacing connections to susceptible items. As was illustrated in the capacitive coupling example of Chap. 10, these connections should be placed where coupling capacitance to the outside world is minimized. This usually means simply increasing distance; however, sometimes effective area may be reduced as well. Certainly resistance or high insulation must be placed between these same interconnects. Interface connectors to the outside world should be recessed.

### 12.11.2 Electrostatic shielding

The use of metal or other highly conductive chassis in equipment designs is an excellent means of avoiding ESD in some stages of production, shipping, and end use. Electrostatic shielding is observed in Fig. 12.38. When metallic conductor B completely surrounds conductor A, then A cannot be influenced by any other outside conductors (or nonconductors, for that matter), such as C and D in Fig. 12.38a. In other

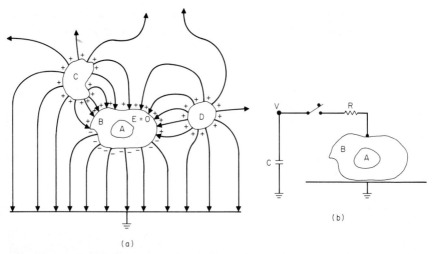

**Figure 12.38** Electrostatic shielding: (*a*) from external fields and (*b*) from direct contact ESD.

words the mutual capacitance between A and other conductors outside of B is zero. A conductive enclosure can be at any potential and still provide shielding; i.e., the shield need not be grounded. Figure 12.38a shows that the field inside the shield is unaffected by flux lines from other charged items, even though the surface charge on the enclosure B is redistributed by induction. Even when there is a direct ESD to the outside surface of the shield as depicted in Fig. 12.38b there is no internal field produced to affect the enclosed A. This 100 percent effective shielding occurs only in the ideal case, but is achieved for all practical purposes by metallic conductors. In actuality a transient field can occur inside a less conductive enclosure dependent upon its surface and volume resistivities and physical dimensions. When shielding ability is marginal or questionable, effectivity tests as described in Chaps. 16 and 17 are advisable.

**12.11.2.1 EMI protection.** Electromagnetic interference protection can be obtained from shielding, so this is consistent with good ESD design measures. Other measures include low-impedance circuitry, a correct grounding discipline, strategic placement, and selection of circuits less vulnerable to slight logic level shifts.

## 12.12 Electrical Design Protection at the Assembly Level

Electrical protective design features at the assembly level can also be cost-effective. The number of nodes or pins requiring protection at the assembly level is much less than required if done at lower levels. There is no easy formula for the best approach. The decision to emphasize protection at higher or lower assembly levels is dependent on factors such as

- End use conditions
- ESD controls in place during processing
- Susceptible item mix
- Maintenance philosophy
- Cost of end item and parts

### 12.12.1 Selection of parts of lower susceptibility

The first step in achieving assembly immunity to ESD lies in appropriate parts selection and control. Selection of parts of lower susceptibilities can be accomplished in several ways. Rarely will a different technological family be available that will be less vulnerable without sacrifice of performance. Sometimes less susceptibility can be found within a generic family from different or even the same manufacturer. In order to exercise the part selection option, one must begin to acquire a repertoire of part susceptibility data. This can be from within one's own facility or other sources as discussed in Chap. 3.

#### 12.12.2 Transient suppressors

Metal-oxide varistor and silicon diode transient suppressors are commercially available and would seem to be an obvious answer to ESD protection at the assembly level. These devices act as low impedances to higher voltage and high impedances to lower voltage. In the high-voltage case the low impedance of the suppressor is expected to force the bulk of the transient voltage drop to occur across the source impedance. Frequently the task is to protect an assortment of susceptible items in an assembly such as a printed-circuit board. Even if the DOD Std classification of these parts is known, the exact failure threshold is usually not known, so a reasonable target protection level is 50 V or even less. Such low voltage levels and the concurrent energy absorption requirements for longer transients require that the suppressor dimensions be large. Thus their inherent capacitances are generally too high to be effectively applied without severely degraded performance at high frequencies.

#### 12.12.3 Cho-Trap

A material commercially available called cho-trap has been cited as having ESD transient suppression characteristics. The proprietary material is said to have nonlinear current voltage behavior. It has been claimed[29] that the material acts as an insulator at low voltages and as a low resistance at high voltages. The material can be custom-formulated as an elastomer to provide versatile application possibilities.

### References

1. Colin Harris, "Input Structure Evaluation Using Specifically Designed Test Structures," *Electrical Overstress/Electrostatic Discharge Symposium Proceedings,* 1987.
2. W. D. Greason, *Electrostatic Damage in Electronics: Devices and Systems,* Wiley, New York, 1980.
3. Charles E. Jowett, *Electrostatics in the Electronics Environment,* Wiley, New York, 1976.
4. T. V. Hulett, "On Chip Protection of High Density NMOS Devices," *Electrical Overstress/Electrostatic Discharge Symposium Proceedings,* 1981.
5. Jack Keller, "Protection of MOS Integrated Circuits from Destruction by Electro-

static Discharge," *Electrical Overstress/Electrostatic Discharge Symposium Proceedings*, 1980.
6. D. J. Wilson et al., "Electrical Overstress in NMOS Silicided Devices," *Electrical Overstress/Electrostatic Discharge Symposium Proceedings*, 1987.
7. R. A. McPhee et al., "Thick Oxide Device ESD Performance under Process Variations," *Electrical Overstress/Electrostatic Discharge Symposium Proceedings*, 1986.
8. R.A. Duvvury, et al., "ESD Protection Reliability in 1 μm CMOS Technologies," *International Reliability Physics Symposium Proceedings*, 1986.
9. R. N. Rountree and C. L. Hutchins, "NMOS Protection Circuitry," *IEEE Transactions*, vol. **ED-32**, no. 5, pp.910–917, 1985.
10. C. Duvvury et al., "A Summary of Most Effective Electrostatic Discharge Protection Circuits for MOS Memories and Their Observed Failure Modes," *Electrical Overstress/Electrostatic Discharge Symposium Proceedings*, 1983.
11. F. De Chiaro, "Input Protection Networks and Fineline NMOS Effects of Stressing Waveform and Circuit Layout," *International Reliability Physics Symposium Proceedings*, 1986.
12. L. W. Linholm and R. F. Plachy, "Gate Protection Using an Arc Gap Device," *International Reliability Physics Symposium*, 1973.
13. F. S. Hickernell and J. J. Crawford, "Voltage Breakdown Characteristics of Closed Spaced Arc Gap Structures on Oxidized Silicon," *Motorola Publication*.
14. Y.-S. Hu et al., "High Density Input Protection Circuit Design in 1.2 Micron CMOS Technology," *Electrical Overstress/Electrostatic Discharge Symposium Proceedings*, 1987.
15. W. Palumbo and M. Patrick Dugan, "Design and Characterization of Input Protection Networks for CMOS/SOS Application," *Electrical Overstress/Electrostatic Discharge Symposium Proceedings*, 1986.
16. T. E. Turner and S. Morris, "Electrostatic Sensitivity of Various Input Protection Networks," *Electrical Overstress/Electrostatic Discharge Symposium Proceedings*, 1980.
17. C. J. Petrizio, "Electrical Overstress versus Device Geometry," *Electrical Overstress/Electrostatic Discharge Symposium Proceedings*, 1979.
18. O. J. McAteer, "Pulse Evaluation of Integrated Circuit Metallization as an Alternate to SEM," *International Reliability Physics Symposium Proceedings*, 1977.
19. R. K. Pancholy, "The Effects of VLSI Scaling on EOS/ESD Failure Threshold," *Electrical Overstress/Electrostatic Discharge Symposium Proceedings*, 1981.
20. D. G. Pierce, "Electro-thermomigration as an Electrical Overstress Failure Mechanism," *Electrical Overstress/Electrostatic Discharge Symposium Proceedings*, 1985.
21. K. L. Chen, "Effects of Interconnect Process and Snapback Voltage on the ESD Failure Threshold of NMOS Transistors," *Electrical Overstress/Electrostatic Discharge Symposium Proceedings*, 1988.
22. R. E. McKeighen et al., "Reversible Charge Induced Failure Mode of CMOS Matrix Switch," *Electrical Overstress/Electrostatic Discharge Symposium Proceedings*, 1986.
23. D. E. Nelson et al., "Design and Test Results for a Robust CMOS VLSI Input Protection Network," *Electrical Overstress/Electrostatic Discharge Symposium Proceedings*, 1986.
24. R. L. Minear and G. A. Dodson, "The Phantom Emitter—an ESD-Resistant Bipolar Transistor Design and its Applications to Linear Integrated Circuits," *Electrical Overstress/Electrostatic Discharge Symposium Proceedings*, 1979.
25. R. L. Minear and G. A. Dodson, "Effects of Electrostatic Discharge on Linear Bipolar Integrated Circuits," *International Reliability Physics Symposium Proceedings*, 1977.
26. J. I. Nischizawa et al., "Field Effect Transistor Versus Analog Transistor (Static Induction Transistor)," *IEEE Transactions*, vol. **22**, no. 4, pp. 185–197, 1975.
27. R. I. Turkman and P. S. Neelakantaswamy," A Novel On-chip Protection Device Using Static Induction Transistor Principle," *Electrical Overstress/Electrostatic Discharge Symposium Proceedings*, 1987.
28. L. S. Avery, "ESD Protection Structures to Survive the Charged Device Model

(CDM)," *Electrical Overstress/Electrostatic Discharge Symposium Proceedings,* 1987.
29. Joseph Pedulla and Paul Malinaric, "Cho-Trap,® A Novel Voltage Transient Protection Packaging Material," *Electrical Overstress/Electrostatic Discharge Symposium Proceedings,* 1986.
30. R. N. Rountree et al., "A Process-Tolerant Input Protection Circuit for Advanced CMOS Processes," *Electrical Overstress/Electrostatic Discharge Symposium Proceedings,,* 1988.
31. R. Hull and R. Jackson, "Analysis of High-voltage ESD Pulse Testing on CMOS Gate Array Technology," *Electrical Overstress/Electrostatic Discharge Symposium Proceedings,* 1988.

Chapter

# 13

# Factory Workplace Considerations and Problem Examples

This chapter is intended to provide insight into possible problem conditions and guidance in the selection of specific control measures. Discussions pertain primarily to uncontrolled conditions, although similar situations can arise in ESD-controlled areas that are not thoroughly planned. The typical factory environment includes countless sources of static electricity. These sources must be sought and recognized and their interface with static-susceptible items must be controlled. Control of this interface is generally referred to as *ESD control* and is covered in Chap. 15. The final part of this chapter discusses some common observances from past factory tours and audits including a few illustrative case histories.

## 13.1 Insulative Materials in General

The modern electronics factory is likely to have many insulative materials to be reckoned with in order to prevent ESD damage. Such materials are commonly used in packaging, furniture, storage areas, automated equipment, and tooling. Factory furniture and fixturing is likely to contain Plexiglas, Teflon, nylon, rubber, and various types of plastics. Packaging materials include Styrofoam, untreated polyethylene, polyurethane, and plastic. Manufacturing processes often involve the use of insulative tapes, acetates (or similar materials) for printed wiring assembly layout templates, insulative tools, and composite materials in automated equipment.

Practically everything in the factory workplace is to be regarded as hazardous from a triboelectric generation standpoint until evaluation verifies otherwise. Verification techniques are discussed in Chap. 16. Major static sources include the following:

- Human body
- Insulative clothing
- Footwear (critical), especially soles
- Flooring, especially carpets and ordinary tile
- Insulative materials commonly used in packaging, furniture, storage areas, and tooling
- Insulative work surfaces
- Certain insulative items essential to the process
- Automated handling equipment

Other sources of static derived in a different manner from the common solid-to-solid contact and separation include the following:

- Video display tubes
- Imbalanced ionizers
- Freeze sprays
- Wave solder machines
- Spray applied finishes and rinses

## 13.2 Contributing Factors

Other factors that contribute to static problems include the following:

- Low humidity
- Cold weather because of associated heavy clothing entering factory, and internal air drying from heating
- Air conditioning's drying effect on internal air
- Use of highly susceptible parts and/or assemblies
- Lack of proper awareness training

## 13.3 Possible Static Problems

This section addresses the problems that are likely to occur in the workplace because of the types of static sources mentioned.

### 13.3.1 Human-body-related problems

The human body has been a prime contributor to ESD problems. The basic problem is ESD damage from human contact with susceptible items. The extent of this problem is aggravated by several factors:

1. The human body has significant capacitance, or charge storing ability.
2. Because of the body's capacitance, the energy level can be appreciable.
3. The high mobility of humans tends to contribute to charge generation and frequent contact with sensitive items.
4. The human is somewhat of a free agent and is difficult to control at all times without sufficient attention to detail in the areas of training and motivation.

### 13.3.2 Clothing-related problems

Insulative types of clothing such as synthetics, wool, and silk tend to generate and hold high static charges. These materials tend to increase the voltage on the ungrounded human body. In the case of grounded personnel, the charge on insulative clothing, being immobile, is not bled off to ground. The immobility of such charge infers that it would be impossible to deliver damaging transient ESD currents to susceptible items. The inference is correct as far as direct contact is concerned; however, charged insulative clothing is hazardous because of possible induction effects. Highly charged clothing can cause failure while being worn if a significant coupling coefficient exists to a capacitance capable of destroying the susceptible item.

A situation that might arise in cold climates where wool sweaters are common is the operator removing a sweater or jacket and placing it in the immediate work area such as on the work surface or stool. The placement of a printed wiring assembly or other sensitive item on top of a highly charged outer garment can result in ESD damage through induction when the item is subsequently handled.

### 13.3.3 Footwear

Footwear is perhaps the most critical item of clothing as far as ESD is concerned. Three pertinent factors are the shoe's capacitance, resistance, and triboelectric relationship with respect to the floor. In the case of specially provided conductive flooring or flooring that coincidentally provides a reasonably low resistance to ground, the conductivity of shoe soles and/or heels can provide a means of personnel grounding, thus preventing static accumulation. Insulative shoe bottoms not only prevent bleedoff to ground but also can increase body potential via triboelectric generation and induction to the wearer's body. Such triboelectric generation can occur even on conductive floors. Whether the flooring is conductive or not, the triboelectric relationship between the floor and shoe materials predetermines charge accumulation tendencies.

### 13.3.4 Flooring

Static electric properties of flooring materials vary considerably. Although triboelectric properties are not directly determined by resistivity parameters, in general the more insulative materials tend to generate higher potentials. Ordinary carpets are notorious static generators and are to be avoided where susceptible items are processed. Even so-called antistatic carpets are often more generative than should be permitted in ESD-controlled areas. Triboelectric propensities of ordinary tile vary over a wide range, but many are highly generative. Ordinary unsealed concrete is usually sufficiently conductive to ground to allow bleedoff of potentials on personnel wearing conductive footwear. Triboelectric generation on unsealed concrete is usually minimal for virtually any common footwear. In practice, however, sealers of various types are generally used. Concrete sealers such as epoxy and other materials are often highly insulative and static-generative. Likewise, wooden flooring is rather neutral in its natural state. The application of varnishes, waxes, and other similar finishes can render wooden floors insulative and generative as well.

### 13.3.5 Furniture-related problems

Potential problem situations can result because of variables in the materials, construction, and installation of items of furniture. Furniture is intended to include items such as workbenches, storage shelves and cabinets, carts, chairs, and stools.

**13.3.5.1 Ordinary work surfaces.** The most critical part of a workbench relating to ESD is the work surface. Work surfaces in areas without ESD controls usually are either of ordinary melamine materials or conductive materials such as stainless steel or copper. Most so-called insulative work station surfaces are not sufficiently prone to static accumulation to cause inductive failures. An exception involving a Plexiglas work surface at an electric power plant is discussed in Sec. 13.2.2. Most ordinary workbenches are covered with a melamine or Formica-type material with surface resistivity ranges that fall between the marginally acceptable high static dissipative range to about $10^{15}$ Ω per square. The major problem with these materials is a resistance to ground that may be too high for effective ESD control. Even where acceptable readings are obtained, resistivity may vary to the extent that some bench locations would have exceedingly high resistance to ground. Second, triboelectric generation of charges on the susceptible item is dependent on the relationship between the work surface and susceptible item materials. Representative results of triboelectric generation of objects on an assortment of ungrounded work surfaces reported by Briggs[1] are summarized in Table 13.1. Grounded surfaces will not prevent triboelectric charging of objects.

**13.3.5.2 Conductive work surfaces.** Conductive work surfaces, generally stainless steel, have been used extensively prior to the emergence of ESD control in the industry. In uncontrolled areas conductive workbenches are not normally grounded intentionally. In such areas the surface may become grounded by placement of electrical apparatus with grounded frames or chassis on the bench. Without added electrical equipment the resistance to ground may still be relatively low depending on installation variables. Even if the electrical resistance to ground is high, a conductive work surface provides a relatively high capacitance to ground because of the large area involved. The impedance to ground for a high-frequency transient is thus very low. This presents a possible charged device model failure hazard if a charged susceptible item touches the work surface.

### 13.3.6 Hazardous items essential to a manufacturing process

Items that are essential to a manufacturing process may be prone to static accumulation. Examples might include tooling, chassis, wiring insulation, solid electrical insulators, insulative part carriers, some printed wiring board materials, or even the parts themselves. Possible detrimental effects from these items must be carefully evaluated. In many cases geometrical and other considerations preclude the occurrences of ESD from such materials even though they are insulative and static-generative.

## 13.4 Static Hazardous Factory Processes

Other sources of static involve certain factory processes that deserve special mention. These processes have been known to cause potential

TABLE 13.1 Results of Triboelectric Generation of Objects on Benchtops

| | Object material—rubbed on benchtop | | | | | |
|---|---|---|---|---|---|---|
| Benchtop material | Ceramic flat pack | Epoxy/ fiberglass PC board | Uralane | Parylene | Rubber RTV | Aluminum |
| Conductive floor tile | 150 V | 3 kV | 800 V | −4 kV | −4 kV | 150 V |
| Carbon polyethylene | 56 V | 3.8 kV | 1.2 V | −4 kV | −4 kV | 16 V |
| Antistatic polyethylene | −14 V | 100 V | 2.6 kV | −3.9 kV | −4 kV | −47 V |
| Melamine (smooth) | 10 V | 1.5 kV | 2.4 kV | −3.9 kV | −4 kV | −128 V |
| Melamine (matte) | −80 V | 200 V | 3 kV | −4 kV | −4 kV | −500 V |
| Benalux (MIL-I-24330) | 25 V | 3.8 kV | 2.5 kV | −4 kV | −4 kV | −1 kV |
| Metafil | −10 V | 3.6 kV | 3.8 kV | −3 kV | −4 kV | 500 V |
| Aluminum | 8 V | 1.4 kV | 2.4 kV | −4 kV | −4 kV | 26 V |
| Stainless steel | 12 V | 1.6 kV | 2.9 kV | −3.3 kV | −4 kV | 0 V |

ESD problems and are often of concern to those instituting ESD controls.

### 13.4.1 Operations involving spraying

Numerous factory processes involve spraying of solvents, paints, or other substances in a liquid or vaporous state. Generally the best way to assess whether these operations are electrostatically hazardous is to take careful measurements of potentials generated during the process.

Many printed-circuit board assemblers apply a conformal coating to the finished product as an insulative protective barrier, normally transparent. The coating is generally applied by either high-pressure spraying or dipping. The dipping process generally does not cause static buildup. The spraying operation tends to generate high static potentials if uncontrolled. Variables affecting static buildup include spray velocity, distance, materials involved in spray and board, and board resistance to ground.

### 13.4.2 Solder removing

Ungrounded solder removing tools have been demonstrated to charge to several thousand volts during the removal process. The charge buildup occurs because of Teflon or other insulative tips prone to triboelectric buildup from the rapid movement of solder and to some extent by the impure air near the point being desoldered. The hazard exists because of the intimate content or close proximity with sensitive nodes being desoldered.

### 13.4.3 Tape application and removal

Insulative tapes are often used in the manufacturing process. Such tapes build up static by triboelectric generation when pulled from the roll. Thus, there is a concern as the highly charged tape is brought near a sensitive item for application. The problem is compounded by further charging as the tape is rubbed during application and during the removal process. Even somewhat benign tape materials such as paper masking tape will show a peak high-voltage generation during rapid removal from the roll. Fortunately, the conductivity of paper tapes is usually sufficient to bleed the charge off to the holder in less than a second at moderate ambient humidities.

### 13.4.4 Video display tubes

Plant modernization has brought about extensive use of video display tubes for various informational purposes. Uncontrolled display tubes generally have high electrostatic field emissions. The hazard presented is no different from that of any high static field, such as from a highly charged insulator. A direct field-induced failure is not extremely likely. The real concern is static accumulation on an item of significant capacitance followed by subsequent discharge through a susceptible item as discussed in Sec. 8.4.1.

### 13.4.5 Ionization

Ionizers are used to neutralize static charge, primarily on insulators. Some manufacturing processes involving movement of insulative materials, such as with refrigerator shells, plastic radios, or VCRs, are prone to static buildup. This accumulated potential can cause nuisance shocks and dust accumulation as well as present a threat to susceptible electronic items. Ionization is generally the best choice to neutralize these charges. Often ionizers are used on a smaller scale to neutralize a single generative material remaining in the process. In any case it is important that the amounts of positive and negative ions remain equally balanced at all times, or static buildup on items in the ionized area can result. This topic is discussed in detail in Chaps. 15 to 17.

### 13.4.6 Automated equipment

The human interface is being reduced if not eliminated from much of the electronics manufacturing cycle through automation. At first consideration this might seem to minimize the ESD problem since the charged human body has been a large contributor to past problems. ESD problems can still result from improperly selected materials and other design and construction imperfections. The primary problem with automated handling equipment is charged-device-type failures. These occur as the devices, or assemblies, become charged during the movement of handling. The amount of charge accumulation is dependent on the materials of the items processed and the contacting handler constituents. A related problem is gradual triboelectric charging of ungrounded metal fixtures until capacitive discharge through a susceptible device occurs after a damage threshold is exceeded.

The modernization and automation trend has brought about increased usage of conveyor belts. Electrostatically these systems can be very similar to a Van de Graaff generator as discussed in Chap. 4. The triboelectric properties determine charge buildup on items stopped, purposely or otherwise, on the moving belt. In addition charges accumulated on the belt can cause inductive charging of conductive items on the belt.

## 13.5 Frequently Observed Conditions

Electronics facilities around the world span the spectrum from no ESD controls to excellent, well-disciplined controls. Where the latter exist, the facility is usually run with forethought, and economics is the primary consideration. In other words, the facility did not go overboard on ESD control implementation because of being sold by an overzealous salesperson. Decisions were based on risk assessment and cost of failures weighed against control measure costs.

Commonly, controls are at an intermediate level with a lack of real commitment. Specific control measures are often instituted because of customer concerns or as a result of an ex-officio ESD crusader with no charter to follow-through on the control program. The remainder of this chapter discusses problematic conditions frequently observed at these facilities. These examples of poor control are intended to reinforce coverage of proper ESD control measures discussed in Chaps. 14 and 15. Chapter 15 explains more fully the reasons why the practices mentioned here are incorrect and suggests reasonable alternatives.

### 13.5.1 Suppliers

Before discussing poor ESD control conditions at an electronics assembly plant, the suppliers must be mentioned. Again a wide spectrum of ESD control would be anticipated at various vendor facilities. The likelihood of poor controls along the way is increased as more intermediaries become involved in the supply chain. Incidences of poor ESD training and practices have been observed at microelectronic distributors, independent test laboratories, and to a lesser extent at device manufacturer's sites.

### 13.5.2 Receiving

As a rule the receiving department workers rarely if ever open shipping packages to the extent of endangering susceptible items. In the rare event that packages must be opened, receiving personnel typically are completely untrained in ESD control measures. Often they have not even heard of static damage to parts.

### 13.5.3 Incoming inspection

Excluding suppliers, incoming inspection is the first place where susceptible items are likely to come into real jeopardy. Personnel are sometimes poorly trained in ESD control but are unaware of their shortcomings in this regard. Static control stations are often limited to one or two partially equipped stations where proper procedures are unlikely to be followed. There is no feedback to this department about static failures during incoming testing or subsequent processes, except in the case of a real line stopper. Therefore the perception of ESD is that it is not a serious problem and the controls in place take care of the problem.

Frequently a positive means of identifying susceptible items is not in place. Workers will respond that they can tell which items are susceptible by the way they are packaged. This leads to other problems: they do not know what constitutes proper packaging; and if a part has no apparent static control packaging of any variety, the assumption is that it is not sensitive. Any effort to show ESD concern whether by a caution label, antistatic, dissipative, or conductive package, or any combination is deemed adequate and indicative of susceptibility. Incoming inspection workers with this attitude fail to see that they are really saying "anything goes."

The problem compounds itself when the frequent occurrences of static-generating "antistatic" labeled tubes, bags, and other containers are considered. Such "hot" antistatic materials may be found throughout the factory if they are not monitored. The concept of static shielding is usually completely misunderstood. Personnel will often discard the usually present outer shield provided by vendors because of the opinion that the inner antistatic package is sufficient. When small numbers of integrated circuits are requested from stores they may be transported in plastic bags or cut DIP tubes rather than in shielding containers.

Proximity to receiving and the variety of sensitive and nonsensitive items received tend to bring insulating packing materials to the incoming inspection area. Discipline is needed to control such notorious static generators, especially when they are in the storage area awaiting possible dispersement to preassembly kitting.

### 13.5.4 Kitting

In the kitting area parts are selected and put into kits sufficient to make up one assembly. The problems begin because of the improper controls at incoming inspection. Parts are likely to be moved to kitting without the proper shielding packages in place. Sometimes insulative materials such as plain plastic boxes, bags, or insulative tapes may be in containers with improperly packaged sensitive parts. A primary source of improper packaging is often the incoming department's attitude that antistatic packaging will protect from such generators. Even worse, if a susceptible item had been received in nonrecognizable or nonexistent static protective packaging, it would have been forwarded to kitting as nonsusceptible. The poor awareness at incoming inspection is likely to influence kitting personnel with some of the same misconceptions. Three common hazards observed include the use of plastic weighing balances to "count" integrated circuits and the use of nonlidded and/or generative tote containers and plain plastic kit setup trays.

### 13.5.5 Assembly areas

The assembly areas are considered most critical because of the amount of handling of susceptible items. This area, being the most visible, is usually the best controlled. Still, common problems exist. Many are housekeeping-related, involving the presence of unnecessary generating materials. Personal items such as plastic cups, thermos bottles, purses, radios, hats, sweaters, and other items of clothing should not be on workbenches. This can be extended to include plastic toolboxes and part bins. An attitude exists that susceptible parts are safe from ESD once placed on a printed circuit or other assembly. Repair stations have similar problems to those of assembly areas.

**13.5.5.1 Wave soldering.** Wave-soldering areas often are uncontrolled, partially because of the belief that parts are safe after installation on printed-circuit boards. The operator is required to be highly mobile, so grounding by wrist strap is inconvenient. The fact that the boards are in movement is reason to be concerned. Insulative rotating brushes are frequently involved with flux removal or other operations. These can be generative. Certain defluxing solutions can be in a conductivity range that is not sufficient to prevent static accumulation. Static-generating varieties of rubber gloves are sometimes used, and boards are frequently transported in ungrounded unlidded conductive trays. Conductive trays will take on the same potential as the person carrying them, which can lead to a failure if another person removes or inserts a susceptible item. Conductive trays can also be charged by induction if placed on charged objects such as insulative chairs, shelves, or other surfaces.

**13.5.5.2 Conformal coating.** Conformal coating operations are prone to static problems, as discussed in Sec. 13.4.1. The operation is frequently observed with ungrounded operators, which adds another possible damage source. The printed-circuit boards are often placed on insulative cushioning materials for the spraying operation. This not only isolates the board from ground, which allows unlimited static accumulation, but it also adds the charge-induction effects from the insulative cushion.

**13.5.5.3 Carts.** Carts are typically left uncontrolled, that is, ungrounded, and often while containing susceptible items that are not properly packaged. To add to this problem, carpet remnants are sometimes placed on the carts to cushion the item being transported. Carts can become charged through the movement of wheels against various types of insulated flooring. In addition, the potential accumulated on the cart pusher is commmon to the cart as well due to hand contact. When the cart is released, a peak potential may remain on the cart and its contents. An ESD failure can result as a person touches a susceptible item, and all the energy stored on the cart and its contents is released to the person, whether grounded or not.

**13.5.5.4 Transient personnel.** Even in factories with good ESD control measures in place there is often a disregard of procedures by transient managers, engineers, and touring customer dignitaries. Oftentimes upper management personnel appear to be ignorant of the ESD problem: that is, they do not practice rudimentary controls when handling products. Engineering personnel have been notorious doubters of the seriousness and extent of the ESD threat. Reasons for this disregard are discussed in Chap. 2. In jest it may be said that to the casual factory personnel it appears that either (1) the engineers know best and the ESD control advocates are ignorant members of a cult that enjoys inconveniencing people, or (2) certain academic degrees held by the engineers render them static-free.

## 13.6  Inspection

It is not uncommon to find ungrounded inspectors handling susceptible assemblies, even though assembly stations are static-controlled. This may be in part tied to the false belief that parts on assemblies are safe. A common specific problem is the presence of plastic microscope covers on the workbench touching or near sensitive items. Perhaps worse, but less common, is the use of plastic templates to inspect for missing or incorrect parts.

## 13.7  Test and Troubleshooting

Test and troubleshooting technicians are generally reasonably well trained in electronics. Static awareness is still often lacking among them, often to the point of regarding ESD as a "MOS" problem. Frequently assemblies are handled by ungrounded personnel. A common problem is utilization of aerosol freon "freeze sprays" to troubleshoot suspected cold temperature failures. Such sprays can generate 10 kV if not of a special antistatic variety. Ungrounded operators removing and inserting printed wiring boards have been frequently observed during system and higher assembly troubleshooting.

## 13.8 Shipping

Shipping departments are often left out of the static awareness training cycle. Principles of proper packaging are not understood. Items returned to vendors for troubleshooting, repair, or warranty replacement are often shipped in nonprotective packaging. The introduction of static electric damage because of incorrect packaging could compound the original problem and confuse the vendor's failure analysis.

## 13.9 Engineering Labs

Engineering laboratories are often exemplary of careless, halfhearted attempts at ESD control or none at all. Typically, engineers and technicians have not regarded ESD as a serious problem and accordingly have not learned of the properties, purposes, and applications of different ESD control measures. Highly sensitive integrated circuits are frequently found in plastic storage cabinets. Sometimes the parts are enclosed in nonshielding antistatic bags. Often there is an apparent disregard for the possible inductive effects from highly generative insulative materials such as plastics, Styrofoam, and polyethlyene found on workbenches. Even when grounded work surfaces are present, susceptible items, particularly printed-circuit boards, will often be found on other nearby ungrounded surfaces. When a relatively high degree of ESD concern is exercised, it is sometimes for the less sensitive items because of misconceptions about part type, technology, or vendor susceptibility levels.

## 13.10 Environmental Considerations

The year-round weather and in-plant environments can have an effect on humidity conditions and thus ESD. The values in Table 13.2 were reported by McFarland[2] showing effects of humidity in a Denver facility. Consider part susceptibility as listed in Table 2.3 for comparison.

## 13.11 Unusual Observances

Potting materials such as room temperature volcanizing (RTV) rubber have long been known to have problems due to breakdowns across

**TABLE 13.2 Effects of Humidity on Static Generation**

| | Electrostatic voltages | | |
|---|---|---|---|
| Means of static generation | 10% RH | 40% RH | 55% RH |
| Person walking across carpet | 35,000* | 15,000 | 7,500 |
| Person walking across vinyl floor | 12,000* | 5,000 | 3,000 |
| Worker at a bench | 6,000* | 800 | 400 |
| Ceramic dips in plastic tube | 2,000 | 700 | 400 |
| Ceramic dips in vinyl setup trays | 11,500 | 4,000 | 2,000 |
| Ceramic dips in Styrofoam | 14,500 | 5,000 | 3,500 |
| Circuit packs as bubble plastic cover removed | 26,000 | 20,000 | 7,000 |
| Circuit packs as packed in foam-lined shipping box | 21,000 | 11,000 | 5,500 |

*Values confirmed by DOD-HDBK-263, dated 5/2/80.
SOURCE: McFarland,[2] with permission of ITTRI/RA.

voids. Such arcing occurs because the dielectric constant differences result in a concentrated field across the void. For this reason great care is taken, such as forming in vacuum to pull out trapped air and other gases. However potting materials with voids can cause triboelectric generation during flexing or pressurizing-depressurizing cycles as reported by DerMarderosian.[3]

## 13.12 Representative Problem Examples

Primary concern for damage brought about by insulative materials in the factory should be directed toward avoidance of induction to susceptible items or to conductive items that contact susceptible items. Discussion of a few selected actual case histories are included to provide insight into hazardous situations from a general viewpoint.

### 13.12.1 Printed wiring assembly "cushioning"

An extremely hazardous situation was observed during a walk through a factory machine shop area enroute to the cafeteria at lunchtime. One of the automated milling machines was "down," and the outside maintenance worker had been called in for repairs. His tool kit and drawings were at the machine, but he apparently was taking a lunch break from his task of troubleshooting. Before leaving he had placed a very large (1 × 2 ft) printed-circuit board assembly on an even larger (2 × 4 ft) sheet of ⅝-in-thick Styrofoam. Later discussions disclosed that the large cushion was carried in the maintenance van to provide cushioning during transportation and during repairs. He discarded the Styrofoam when informed as to how easy it is to induce more than 5 kV on a board by placing it on a sheet of Styrofoam that has been rubbed by hand.

### 13.12.2 Upsets of electric utility plant control system

Several years ago I was called on to investigate troublesome and costly control system upsets occurring at an electric power plant in the Midwest. The newly installed sophisticated computerized control system monitored all processes within the large power-generating facility. Payment for the computerized system had been forestalled due to unexpected problems at one of two similar control rooms. For no apparent reason the system would lock up in a nonoperating state several times a day. Normal functioning could be restored, but only after rebooting the system. Several minutes were required to reconfigure to the lost mode of operation. This was not only an inconvenience, but it could be critical during a plant malfunction. Static electricity was suspected by the on-site field engineer owing to the nature of the problem and the human interface via keyboard in the control room. An extensive static control program was being considered with no real determination of the exact circumstances relating to the upset occurrences.

A few initial experiments revealed that human body potentials were too low to affect the equipment even with several types of footwear conducive to static. The few insulative materials present were found to be nondamaging as well. Upon inspection one significant difference between the two control stations was cause for concern. The problem control station had a ⅜-in Plexiglas work surface about 2 in from the keyboard, whereas the other station had a statically neutral wooden work surface. Rubbing the Plexiglas by hand would rather easily generate 10 kV on the surface. This potential had no effect on the keyboard, however. After rubbing a spot about 3 in from the keyboard, I quickly scanned for an item with a particular material makeup. Reflecting on the principles of induction, I knew I needed first of all an item with a conductive portion that could have a potential induced on it from the Plexiglas. Second, the conductive portion had to be sufficiently large to have the capacity to deliver significant energy. A mental image showed that a conductor such as an aluminum ash tray would have done nicely except for one vital factor. If I tried to move it to the point of contact with the keyboard, the static potential would discharge to my hand upon touch. Therefore I needed an item that contained the conductive portion but also had an insulative part to serve as a handle in order to move it. There, right before my eyes, I finally focused on an 8- by 11-in clipboard with a rather large metal clip.

I picked up the clipboard and placed it over the rubbed Plexiglas.

The metal clip measured 4 kV on my hand-held static field meter. I reached down to the insulative "board" and moved the clip over to contact the edge of the keyboard. The control system immediately became upset. The on-site field engineer smiled broadly as his problem was solved in less than a half-hour. After recommended corrective action of removing the Plexiglas, there were no further upset problems at this facility.

### 13.12.3 Acetate printed wiring layout templates

A frequently observed practice is to use a transparent acetate or similar material with an imprinted template showing part locations on a printed-circuit board assembly. Such templates often are prone to static accumulation. If a charged template is placed on a printed-circuit board, a static discharge can occur when the board is touched because of simple induction. If the board is insulated from ground and the template is subsequently removed, a charge of opposite polarity can remain on the board, presenting a double jeopardy by compound induction, as discussed in Chap. 4.

### 13.12.4 Masking tape induced failures

The question of whether relatively small insulators have sufficient capacity to deliver destructive energy by induction or sufficient fields to cause oxide breakdowns often arises. Typical lengths of insulated tapes used in the industry are smaller than 6 in, thus having little apparent energy-delivering capacity when compared with that of the human body, for instance. However, the intimate contact of tapes with susceptible items or connecting conductors increases the concern because of the direct inductive coupling possibility.

Experiments reported by Baumgartner[4] of Lockheed provide a good deal of insight into the proper concern for the hazards of masking tape in the workplace. Eight different tape types were used. Five were insulative with surface resistivities of $>10^{14}$ $\Omega$ per square, and three were aluminum of different varieties. The aluminum tapes had discontinuous resistivities that measured greater than $10^{13}$ $\Omega$ per square at 10 V and less than 10 $\Omega$ per square at 25 V. Table 13.3 lists the effects of application, finger rubbing for adhesion, and removal of eight different tapes with an aluminum plate. Figure 13.1 shows the equivalent circuit of the experimental printed-circuit board used to test the effects on discrete devices. Associated parasitic and mutual capacitances, as well as intertrack resistance $R_1$, is indicated in the diagram. Similar tests on the board with no devices installed resulted

TABLE 13.3  Effects of Application, Finger Rubbing for Adhesion, and Removal of Eight Different Tapes with an Aluminum Plate

|  | Voltage, kV | | |
| --- | --- | --- | --- |
| Tape | Applied | Rubbed | Removed |
| Cellulose acetate | >2.0 | 1.0 | >2.0 |
| Teflon (black) | >2.0 | 0.9 | 1.2 |
| Polyester (green) | >2.0 | 1.1 | 1.3 |
| Teflon (white) | >0.9 | 0.3 | >2.0 |
| Mylar (tan) | >2.0 | 0.5 | >2.0 |
| Aluminum (metalized) | 0 | 0 | 1.3 |
| Aluminum (embossed) | 0.1 | 0 | 0.2 |
| Aluminum (conductive) | 0 | 0.1 | 0.2 |

SOURCE: Baumgartner,[4] with permission of the EOS/ESD Association.

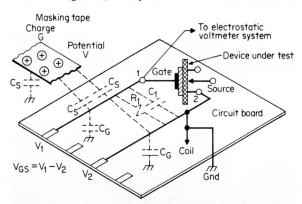

**Figure 13.1** The equivalent circuit of the experimental printed-circuit board used to test the effects of tape on discrete devices. (*From Baumgartner,[4] with permission of the EOS/ESD Association.*)

in 750 to 1500 V during application, 250 to 500 V during rubbing, and approximately 250 V during removal. Calculated energies were from 0.37 to 13 µJ. When the taping operations were performed with rapid motions, a benchtop ionizer at a distance of 2 ft had little effect on decreasing these peaks, as shown in Fig. 13.2.

The effect on MOSFET gate-to-source resistance was evaluated using the test board after application, rubbing, and removal of all eight tape types. Two of the device types, the SD212D MOSFET and the

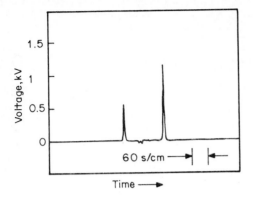

**Figure 13.2** Voltage peaks produced by Teflon tape during application, rubbing, and removal from an aluminum surface while in the presence of a benchtop ionizer.

3N163 MOSFET, were field-sensitive, having unprotected oxide breakdowns at 60 and 160 V, respectively. The third MOSFET device type tested, the 2D213, contained an input protection diode so it might more appropriately be categorized as energy-sensitive. Results for all three device types are given in Table 13.4. A resultant damaged gate of a 3N163 device is shown in Fig. 13.3. From these tests one can conclude that even though inductive coupling from the insulative tape is involved, the hazard seems to be primarily to field-sensitive devices.

### 13.12.5 Verification of hazard from garments

Speaking from my personal experiences, I recall when several blends and combinations of Dacron and nylon garments were in use at a microelectronics facility. After experiencing significantly high failures from static electricity I became concerned about garments. After instituting the normal good practices of ESD control such as grounded personnel, grounded work surfaces, protective packaging, and purge of notorious generators, a remaining concern was the fact that highly generative synthetic smocks were being worn.

Upon initial consideration, a highly charged smock (to a potential of 30 kV) was obviously not as hazardous as an equally highly charged conductor. The insulative smock contained charge that was virtually immobile, and it is impossible to deliver any significant concentration of this charge, and thus energy, to a sensitive item. There is, however, a static hazard from smocks that is twofold: voltage-sensitive parts can be damaged by potentials determined by the electrostatic field of the charged smock and the smock could induce a charge separation on a conductor, possibly causing sufficient current to destroy an energy-sensitive item.

TABLE 13.4 Gate-to-Source Resistance Changes due to Taping Operations

| Device | Resistance, Ω | |
|---|---|---|
| | Before EOS | After EOS |
| SD212 | | |
| No. 1 | $-2.0 \times 10^{14}$ | $-2.5/3.0 \times 10^{9}$ |
| No. 2 | $-2.0 \times 10^{14}$ | $-1.0 \times 10^{9}$ |
| No. 3 | $-1.5 \times 10^{14}$ | $-3.0/4.0 \times 10^{9}$ |
| SD213 | | |
| No. 1 | $+1.0 \times 10^{13}$ | $+1.0 \times 10^{13}$ |
| | $-0.7 \times 10^{10}$ | $-0.7 \times 10^{13}$ |
| No. 2 | $+0.7 \times 10^{13}$ | $+0.7 \times 10^{13}$ |
| | $-0.7 \times 10^{10}$ | $-0.7 \times 10^{10}$ |
| No. 3 | $+6.0 \times 10^{14}$ | $+6.0 \times 10^{14}$ |
| | $-1.0 \times 10^{10}$ | $-1.0 \times 10^{10}$ |
| 3N163 | | |
| No. 1 | $-2.0 \times 10^{12}$ | $-1.7 \times 10^{9}$ |
| No. 2 | $-1.0 \times 10^{14}$ | $-1.0 \times 10^{9}$ |
| No. 3 | $-2.0 \times 10^{14}$ | $-3.0 \times 10^{12}$ |

SOURCE: Baumgartner,[4] with permission of the EOS/ESD Association.

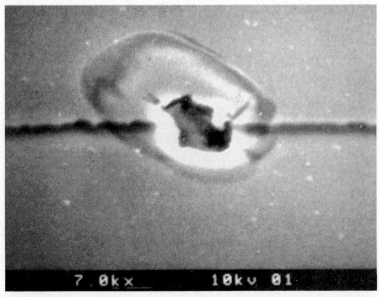

**Figure 13.3** Damaged gate of a 3N163 MOSFET caused by taping operations. SEM magnification 7000X. (*From Baumgartner,[4] with permission of the EOS/ESD Association.*)

#### 13.12.5.1 Destruction of part by charged garment.
Destruction of a bipolar operational amplifier was accomplished by waving a 30-kV smock to the point of touching a 6-in length of wire with a crocodile clip on the end tied to a sensitive input. $V_{SS}$ of this device was tied to ground during this experiment. For comparison, this type of custom operational amplifier was sensitive to about 400 V from the human body model.

### 13.12.6 Voltage suppression

The relationship between charge, voltage, and capacitance, that is, $Q = CV$, results in a sometimes puzzling phenomenon called "voltage suppression." This term is used to define the reduction in voltage that occurs when a charged item approaches ground or some other relatively large object. The resultant increase in mutual capacitance means the voltage must decrease for a given charge.

Voltage suppression occurs to the charged human body as well as to inanimate objects such as tote boxes, bags, trays, or the sensitive item itself. A classic example can occur with seated operators. Often the chairs or stools at workstations have a footrest provided, as part of the bench if not the seat itself. A seated operator might have a capacitance with respect to the floor of 200 pF with the soles and heels of both shoes against the floor. Ordinary insulative shoes are assumed in this example. Suppose the operator's wriststrap was not functioning and the body potential was 100 V. The charge in this case would be equal to $CV$ or 20 µJ. If both feet are raised or on the footrest the capacitance might reduce to as low as 60 pF. Assuming that no additional charge was generated by the movement and the original 20 µJ remained, the body potential would increase to 333 V, obtained from $Q/C$. This may illustrate to some extent why static failures are erratic unpredictable occurrences. Raising one foot would increase the voltage to some intermediate value between 100 V and 333 V.

### References

1. Charles Briggs, Jr., "Electrostatic Conductivity Characteristics of Workbench-Top Surface Materials," *Electrical Overstress/Electrostatic Discharge Symposium Proceedings*, 1979.
2. W. Y. McFarland, "The Economic Benefits of an Effective Electrostatic Discharge Awareness and Control Program—an Empirical Analysis," *Electrical Overstress/Electrostatic Discharge Symposium Proceedings*, 1981.
3. A. DerMarderosian and L. Rideout, "The Generation of Electrostatic Charges in Silicone Decapsulants during Cyclic Gaseous Pressure Tests," *Electrical Overstress/Electrostatic Discharge Symposium Proceedings*, 1979.
4. G. Baumgartner, "ESD Analysis of Masking Tape Operations," *Electrical Overstress/Electrostatic Discharge Symposium Proceedings*, 1988.

# Chapter 14

# ESD Control Management

## 14.1 Introduction

ESD control, an important issue today, is likely to become paramount in future years, perhaps for decades. Attainment of technological advances in the areas of superconductivity, higher speeds, and less power consumption will be pursued in spite of any static-susceptible materials. It appears, therefore, that increased future ESD susceptibility is likely. On the other hand, the possibility of a technological breakthrough that would remove or significantly reduce the ESD problem is remote.

This chapter discusses the establishment of an ESD control program, determining the extent of control measures and managing the program. The extent of an ESD control program is a matter of economics and other important variables that are facility- or operation-dependent. Therefore each facility or operation must tailor its program to its own needs. The task is difficult, as the extent of ESD problems and/or potential problems is often masked and not properly measured.

## 14.2 The Control Problem

Much of the literature on the subject of ESD abounds with well-intentioned guidance from zealots who would have one believe that Chicken Little was on the right track although he hadn't heard yet of ESD. Such guidance would (and has, in many cases) lead to the conclusion that every available control measure is necessary, perhaps in redundant multiples. The preponderance of this type of information does not serve well in convincing the many doubters that the problem is real. Historically, it has been much easier for staff members with indifferent or negative ESD control attitudes to gain influence with management. After all, what manager wants to be told that a serious problem exists throughout the entire facility and that the manufacturing process in all locations must be changed? The facts that doubters have outnumbered believers and that zealots occasionally stretch the truth counter our love for both majority rule and honesty. The end result is that many managers react to the customer's demands for ESD control with minimal effort other than a facade of meeting the seemingly unnecessary requirements.

How then should a concerned individual tailor an ESD program to best meet the needs of a facility or operation? How does one find a proper balance between the extremes of "every control" and "no control" measures?

## 14.3 ESD Control Funding

The potential for static problems at most facilities falls someplace between the extremes held by the doubters and the zealots, but management often has little solid information from which to assess the true situation. The cost of extensive failure analysis to get the clear visibility desired could cost more than the ESD control measures under question. Many justifiable reasons can be cited for the lack of good visibility of the extent of losses from static electricity. Table 14.1 lists several such factors that were discussed in Chap. 2 on ESD awareness. The natural inclination drawn from these considerations is to completely doubt or at least minimize the estimated extent of potential ESD problems.

## 14.4 Relating to ESD Experiences of Others

Many look to the literature for ESD experiences of other companies as justification for ESD controls to be instituted at their own facility. The

**TABLE 14.1 Reasons for Minimization Estimate of ESD Problem Magnitude**

Most failures occur below the threshold of human sensitivity.

Static often affects many part types with no apparent failure trend.

ESD failures, likely at first electrical test, are often not analyzed.

Under cursory failure analysis static damage is usually invisible.

Experience has taught many that improper practices are harmless.

Static failures can occur through subtle rather than direct paths.

many case histories in Chaps. 10 and 13 clearly indicate that others have had extensive and costly static problems. The approach of relying entirely on the experience of others, therefore, tends to direct one to the most thorough and possibly extreme measures. An objective determination can be made only when self-assessment predominates, that is, analysis of circumstances related to one's own facility and product.

## 14.5 Determination of ESD Control Needs

In recognition of the opposing opinions of how big the ESD problem is, it may appear difficult to decide the optimal amount to dedicate to controlling or eliminating it. Pertinent considerations must be put into perspective in order to provide guidance in this regard. There is no universal set of controls that is best suited for all. Each application must be assessed of its own individual factors to be economically sound, yet effective.

### 14.5.1 Known susceptibility level extremes

Operations with extremely few static-sensitive items can minimize the workstations and other areas requiring control by segregation of processes involving those items. Operations with a high percentage of items having known histories of extreme ESD sensitivity, on the other hand, will require extreme control measures throughout. In either of these cases appropriate allocation of funds for control measures can easily be justified to management. Thus ESD control decisions are straightforward for those with known item sensitivities at either extreme. However, without prior knowledge of susceptibility levels, the expenses required for categorization testing are described in Chaps. 8, 16, and 17 as frequently being prohibitive.

### 14.5.2 Unknown or moderate susceptibility levels

Table 14.2 is intended to provide guidance in deriving the appropriate level of controls necessary for a cost-effective program tailored to a given application. It is apparent from the table that precise quantification of failures and sensitivity levels is not the only factor affecting the degree of controls necessary. Those considerations listed must be evaluated by management to arrive at a designated ESD control program with cost effectiveness. The intent of Table 14.2 can be summarized as follows: *Each measure instituted must be judged on its merits as compared with the risk of ESD failure without the control as well as the impact of failure.*

TABLE 14.2  Considerations to Derive Proper Extent of ESD Control Measures

| Consideration | Result | Effect | Tendency of control extent |
|---|---|---|---|
| Susceptibility levels* | Class 0 | Acute degree of failures | Rigorous |
| | High | High degree of failures | Thorough to rigorous |
| | Low | Few failures likely | Minimal to none |
| | Moderate | Moderate failure extent | Moderate |
| | Unknown | Unknown | Reactive to customer demands |
| Percentage of susceptible items | High | High degree of failures | Thorough to rigorous |
| | Low | Few failures likely | Thorough in segregated area |
| | Moderate | Moderate failure extent | Moderate to thorough |
| | Zero | No ESD failures | None |
| | Unknown | Unknown | Reactive to customer demands |
| Susceptible item cost | High | High cost of failures | Thorough to rigorous |
| | Low | Low cost of failures | Minimal to none |
| | Moderate | Moderate failure cost | Moderate to selectively thorough |
| | Unknown | Unknown cost effect | Reactive to customer |
| Criticality of mission | Space | Severe consequences | Rigorous controls warranted |
| | Life | Severe consequences | Rigorous controls warranted |
| | Military | Serious consequences | Thorough to selectively rigorous |
| | Commerce | Moderate consequences | Moderate to selectively thorough |
| | Consumer | Trivial consequences | Minimal to no controls |
| Customer rules | Severe | Customer dissatisfaction | Thorough to rigorous |
| | Moderate | Customer dissatisfaction | Moderate |
| | Minor | Customer dissatisfaction | Minimal |
| | None | Customer dissatisfaction | None |

*Note:* Susceptibility categorization methods and levels may vary at different facilities. The following are typical susceptibility ranges as determined by the standard 100-pF, 1.5-k$\Omega$ HBM test circuit:

Class 0 — < 150 V
High — < 2000 V
Moderate — 2000 V ≤ 4000 V
Low — > 4000 V
Insensitive — > 16,000 V

**14.5.2.1 ESD control program levels.** At the outset it is necessary to define the level or degree of the overall ESD control program to be implemented for a given operation. Before attempting to discuss this determination several definitions of ESD control level categories are in order. The following are intended to give a cursory overview of the control program categories, with greater details to be found in Chap. 15:

1. *Minimal:* A minimal ESD control program consists of a few selected control measures usually instituted at work areas having a known ESD problem history. The extent of these controls beyond personnel grounding is generally determined by the extent and cost of continued ESD failures.
2. *Moderate:* A moderate ESD control program consists of the four

basic good practices of personnel grounding, grounded work surfaces, purge of static generators, and protective packaging. Details of these control practices are discussed further in Chap. 15. The controls are implemented wherever susceptible items are handled, and documented procedures are maintained.
3. *Thorough:* A thorough ESD control program consists of all the measures of a moderate program with supplemental measures and better discipline. Supplemental measures might include the use of humidity control, ionizers, ESD awareness training, and some redundancy, such as conductive flooring in addition to wrist straps. Periodic audits and calibration checks of ESD control equipment would be conducted and documented.
4. *Rigorous:* A rigorous ESD control program consists of a well-planned and -documented set of regulations that are adamantly followed. The control measures of a thorough program would be implemented with additional measures, greater redundancy, attention to detail, and even more stringent discipline.

### 14.5.3 Criticality of mission

Criticality of mission is one item listed that can be the sole criterion to determine the degree of controls necessary. The possibility of a latent failure due to ESD was substantiated in Chap. 11. For each application the criticality must be weighed carefully through considerations of cost of failure in terms of human injury, financial loss, and customer image.

The consequences of such a failure in the case of critical space or life-dependent applications warrants thorough to rigorous ESD controls. This does not mean utilization of unnecessary or ineffective measures, although redundant control materials and well-disciplined checks and balances are often required. Chapters 15 to 17 discuss the means of selecting and evaluating control measures and materials.

End usage with comparatively trivial consequences of failure such as in computerized toys or other electronic gadgetry does not warrant thorough controls. In such an endeavor ESD controls could be completely avoided unless justified by other factors in Table 14.2. Some measures are generally found to be cost-effective in increasing factory yields by eliminating excessive in-process failures.

Military applications have varying degrees of criticality, but in general are of a serious nature. The overall control level would be anticipated to be in the thorough to rigorous range, with a leaning toward the latter.

Missions of "commerce" items are those products used in commercial operations where failures would have consequences that are moderate to high in terms of cost and other detrimental effects. For example, digital control circuitry might be used in applications that range from washing machines to the monitoring of power plants or large production facilities. Thus, the control program will consist of measures that are in the moderate to thorough category range, with a tendency toward the former.

### 14.5.4 Customer requirements

Customer requirements might also dictate the extent of ESD control. Usually customer requirements are related to the customer's perspective of criticality of mission. Controls imposed merely to meet a customer specification, however, are often not enforced with the proper commitment. They can be viewed as a stopgap until the contractor becomes inherently concerned about ESD.

### 14.5.5 Assessment of other factors affecting ESD control

Those contractors who do not have a critical mission or customer-defined requirements are left with the other considerations to weigh. Of these, high cost of failures seems to be the easiest to resolve. In practice even the champion doubters can be seen taking great care to wear a wrist strap when handling a $100,000 hybrid or very large-scale integragion (VLSI) device.

The considerations remaining to be evaluated are

1. Susceptibility levels of items processed and of items delivered
2. Percentages of susceptible items of each classification
3. Cost of susceptible items
4. Criticality of mission other than critical and trivial
5. Customer regulations other than thorough or rigorous

The first of these five items is the most important. After all, the second and third considerations are dependent on having classified items by their sensitivity. Once susceptibility levels have been determined, relative percentages and costs of each are easily determined. The fourth and fifth considerations tend toward neutrality, again pointing toward classification of susceptibilities as the primary task at hand.

#### 14.5.5.1 Determination of susceptibility levels.
Susceptibility levels can be determined to varying degrees of accuracy by several methods:

1. Rigorous testing as per Mil Std 883, Method 3015
2. Data base references such as RAC Vzap 2

3. Mil standard 1686A, Appendix B
4. Vendor supplied levels
5. Information supplied by peers
6. Generic part type generalization
7. Engineering analysis and estimates
8. Information from literature

Data by all but method 1 are rather easily obtainable and should be judiciously utilized in lieu of actual testing. The present aim is to reach a general decision as to the extent of ESD controls needed. Subsequent refinement of the categorization may be deemed necessary for decisions related to specific ESD control options. Therefore a great deal of precision can be sacrificed for the initial determination of whether controls are to be minimal, moderate, thorough, or rigorous. Problems associated with methods 1 to 3 are discussed in greater detail in Chap. 16. Nevertheless, a cursory but reasonably accurate categorization can be accomplished by referring to available data bases such as Vzap 2, DOD Std 1686, Appendix B, and additional generic extension of categories. Special cases for new technologies or other items not included in these two references can be handled by vendor or peer contacts, engineering analysis, or actual susceptibility testing.

A cautionary word on engineering analysis points out that historically design engineers have tended to underestimate the susceptibility of generic part types. It is anticipated that this tendency will change realistically with time and increased knowledge of ESD within the industry. In the meantime there is no real substitute for actual testing of the part in question to resolve specific sensitivity questions.

### 14.5.6 Summary of final determination of extent of controls

The control level determination outlined in Secs. 14.5.1 to 14.5.5 is summarized in Table 14.3. The information is intended only as a guide for management decisions based on the factors listed. This tabulation clarifies that many ESD control level decisions are easily determined, whereas a few cases will be judgmentally decided.

### 14.6 ESD Control Management

Of foremost importance is selection of a person to oversee all ESD control-related activities. Table 14.4 lists such activities and the disciplines or specialties that need to be involved. The table indicates the

TABLE 14.3 Summary of Determination of Extent of ESD Control Measures

| Mission criticality | Customer requirements | Susceptible levels | Percentage susceptible | Susceptible item cost | Recommended controls |
|---|---|---|---|---|---|
| Any | Any | None | 0% | N/A | None |
| Any | Any | Class 0 | >0% | N/A | Rig.* |
| Space | N/A | Susceptible | >0% | N/A | Rig. |
| Life | N/A | Susceptible | >0% | N/A | Rig. |
| Any | High | Susceptible | >0% | N/A | Thor. to Rig. |
| Any | Any | Susceptible | >0% | High | Thor. to Rig.* |
| Military | Any | High | >25% | N/A | Rig. |
| Military | Low | Low | N/A | N/A | Min. to Mod. |
| Military | Low | Moderate/High | >25% | N/A | Mod. to Rig.* |
| Military | Low | Moderate/High | >50% | N/A | Thor. to Rig.* |
| Military | Low | Moderate | <10% | Low/Moderate | Mod. to Thor.* |
| Military | Low | High | <10% | Low/Moderate | Mod. to Rig.* |
| Military | Moderate | Low/Moderate | >0% | Low/Moderate | Mod. to Rig.* |
| Commerce | Low | Moderate/High | >50% | N/A | Thorough |
| Commerce | Low | Low | >50% | High | Thorough |
| Commerce | Moderate | Moderate/High | >10% | Low/Moderate | Moderate |

*Note:* The more stringent controls are adapted selectively at locations where the need exists.

need for an ESD control director to coordinate all activities and to be the focal point for the more controversial aspects of the program. A recent industry survey indicated that about one-half of the large electronics and defense contractors are beginning to recognize this need.

For larger facilities one or more full-time ESD coordinators may easily be justified, while smaller sites may support the part-time efforts of a person for this effort. The remainder of this chapter is intended to help guide the ESD coordinator in dealing with the many tradeoff decisions in management of an effective and economical control program. In those cases where the ESD control director has limited knowledge of the subject, additional support through qualified ESD consultants is highly recommended because of the many subtleties requiring specialized knowledge.

The diverse areas requiring leadership of a competent ESD control director are indicative of the broad technical base required. Unfortunately managers sometimes erroneously see this assignment as an extremely narrow discipline and deal with it accordingly. After all, the problem is seen as just one of many, and prevention should be simply a matter of committing funds for some control materials such as wrist

TABLE 14.4  ESD Control Activities with Participating Disciplines

| Activity | Primary discipline | Secondary disciplines |
|---|---|---|
| Design protection | Electrical design | Mechanical design; reliability; parts<br>ESD control director |
| Susceptibility classification | Electrical design | Reliability and parts engineering<br>ESD control director |
| ESD control Policies | ESD control director | All disciplines and upper management |
| Control equipment and materials selection/evaluation | ESD control director | All disciplines as applicable |
| ESD awareness training | Training department | All disciplines and ESD control director |
| ESD control Problem resolution | ESD control director | All disciplines as applicable |
| ESD audits | ESD control director | Manufacturing, QA, other area managers |
| Part vendor control | Parts engineering | ESD control director, QA, reliability |
| Subcontractor control | Quality Assurance | ESD control director, reliability |

straps, work surfaces, and special packaging. As general knowledge increases on the subject, ESD control is likely to emerge as a recognized specialty after growing pains are endured as with quality assurance and/or reliability engineering in past years.

## 14.7  Convincing Upper Management

Chapter 2 covered methods for convincing management and others of the existence of the ESD problem. Detailed presentation of the facts uncovered in Secs. 14.1 to 14.6, with emphasis on cost factors, will show that the recommended level of ESD control has been thoroughly researched. This is usually sufficient to convince management that the controls are warranted.

## 14.8  Establishing the ESD Director Position

The discipline of ESD control as a career specialty is a definite need that has arisen with the emergence of the high incidence of ESD susceptibility in modern electronics. Yet, since the need is new, managers are likely to raise several questions concerning the establishment of positions to meet this need:

1. Why can't our line supervision manage ESD control?
2. If a specialist is required, should it be a full- or part-time position?
3. What level of the career ladder should the position of ESD director be?
4. What are some desirable qualifications for this position?
5. Through which functional group manager should the ESD director report?
6. What additional personnel support will be required?
7. Is the cost in salary justified?

### 14.8.1 ESD control by line supervision

Line supervision plays a major role in attaining effective ESD control. They must ensure that their personnel are properly trained, the ESD-related documentation is current and strictly followed, and the ESD control materials are monitored and calibrated as needed. However ESD is just one of many problems encountered in the day-to-day endeavors of line supervisors. They cannot be expected to keep abreast of the fast-moving advances in the field of ESD control. The establishment of requirements as well as selection and proper monitoring procedures for control materials are best left to a specialist in a functional support group.

### 14.8.2 Full/part time specialists

Support from a specialist(s) is needed for continued maintenance of any ESD control program. The amount of dedicated effort depends on many of the factors addressed in Secs. 14.1 to 14.6 coupled with the size of the facility and cost of controls.

**14.8.2.1 Cost of controls as basis for ESD personnel.** To accurately determine the extent of ESD problems is difficult or impractical for most situations. Regardless, the industry has moved toward institution of ESD controls as driven by customer demands, desire for peace of mind, yield enhancement endeavors, or other dictates. The cost of controls is often substantial, especially when done without dedicated ESD personnel. Annual expenditures for such items as protective bags, tote boxes, wrist straps, and work surfaces can easily reach millions annually for large plants with extensive use of integrated circuits. The amount allocated for these and other controls is based on criticality of mission, customer demands, item susceptibility, and cost of failures. Since these important factors were accounted for in determining the extent of controls, they need not be reexamined. A direct correlation to personnel needs is inherent if they are based on the cost of already justified ESD controls. A suggested minimal level of effort is 10 percent of the cost of controls.

### 14.8.3 Position level of ESD director

The ESD director needs to be a management level position in order to

1. Communicate effectively with management in other disciplines
2. Direct the efforts of management level ESD control committee members
3. Initiate timely corrective action measures
4. Manage other ESD specialists
5. Coordinate ESD activities of other disciplines
6. Deal effectively with customer management in resolving ESD conflicts

### 14.8.4 Position qualifications

Ideally the candidate would possess the following characteristics:

- A degree in electrical engineering or physics
- Several years of hands-on experience in areas of electronics such as failure analysis, quality assurance, and/or ESD studies
- Broad knowledge of the ESD control field
- Professional recognition within the EOS/ESD community
- Management experience

A candidate deficient in ESD control knowledge or management experience should be considered only if adequate support can be provided until the deficiency is corrected. Such support can be obtained from other employed personnel or from outside consultants. If more than one person is to be dedicated to ESD control, then complementary qualifications can be selected to best fulfill the needs.

### 14.8.5 Functional group for ESD director

Likely candidates for the position of ESD control director are those champions of ESD control who have emerged within the industry to fill this critical need. Such individuals are typically from disciplines where static problems have been encountered firsthand such as failure analysis, quality assurance, test engineering, reliability, manufacturing, and others. Many have had difficulty convincing their immediate managers of the need for their continued full-time efforts to direct ESD activities. The perspective of the managers is understandable. Whatever functional group is involved, ESD control management has not been historically within the group's area of responsibility. Furthermore, the individual middle manager is not likely to see the need for a full-time director. If the need is seen at all, there is no apparent incentive for the group manager to undertake the task, possibly at the expense of other responsibilities. The decision to devote personnel to ESD control is generally best made at a higher management level.

**14.8.5.1 Attaining upper management decision on ESD personnel assignments.** Upper management must be made to recognize the need for devoted ESD personnel by effective presentation of pertinent data. Generally, upper management commitment comes from a two-step process: being made aware of the severity of the ESD problem as discussed in Sec. 2.4.2; and subsequently becoming convinced of the need for a centralized ESD coordinating activity. The awareness presentation, done properly, will usually bring about the institution of control practices. In most cases the control program is established with temporary, part-time committee efforts. Without dedicated attention, costly mistakes are likely to be made.

The second step of the upper management commitment process is best accomplished through a one-on-one meeting with the appropriate manager. At this point the decisions are going to evolve primarily around long-term economics, and the manager will not depend on staff support for technical matters. If ESD control infractions were a frequent target of customer audits, the subject may have been a "thorn in the side" and the manager may already be primed to listen. A suggested agenda follows:

1. Problems cited in past customer ESD audits
2. Cost of ESD controls
3. Decentralized ESD control decision making
4. ESD control activities
5. Lack of detailed knowledge resources being applied to control decisions
6. Cost versus knowledge versus effectiveness relationship
7. Examples of costly ESD control errors
8. Future ESD concerns
9. Relationship to customer image and future business
10. Recommendations with return on investment (ROI) estimate

The preceding items can be organized and presented in several ways, but they could make a good justification for a centralized ESD activity. Examples of presenting some of the items are discussed in ensuing paragraphs. With upper management support the assignment of ESD control to a functional group would depend on the organizational structure. Candidate groups would include operations, quality assurance, reliability engineering, parts and materials, or manufacturing.

## 14.9 Additional Personnel Needed

Additional ESD specialists reporting to the ESD control director may be warranted based on the analysis of cost of controls. These personnel could take on such endeavors as ESD audits (both internal and at subdivisions, subcontractors, and vendors), evaluation of control materials, problem investigation and resolution, and ESD-related research. Since ESD control is virtually everybody's job, each functional group and off-site facility will be required to have at least one management level member of an ESD control committee chaired by the ESD control director. Such a committee might meet weekly during initial control policy generation or for problem resolution. Once a program is successfully under way, the frequency of meetings could taper off to semiannual occurrences.

## 14.10 Application of Knowledge to ESD Control Decisions

Many subtle problems can occur due to static electricity. ESD control products are sometimes constructed with improper materials and/or ineffective designs. The state of the art is rapidly changing. New problems with ESD control equipment arise frequently. As already discussed several times, widespread misconceptions about ESD and ESD control are the norm rather than the exception. Control product vendors and customers cannot always be relied on for proper guidance in ESD problem resolution. Control measures are expensive, and improper application can be even more costly. The point is that the best available knowledge base must be behind ESD control decisions. This will happen only when upper management is made aware of the many pitfalls and possible overexpenditures that can result without a concerted effort in this regard. This may be the best argument for the establishment of an ESD control director position.

Figure 14.1 is a hypothetical curve indicating the growth of overall ESD awareness from a time of several years ago, then $(T)$, until now $(N)$. Overall ESD awareness means the attainment of a belief that the ESD problem is real and significant on the part of the vast majority of

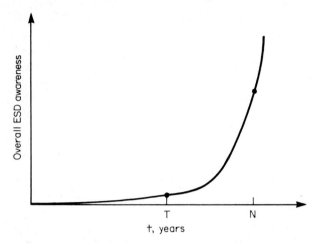

**Figure 14.1** Hypothetical curve indicating the growth of overall ESD awareness from a time of several years ago, then ($T$), until now ($N$).

those in the electronics industry. For the most part it must be agreed that widespread publicity has brought about this awareness growth. Figure 14.2 is a hypothetical curve showing the growth of ESD knowledge among this same electronics populace. Because of the increased awareness and customer demands, ESD controls, and thus expenditures, have grown considerably. For those facilities without dedicated ESD specialists, the expenditures are decided by various members of the general populace, that is, people from various disciplines requiring control measures, but without detailed knowledge of the subject. Assuming control decisions by such personnel with a knowledge base according to Fig. 14.2, the expenditures and control effectiveness are likely to follow a relationship similar to that shown in Fig. 14.3. With

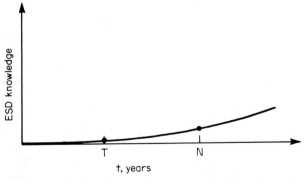

**Figure 14.2** Hypothetical curve showing the growth of ESD knowledge among the general electronics industry populace.

ESD Control Management 377

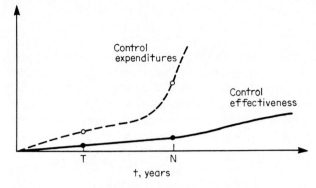

Figure 14.3 Likely growth of ESD control expenditures and effectiveness assuming decisions made by persons with ESD awareness and knowledge as shown in Figs. 14.1 and 14.2.

a knowledgeable centralized ESD control organization behind each decision, the knowledge, expenditures, and effectiveness are apt to follow a much more cost-effective pattern as shown in Fig. 14.4.

## 14.11 Return on Investment from Assigned ESD Personnel

The recommended approach in this chapter is to base funding for ESD on the cost of controls as determined after careful analysis of all per-

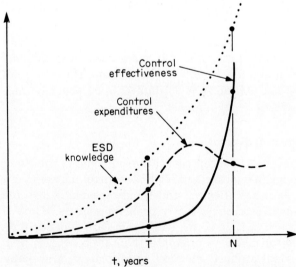

Figure 14.4 Anticipated relationship of control expenditures and effectiveness where decisions are directed by a centralized specialty group with a good knowledge base.

tinent factors. A minimum of 10 percent of the cost of controls was suggested. For justification let us compare events that might be occurring on a given day at a hypothetical electronics manufacturing facility first without, and then with, an ESD director. The plant in question delivers large systems requiring the annual use of about 5 million integrated circuits to produce about 45,000 printed-circuit boards. Customer requirements have already demanded an ESD control program in place. With all the processing of susceptible parts and assemblies at a given time the following are in use:

- Over 200,000 antistatic bags and about 20,000 metallized bags
- About 6000 conductive and 48 antistatic tote boxes
- 800 "static controlled work stations" in several locations, including "grounded" work surfaces and wrist straps
- About 3000 DIP tubes of conductive, dissipative, antistatic varieties and a few ordinary plastic varieties
- Some special tiles, floor mats, and carpets
- Some special chairs and chair covers
- Conductive trays
- 2500 special smocks and other garments
- 500 ionizers
- 300 wrist-strap checkers

Other items available include: 104 static field meters of various types, 20 resistivity meters, and assorted other items including waxes, sprays, lotions, and creams. In addition to the items in use there are in stores about 5 million antistatic, 300,000 metallized, and 200,000 static dissipative bags and a few hundred wrist straps. The cost of all these items, including maintenance, is estimated to be about $3 million annually.

### 14.11.1 With no ESD control director

The following are believed to be a representative outcome in the event of no assigned knowledgeable focal point for ESD control matters. Virtually all areas of a plant are affected, and the discussion is directed toward some of the major and most costly problem sources. Some of the hypotheses are almost sure to occur, while some are dependent on too many interrelated variables to estimate their probabilities. Primary concerns are control product selection, qualification, and maintenance; awareness training; handling procedures; problem occurrence, detection, and resolution; schedule slippage; overall ESD control policy; and cost factors.

**14.11.1.1 Control item selection.** Control items would be selected using the following criteria:

1. ESD control items will be selected to fulfill perceived customer requirements, from vendor sales pitches, and through misconceptions.
2. Personnel with virtually no knowledge of the subject order antistatic bags, totes, or even shunts and work surfaces of a certain color, with no other stipulation.
3. Antistatic shunts and work surfaces ordered (even if within the antistatic resistivity range, the materials are not sufficiently conductive for these applications).
4. Ignorance of the differences between conductive, dissipative, and antistatic materials may lead to an assortment of products perceived as equivalent, when each has its own proper application.
5. Generally items are purchased from local product representatives. These representatives often carry ESD items as a sideline and are not knowledgeable of the product or need.
6. Often assorted items such as chairs; conductive tiles or mats; antistatic carpets; antistatic waxes, creams, and solutions; and ionizers are purchased with improper application intended.

**14.11.1.2 Control product qualification and maintenance.** Control product qualification and maintenance would consist of the following:

1. There are no qualifications, incoming tests, or maintenance performed. Products are perceived to be cut-and-dried visible measures that will eliminate static problems and will not degrade.
2. With no qualification or incoming tests, static-generating "antistatic" materials may be received and used, possibly for several years.
3. Some do not check wrist straps; some check monthly or less frequently, and sometimes improperly.

**14.11.1.3 Awareness training.** Awareness training would be limited to the following:

1. ESD training consists of viewing one of several available videotapes on the subject, with about 5 to 10 min of additional discussion.
2. Misconceptions about static principles and control product usage is evident on the part of factory operators, management, engineering, and sometimes the awareness instructors.

**14.11.1.4 Handling procedures.** Handling procedures reflect improper control item selection and the lack of ESD awareness. The many myths discussed in Chap. 2 are probably widely believed and will not be dispelled until proper training takes place. Several of the following poor practices inviting ESD failures were discussed in the factory examples of Chap. 13:

1. Susceptible items are carried in unlidded conductive trays. This common error can lead to static failures by induction if the tray is placed on an insulator and because the tray is at a common potential with the carrier.
2. Metallized bags and antistatic items are used beyond their effective useful life.
3. Antistatic and conductive DIP tubes are interchanged as if equivalent.
4. Receiving inspection interprets a static caution label, or the word *antistatic* on an incoming product as indicative of proper packaging as well as susceptibility. The lack of such visible notation is taken as an indication of nonsusceptibility. (In fact generating packages are often labeled antistatic.) The classification of and subsequent handling of susceptible items should not be decided on the basis of its incoming package from a particular vendor.
5. Improper incoming packaging is maintained throughout much of subsequent processing.
6. Sometimes the incoming packaging integrity is immediately compromised, as operators do not recognize the external shielding provided by vendors and retain only the internal antistatic DIP tubes or bags.
7. Ionizers may be looked on as a panacea and improperly used in lieu of personnel grounding.
8. Static control may be overdone by utilization of layers of redundant measures as well as separate ground systems with large copper ground buses to bleed away less than a milliampere. Such procedures are sometimes based on the belief that most integrated circuits fail at static potentials of 10 V or so.
9. The improperly selected control items as discussed in Sec. 14.11.1.1 are used throughout the facility.
10. The list could go on, but errors of the types listed are sufficient to make a case for cost avoidance.

### 14.11.1.5 Problem occurrence, detection, and resolution

**14.11.1.5.1 Occurrence.** The examples discussed in Chap. 13 are illustrative of many failure incidents that might occur at such a facility. Sometimes piecemeal static control efforts are worse than none. For example, failures can result from induction effects on unlidded conductive trays or from unbalanced ionizers. Open or intermittent wrist straps that are not checked properly if at all could result in failures. An ungrounded operator is potentially more dangerous at a grounded than an ungrounded workstation. Improper grounding could result in higher probability of failure but also presents safety concerns.

Consider the tremendous consequences of ineffective ESD control materials. Since these items are not likely to have been selected properly, it is extremely likely that some are ineffective and some may actually be causing problems. Suppose the antistatic bags were found to generate static. Not only has the cost of about $1 million dollars been wasted, but the failure incidence may have increased as well. The same fears relate to DIP tubes, for which static generation has been a common problem when not monitored. Improperly selected garments may also be generative or sources of conductive contamination that could result in shorts. With such poor operator (and supervision) training, it would not be surprising to find unnecessary static generative materials at the workstations. Without a director or knowledgeable ESD focal point the controls in place are most likely to fall behind the state of the art, be ineffective, and lead to a compounding of present and future problems.

**14.11.1.5.2 Detection.** Problems with such products are most likely to be detected by major customers at their incoming inspection. This will lead to possible conflicts, poor image, and loss of future business. If such problems are detected in house it is likely the result of costly failures caused by ineffective control measures. Difficulties in identifying ESD as the cause of failure was covered in detail in Chaps. 9 and 10 and latent failures in Chap. 11.

**14.11.1.5.3 Resolution.** The tendency is to "resolve" ESD-related problems by committees of many manufacturing and quality assurance management personnel rather than under the direction of a person knowledgeable in ESD control. This can result in corrective action taking months rather than days without the influence of a knowledgeable information base. The poor control scenario described is likely to result in many problems, and such meetings are likely to be held on a continual basis, involving different people. The meetings may be at the same or different locations and may address the same or different problems. The cost factor is further compounded as these management attendees assign a second tier of underlings to investigate specific aspects of the problem; that is, reinvent the wheel. Delivery schedules can be upset from awaiting problem resolutions. Finding solutions to problems with ESD control materials and/or procedures can be as time-consuming as arriving at corrective action for hardware failures.

**14.11.1.6 Cost factors.** The factory ESD control policy varies from area to area, and a wide variety of control items have been selected with little knowledge of important parameters. If one were to account for the cost of the ineffective materials purchased, resulting ESD failures, lost future business, and poorly directed meetings of highly paid individuals discussing these problems, it might be easy to justify several ESD specialists.

### 14.11.2  With ESD control direction

A cursory estimate of the return on investment can be gotten from modifying assumptions of Sec. 14.11.1 to what is more likely under proper direction. Keep in mind that it makes little difference if a contractor has knowledgeable ESD experts employed at tasks in areas other than ESD. Without such a focal point other company employees are left to their own wits as to how best to control static electricity. The cost of controls at $3 million per year would justify $300,000 for several ESD control people. An ESD control director, with three or four full-time support personnel, could have done the following:

1. Ensured proper specification of each control item's electrical parameters
2. Operated from a base of knowledge of the product and the reputable vendors
3. Qualified control items for initial acceptance for usage
4. Overseen incoming inspection and evaluations on the product
5. Taken timely corrective action in the event of any problem
6. Established a proper training program
7. Seen that measures are consistent with the state of the art
8. Adopted an overall ESD control policy

## 14.12 Establishing an ESD Control Program

The guidelines contained in this chapter are intended to be utilized by individuals attempting to establish a realistic control program independent of customer-dictated regulations. Needless to say customer requirements must be met or, if inappropriate, renegotiated. The approach that follows has proven to be successful after other approaches have failed. It is assumed that the control plan is being initiated by a champion of the cause, hopefully the ESD control director. This individual should develop a plan that will involve other appropriate disciplines in decision making and is therefore likely to win broad acceptance and cooperation. The items that follow should be included as befits the needs of the facility:

1. Outline for development of program developed
2. Extent of ESD control determined by criteria per Sec. 14.5
3. Results presented to upper management
4. Funds committed for program development
5. ESD champion designated (ideally ESD control director)
6. ESD control committee formed
7. ESD control policy and top-level requirements formulated
8. Lower-level operating procedures developed
9. ESD awareness training initiated
10. Periodic ESD control committee meetings held

### 14.12.1  ESD control committee

The ESD director, regardless of the extent of knowledge of the subject, must enlist the help of other disciplines in formulating a top-level requirements document for the company or plant. There are many details, perhaps involving factors other than ESD, that are affected by ESD control decisions. Management personnel from the work areas and disciplines involved are best suited to consider tradeoffs involving human comfort, economics, aesthetics, psychology, safety, and discipline matters. For these reasons and for the enhanced likelihood for acceptance and future cooperation, an ESD control committee made up of management-level members from all appropriate disciplines is recommended. The ESD control director is normally best suited to chair such a committee. The purpose of the committee is to formulate the ESD control policy and top requirements document, review lower level operating documents, help resolve future problems, and provide a forum for continued communications.

### 14.12.2  Requirements documentation

A top-level ESD control document is needed to describe a range of acceptable parameters for all ESD control products and procedures to be implemented throughout the facility.

This document is to be developed with the aim of being useful to those formulating lower-level working or operating procedures. Since each operation has a different set of variables to consider, latitude for differences in control measures must be provided. This will become more apparent after more detailed discussion in Chap. 15.

Each operational group will tailor a working-level document written in terms specific to its area. Technical terminologies will be dropped in favor of descriptive terms familiar to the operators, such as #3 blue tote boxes, or pink-poly envelopes. Committee review will ensure compliance with the top-level document. ESD awareness training should be given to all personnel including management prior to instituting each operating procedure. Additional refresher and makeup training sessions must be required to assure effective discipline.

## 14.13 Periodic ESD Committee Meetings

The ESD committee should meet weekly during the urgent task of developing the top-level requirements document and initial operating-level specification reviews. With the completion of these tasks the committee, or subcommittees thereof, should be required to meet only occasionally for problem resolution. The committee should at a minimum schedule semiannual meetings for communications and continued interest in ESD control. Many have found it worthwhile to name ESD facilitators in each working group. These people are designated to take a special interest in ESD and help in on-the-job training and in self-monitoring controls in their work area.

# Chapter 15

# ESD Control

## 15.1 Introduction

ESD control is the means used to control the interface of susceptible items with sources of static. Control of this interface can be by compartmental segregation, shielding, or maintaining distance; eliminating generative materials by antistatic treatment, replacement, removal, or ion neutralization; design hardening; and strict grounding and other disciplined procedures. This chapter discusses the uses of ESD control measures and some common problems likely to be encountered. The specifics of an ESD control program must be evaluated on economic and effectivity merits. Therefore each facility or operation must tailor control to its needs using information gathered in the analysis outlined in Chap. 14.

## 15.2 Tailoring Factory Controls Based on Susceptibility

In the event that some of the major factors such as customer requirements, overall product costs, and reliability requirements are high, then ESD control measures might be based on susceptibility levels. Kirk[1] described just such a progressive approach to ESD control. Susceptibility levels were based upon DOD Std 1686 with the added "0 class" of devices failing below 171 V. Table 15.1 lists the controls that are added for each susceptibility level. With this approach a single top ESD control document can cover many departments or facilities.

An important point is to be made before getting into the details of ESD controls, primarily as they are applied in a manufacturing facility. ESD control affects the entire life-cycle costs of a product. For this reason appropriate attention and measures must be applied to areas of

TABLE 15.1 Progressive ESD Control System Based on Sensitivity Classes

| Class | Protection method | | |
|---|---|---|---|
| | Facility | Process | Personnel |
| 3 | | Sandblasting not permitted | |
| 2 | Caution sign at entrance<br>Conductive floors<br>Conductive chairs<br>Conductive carts<br>Grounded work surfaces<br>Grounded shelving<br>Grounded soldering tools<br>Grounded cleaning station | Protective packaging<br>Caution labels<br>Antistatic cushioning<br>Ionizers during potting<br>Grounded tools<br>Handling procedures<br>Troubleshooting procedures<br>Burn-in and life test* | Conductive footwear |
| 1 | | Nonconductors kept at 2 in distance<br>Environmental tests<br>Ionized air at receiving, inspection, and test | Special smocks and gloves |
| 0 | | Shielding bags<br>Ionized air | Wrist strap |

*At receiving inspection:
Class 0 — 0 to 170 V
Class 1 — 171 to 1000 V
Class 2 — 1001 to 4000 V
Class 3 — 4001 to 15,000 V

life cycle going well beyond production facilities and processes. Table 15.2 lists major areas where attention to ESD control is required.

It is strongly recommended that consideration be given to segregating susceptible items from nonsusceptible items where practical. In general ESD control measures are costly, and separation can reduce capital expenditures, control monitoring, and confusion.

TABLE 15.2 Life Cycle Areas of ESD Control Impact

Design, both conceptual and hardware
Parts, materials, and processes
Production, including quality assurance inspections and test
Purchasing of hardware and ESD control equipment
Vendor and subcontractor controls
Installation setup, repair, and maintenance
End use

## 15.3 Primary Controls

In its simplest form a static control program can be condensed to four primary control measures or good ESD practices:

1. Personnel grounding
2. Grounded work surfaces and equipment
3. Purge of static hazardous materials (static generators)
4. Protective packaging for susceptible items when transported or stored away from the work area with the first three controls

Primary controls are those measures essential to prevention of ESD failures. In general, they are applicable in some degree to any control program. The first three items define a controlled workstation, whereas the fourth, protective packaging, provides for transport and storage under otherwise uncontrolled conditions.

### 15.3.1 Complications involved with the four primary controls

As presented in the preceding paragraph, static control might appear to be rather simple. However, several details are taken for granted in this simplification, such as knowledge of which items are susceptible, trained operators, and full commitment to these four controls with no room for error. This requires good understanding of all the materials involved, possible inherent problems with them, or problems from improper usage. Nevertheless, all static control programs should begin with an attempt at accomplishing these four controls. The remainder of this and the next chapter deal with the many complications that arise along the way. To understand why the list of controls seems to grow endlessly consider a few additional items.

Personnel grounding is not as easy as it appears at first glance. What about: safety measures, grounding operators that must move about considerably, clothing, transients in the area, reliability of grounding apparatus, and even perceived psychological effects? Work surface and equipment grounding raises similar questions on safety concerning the ideal range of conductivity and best method of grounding. Ground loops created by indiscriminate grounding might bring about parametric electrical problems in certain equipment. How do you ground insulative equipment?

Purging static hazardous items is desirable, but it is usually impractical to remove every such item from the work area. Other means of controlling static generative materials must be pursued. Protective packaging comes in many types with varying degrees of effectiveness and even different purposes. Therefore protective packaging must be selected on the basis of many characteristics associated with the product, its environment, and packaging options available.

## 15.4 Secondary Controls

Secondary controls are defined as those controls required to support or ensure effectivity of the primary controls. These include written procedures, training, self-checks and audits, vendor/subcontractor surveillance, and occasionally additional control items. Any additional control items would be used to eliminate deficiencies in fulfillment of primary controls for a given application. Measures included might be extraordinary procedures, selective ionization, replacement materials, topical antistatic treatment, special garments, flooring, and floor covers or treatment.

## 15.5 Tertiary Controls

Tertiary controls are measures that are helpful but not essential, redundant, of limited or controversial effectivity, or regarded as extraordinary but desirable efforts. Desirable items that may be regarded as nonessential to a given operation might include rigorous sensitivity classification, hardware/drawing marking, motivational posters, disciplinary measures, qualification of control materials, corporate policies, ESD as a design review item, ESD research and development, professional association activities, and controls initiated for the sake of image enhancement in the customer's eyes. Redundant items might include conductive flooring, humidity control, and workstation ionization. Perhaps the most controversial item is room ionization.

## 15.6 Importance of Controls from All Three Categories

Separating ESD controls into primary, secondary, and tertiary categories is intended to help put the ever-growing variety of available measures into a relative perspective. This is not to indicate that secondary, or even tertiary controls are unimportant or to be disregarded. Secondary controls are defined as being required to support the primary controls. Thus secondary controls play an essential part in any control program. For that matter, the categories designated herein are not steadfast. Sometimes a secondary or tertiary control measure might be effectively shifted to primary importance due to individual circumstances. An example would be the use of an ionizer to neutralize an insulative item that could not be purged. The same ionizer would be tertiary only when used as a redundant measure.

## 15.7 Primary Control Considerations

The following paragraphs discuss considerations and problems associated with implementation of primary control measures.

### 15.7.1 Personnel grounding

Most ESD control professionals agree that the first item of defense against ESD is personnel grounding. This is usually accomplished by use of a wrist strap connected to ground. Where conductive flooring is in place, conductive footwear is also necessary to assure that the body is electrically common with the flooring. Although ionization is discussed in detail in subsequent paragraphs, a cautionary note is made at this point that ionization should not be regarded as an effective personnel grounding apparatus.

**15.7.1.1 Wrist straps.** Personnel grounding wrist straps are available with several band varieties including conductive plastic, bead chain, Velcro with metal skin contacter, expandable watch band, and conductive fiber. The strap itself is usually of multistrand wire but is also available in solid conductive plastic and nylon varieties.

**15.7.1.1.1 Safety considerations.** Most varieties of wrist straps include a series resistor, typically 1 M$\Omega$, for safety purposes. The reason for this resistor is to limit current through the body in the event of accidental contact with 120 V ac. The hypothetical scenario is that an operator touches 120 V with one hand while the other hand is tied to ground via the wrist strap. With no resistance electrocution could take place. To limit current to a safe level of 0.5 mA maximum, the resistance must be at least 240 k$\Omega$. The industry standard has been placed at 1 M$\Omega$, thus limiting current to 0.12 mA.

**15.7.1.1.2 Problems with wrist straps.** The most notable problem with wrist straps of all types has been open circuit (or discontinuity), both intermittent and permanent. For this reason most users have seen fit to buy wrist strap checkers to monitor continuity. Some have gone so far as to purchase instrumentation that does continuous monitoring and sounds an alarm when an open circuit or high resistance condition is detected.

Solid conductive plastic-type wrist straps were used extensively in the early seventies. These are now regarded as outmoded because of a number of inherent problems. They were prone to intermittent open circuits at sites of creasing in the ribbonlike material. In addition the series resistance provided is the integral resistance of the ribbon material itself, typically about 270 k$\Omega$ for a 4-ft wrist strap. Since these straps are uninsulated, a much lower series resistance can result when the strap touches ground at a short distance from the wrist. These straps thus compromise human safety, so their usage should be discontinued and dropped from any further consideration.

Intermittent skin contact of the loosely fitting bead-chain-type wrist bands has been reported. Probabilities are that loss of contact would exist for only a fraction of a second during which time appreciable static accumulation would be rare. Thus the bead-chain-type may be a reasonable and relatively inexpensive choice for most applications. They should not be used where susceptibility levels are less than 100 V or where severe ESD control requirements are imposed.

The metallic expandable watch-type bands present an electrical safety hazard if used around live voltages. A potential difference across two points on a highly conductive band could conduct high current resulting in a burn injury to the wearer. Since this type of band has been extremely popular, several manufacturers now produce an expandable band with an insulative exterior. This type of insulative band should be carefully evaluated for integrity and life of the insulator under worst-case simulated life conditions.

Velcro bands have been reported to sluff and contaminate electronic products. Many bands were made in a manner that skin contact was accomplished by a metal contacter just opposite the strap-to-band connection. These tended to lose continuity when the straps were stretched, especially if loosely fitting. This problem has been overcome by placing a second metallic contacter on the opposite side of the band or conductive fibers around the inner surface.

Expandable cloth bands are commonly used, usually with inner conductive fibers and a metallic contact opposite the strap connection. Looseness of fit or discomfort brings up the issue of size. Bands are now available in several sizes, typically small, medium, and large, from several manufacturers. Adjustable bands to fit all sizes have recently been developed.

The outer layer of the wrist often has a "dead skin" layer that can act as an open circuit. Dry skin can add to the problem of good electrical contact from band to wrist. This is particularly true in dry atmospheres. Antistatic creams are usually sufficient to overcome this problem. A similar problem of body hair acting as an insulator is also helped by antistatic cream. A good practice is to instruct operators to wear the wrist strap in such a way as to place the contact point at the underside of the wrist where hair is minimal.

**15.7.1.1.3 Recommended wrist straps.** The best assurance for good wrist straps is to select those that have been certified for compliance with EOS/ESD Std 1. When these are not available, user evaluations are advised. Some vendors give replacement warranties for periods of 1 year. This may be regarded as an indication of improved reliability.

**15.7.1.1.4 Wrist-strap connection to ground.** Many problems have occurred through improper connection of wrist straps. Wrist straps should never be connected to the work surface as a path to ground unless the surface is metal. Most work surfaces add far too much resistance for effective personnel grounding. Using alligator clips for connection to a work surface increases the problem through excessive contact resistance. The strap connection point should never be left to chance. In general, a separate hard wired path to a connecting socket or bus for the strap connection should be provided.

**15.7.1.2 Conductive flooring/footwear.** Personnel grounding can also be accomplished through conductive flooring used in conjunction with conductive footwear. Conductive flooring can be achieved by using special conductive tiles, mats, unsealed or conductively sealed concrete, paints, or other conductive floor coverings or finishes.

**15.7.1.2.1 Applications for conductive flooring/footwear.** Conductive flooring options are commonly used in applications such as:

1. Processes where operator mobility precludes effectivity of wrist straps, typically selectively located mats
2. New or renovated buildings where installation of conductive flooring is not a major expenditure
3. High reliability or other critical static controlled areas
4. Antistatic carpets for computer stations in order to limit voltages below thresholds that would cause upsets
5. Special finishes applied to ordinary or conductive tile to inhibit triboelectric generation

**15.7.1.2.2 Problems with conductive flooring/footwear.** The following paragraphs discuss some of the numerous problems that have been reported with various types of conductive flooring and footwear.

**15.7.1.2.3 Conductive tile problems.** The most common complaint about conductive tile has been the fact that ordinary floor maintenance, which includes waxing, renders the tile nonconductive and static-generative. This is not a fault of the tile but of poor discipline or lack of control of janitorial services. The second most frequent complaint is related; lack of the protective wax layer leaves the tile prone to marking from shoe soles and perhaps some of the conductive carbon particulate within the tile itself.

The most important issue with conductive tile is the resistance to ground. Improper installation can easily result in high resistance to ground and loss of effectiveness. The other concern is for safety if the resistance to ground is too low. Thus, qualification testing and periodic monitoring during usage is essential to the use of conductive flooring.

**15.7.1.2.4 Conductive mat problems.** Conductive mats are commonly black and can accumulate sufficient surface dirt and contamination to cause excessively high resistance with no visible indications. This problem and safety concerns justify periodic resistance to ground monitoring. Mats are now available in grid patterns as well as solid configurations. The grids allow much contamination to fall through the void areas and offer slip or skid resistance as well.

**15.7.1.2.5 Antistatic carpet problems.** Antistatic carpets are a good choice if the need is to limit human body potentials to a maximum of about 2000 V. Accordingly, they are commonly used in computer operations where upsets are to be avoided. This 2000-V limit is usually intolerable in manufacturing operations where integrated circuits and other susceptible items are to be handled. The most common problem has been misunderstanding and misapplication of these carpets in manufacturing operations.

**15.7.1.2.6 Floor finish problems.** Special floor finishes or waxes for ESD control typically have antistatic rather than conductive features. Usage for treatment of ordinary tile seems appropriate to inhibit triboelectric generation caused by contact-separation of the shoe and floor. However, many have found the antistatic effect to be short-lived. Walking has been shown to wear away the effective overcoat in just a short time, rendering the tile static-generative. This points out the need for periodic monitoring and perhaps qualification tests of floor finishes under consideration.

Another possible usage is a replacement wax for conductive tile. This has been reported to be effective in most cases. One caution is to keep in mind that the antistatic treatment will increase the resistance to ground and must be controlled. This again points out the value of periodic monitoring and qualification testing.

## 15.7.2 Equipment grounding

Equipment grounding refers to electrically connecting exposed conductive surfaces to ground. This has been a common practice for much electrical equipment for safety reasons. Thus the chassis of oscilloscopes, curve tracers, power supplies, temperature chambers, etc., are likely to be tied to ground internally when received. This is desirable also for electrostatic discharge prevention in that every conductive item including the human body can be at a common electrical reference and potential differences leading to ESD are avoided.

**15.7.2.1 Grounded work surfaces.** The first order of business concerning equipment grounding is to provide a grounded work surface. This is normally a bench top cover connected to ground. The first requirement is that the surface in use have sufficient conductivity to be groundable. The primary purpose for a grounded work surface is to effectively ground conductive items contacting the surface. Thus much other equipment becomes grounded by virtue of this work surface. It also provides a redundant path to ground for the operator at all times when the surface is touched. An additional benefit is assurance that the surface will not acquire a triboelectric charge.

Antistatic materials are not generally recommended for work surfaces because of the excessively high resistance, dependence upon humidity, and limited life. However it is conceivable that these materials might be found desirable in the event of frequent charged device failures. If antistatic work surfaces are used, frequent monitoring and topical antistatic treatment is recommended to maintain an acceptable upper limit on resistance to ground.

Static dissipative materials in the range of $10^7$ to $10^9$ are recommended for most applications. This range does not have the extreme sparking problems of conductive materials yet is of sufficient conductivity to effectively bleed off potentials.

Conductive work surfaces such as stainless steel are often favored because of durability, rapid charge bleed off, and other factors. An undesirable feature is the propensity for sparking when a sufficiently charged conductive item is brought near or in contact with the surface. Thus if conductive surfaces are used, extreme care must be taken to prevent charge accumulation on susceptible items to avoid charged-device-type failures.

**15.7.2.1.1 Further charged device failure considerations.** Bossard et al.[2] calculated the ratio of average power density to the Wunsch Bell density as a function of $R_p$ and $C_{device}$. The model used an assumed 2-nC charge, $L_d$ = 10 nH, and $R_d$ = 25 Ω for a junction area of 160 μm$^2$. Figure 15.1 shows the results for five different values of capacitance. The authors assumed a safe level was obtained when the power ratio was 0.001 or less. This relates to a work-surface resistivity of about $10^5$ Ω minimum. The authors cautioned that the resistivity must be

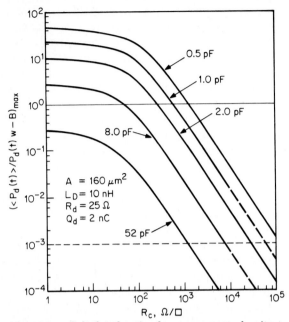

**Figure 15.1** Calculated ratio of average power density to the Wunsch Bell density as a function of $R_p$ and $C_{device}$ values. The model used an assumed 2-nC charge. $L_d = 10$ nH and $R_d = 25$ Ω. (*From Bossard,[2] with permission of ITTRI/RAC.*)

distributed rather than a lumped element. For instance a metal table top with a capacitance of >10 pF would effectively shunt any series resistance that might inhibit CDM failures.

I would like to add that conductive table tops need not be completely ruled out, as they have been used effectively by numerous manufacturers. However, the CDM threat with such surfaces must be addressed. A key factor is the requirement to strictly limit possible charge accumulation on susceptible items processed on conductive surfaces.

**15.7.2.2 Grounded miscellaneous equipment.** Miscellaneous conductive equipment, especially if likely to contact susceptible items, should be grounded. Storage shelving in particular can be a source of problems. Even where the more notorious generative materials such as Plexiglas and insulative foams are avoided, varnishes and finishes can be a problem source. The tips of soldering irons are to be grounded, mainly to prevent ac leakage, which can be a safety hazard as well as a cause of EOS. Vacuum-type solder-removing tools can generate several thousand volts at the tip. Tools are available with conductive tips that are either hard wire grounded or connected back to the hand of the supposedly grounded operator. The hard wired variety is recommended. Other items to be considered for grounding include work area laydown tables or shelving, temperature chamber trays or platforms, and subassembly or assembly chassis being manufactured or repaired.

### 15.7.3 Grounding considerations

Grounding of personnel and equipment is accomplished in various ways throughout the industry. Many of the variations devised are costly, ineffective, and sometimes unsafe.

**15.7.3.1 Separate ESD control grounds.** Many installations have been observed with so-called separate ESD grounding systems. These systems are generally shown with great pride, perhaps proportional to the tremendous cost involved. The initial expenditure is for separate buried grounding plates. This is followed by extensive connections through large copper buses, sometimes incorporating the structural steel beams of the building in the path. Then all "ESD controlled equipment" is carefully tied in with heavy gauge copper cables to this separate ground. It is extremely unlikely that such a ground system could be controlled and monitored sufficiently to keep it forever separate from third-wire common ground. As a matter of fact, one probably need not look far to find connections between the separate and common grounds. Often cables from chassis to the separate ground are in plain sight, while the chassis was already tied to common internally. Electrical items placed on conductive work surfaces are often tied to common ground to avoid leakage potentials on the chassis.

Why was the separate ground desired in the first place? Many assume that a very good low-resistance ground is needed for ESD control. This is obviously not the case since 1 M$\Omega$ is typically placed in series with the wrist strap, and grounded work surfaces may be as high as 100 M$\Omega$ above ground. The recommendation is that all grounding for ESD control be to third-wire common. Connections should be reliable; not dependent on conduction through enamel, loose screws, or clip leads. If third-wire common is found to be a poor ground, then it is essential that it be corrected for safety reasons.

**15.7.3.2 Conductive work-surface grounding.** Most static control manuals and guides, including DOD Hdbk 263, indicate that a resistor (typically 1 M$\Omega$) be installed between the work surface and ground. The reason given is to limit the current in the event of the operator accidentally touching 120 V ac with one hand while touching the work surface with the other.

In the case of highly conductive surfaces such as stainless steel, a different hazardous circumstance is believed to be more likely. Electrical apparatus is commonly used on such work surfaces. Leakage of 120 V ac might reach the work surface via frayed wires, floating chassis, or other sneak paths. With a high resistance between the conductive surface and ground, the surface will be held at 120 V. This presents a serious hazard in the event of an operator touching the surface and some point tied to ground such as the chassis of an electrical instrument. For this reason the safer option is believed to be a direct connection between the work surface and ground.

**15.7.3.3 Ground fault interrupters.** A ground fault interrupter (GFI) is an inexpensive safety item that detects the current imbalance brought about by a short circuit or leakage path. Such a condition causes current that would normally be conducted on the return wire to be diverted to third-wire common or another ground. Most GFIs will detect such a current at a level of a few milliamperes and will shut off power within several milliseconds. Ground fault interrupters are recommended for all ESD control personnel grounding systems, especially installations using conductive work surfaces.

Large capacity ground fault interrupters can be installed for each circuit breaker, or smaller units can be installed for one or two workstations. The biggest reported problem has been the nuisance factor caused by detection of cumulative low-current leakages or ground noise transients. The sensitivity allows circuit-breaker-type GFIs to shut down frequently when no real hazard exists. The best option may be to install GFIs at every station. Frequent shutdowns of an individual station are likely to be indicative of an electrical problem or poor ground and should be investigated and corrected.

### 15.7.4 Control of static hazardous materials

Static hazardous materials are those insulative materials prone to static accumulation and retention. Some of the more notorious include Styrofoam, polyethylene, Plexiglas, cellophane tape, silk, wool, jersey, Dacron, nylon, rubber, Teflon, acetate, and plastics in general.

**15.7.4.1 Purge of unnecessary hazardous materials.** A static control program should prohibit indiscriminate placing of generative materials at a static controlled workstation. All controlled stations should be purged of such materials to the extent practicable. Common hazardous items include plastic toolboxes or part bins; plastic cups or glasses; thermos bottles; items of clothing such as hats, gloves, sweaters, or even umbrellas; cellophane tape; and Styrofoam or rubber cushions or pads.

**15.7.4.2 Control of necessary static hazardous materials.** Frequently one or more static hazardous items will be required in a process and cannot be readily purged from the operation. These might include Mylar tape, insulative items that are part of the assembly, necessary electrical insulators, insulative instruments, etc. Hazardous items require strict control to avoid ESD failures during processing. Several options must be evaluated as applicable: replacement with nonhazardous item, treatment with topical antistat, neutralization by bench top ionizer, or following the safe distance rule. Ionizers are discussed in Sec. 15.8.5.1. The safe distance rule means that personnel must be trained to keep a separation of a predetermined distance between the static hazardous and susceptible items.

**15.7.4.3 Safe distance rule.** The appropriate safe distance to be maintained between static generative and susceptible items is somewhat controversial. DOD Std 1686, May 1980, requires a distance of 1 m from the periphery of the work area. Most manufacturers find it impractical to enforce this rule over such an extensive area. In actuality, the appropriate distance should be based on several factors, such as:

1. The amount and type of generative materials likely to be encountered
2. Criticality of the end use
3. Cost of failures to determine acceptable degree of risk
4. Other controls in place

By far the most important considerations in deciding on a safe distance are variables relating to expected or known insulative items in the work area. Most important of these are the geometrical shape and material makeup of the items in question in relationship to susceptible items being processed. For instance, the material may limit voltage accumulations to relatively low levels. Irregular or small geometrical shapes further limit inductive coupling to susceptible items. My personal investigations have found that 15 cm (approximately 6 in) is reasonably safe for most common insulative items found necessary in static controlled work areas.

**15.7.4.4 Insulative solvents.** Cleaning agents for static control materials must be selected that do not leave insulative films. Spray electrification can occur with the use of high-resistance solvents. Perhaps even more important in this regard are solvents used in the processing of susceptible items. Addition of alcohol or other additives is often done to increase conductivity of solvents to eliminate static electrification from such fluids.

### 15.7.5 Protective packaging

The preceding controls are sufficient for protection of susceptible items while at a static controlled workstation. This section discusses provisions for protection when a susceptible item is removed from the static controlled station.

**15.7.5.1 Static shielding packaging.** For practically all operations potentially damaging electrostatic fields or actual contact with charged items is likely to be encountered when away from static controlled stations. Charge sources might include ungrounded personnel, clothing, packaging materials such as Styrofoam and polyethylene, and cathode ray tubes at video or computerized stations, carts, and other items. Therefore, protection of susceptible items must be provided by assuring at least one level of electrostatic shielding. Electrostatic shielding is provided by a conductive layer surrounding the item. This may be in the form of a conductive chassis with conductive protection over all connectors, or conductive boxes, bags, DIP tubes, or storage bins. Static shielding materials consist of numerous options from solid metal, metal foil, metallized plastics, carbonized plastics, carbonized or metallized paints, and other means of attaining conductivity.

**15.7.5.1.1 Static shielding boxes.** Static shielding boxes are available in several conductive materials including solid metal, aluminum foil layered, conductive plastics, and carbon-layered cardboard. Any of these materials can be effective if configured properly. The most common problem with static shielding boxes often goes undetected, that is, electrical discontinuity between the box and its lid. This usually occurs because of improper design of the box. Some static control manufacturers have a history of touting the characteristics of the material in their product with little or no regard for its final configuration or end use. Sluffing of conductive particulate is a concern when using carbon-filled plastic or conductively coated cardboard. Abrasion of the conductive paints or other cardboard coatings can reduce shielding effectiveness.

**15.7.5.1.2 Homogenous material shielding bags.** Static shielding bags come in several varieties and combinations of materials. The most common shielding bags of single homogenous material are aluminum foil and carbon-impregnated plastic bags. Nontransparency of both aluminum foil and black carbonized plastic bags is seen by many as a disadvantage for in-plant processing. Shielding effectiveness of most black carbonized plastic bags is limited but sufficient for most applications. Aluminum foil provides excellent shielding but is easily torn and thus has a short usage life. Conductive particulate contamination is a concern for both types of bags.

**15.7.5.1.3 Multilayer shielding bags.** The nontransparent aluminum foil lined type Mil B 81705-3 bags with antistatic Tyvek exterior offer excellent shielding and durability. These bags are commonly used as used as shipping bags where transparency is of lesser importance. Multilayer metallized bags are extremely popular because of the excellent shielding and transparency. These bags are fabricated in several variations by different manufacturers. Figure 15.2 shows a representative sketch of layers to be encountered. Typically the metal layer is of vacuum-deposited aluminum or nickel. The inner antistatic layer is provided to inhibit triboelectric charging due to movement of the contents within the bag.

The only negative aspect of the type Mil B 21705-3 bag is lack of transparency. The transparent feature provided in the metallized bags is due to the thinness of the conductive layer. Unfortunately, thinness is directly correlated to fragility. The metal layer is easily damaged by the flexing from normal handling. For this reason most vendors recommend discarding the bags after one usage. Users are left with the task of defining *one usage*. An item going through many handling processes may require several bags for that one usage. Users must monitor and check the effectivity of bags to determine the most cost-effective replacement cycle. Bags that are obviously crinkled, however, should be replaced, as it is highly probable that the conductive layer is discontinuous. In particular the bottom crease is critical to the effectivity of the conductive layer.

Triboelectric generation by the internal "antistatic" layer has also been a problem. This is believed to be due to topical antistatic treatment of the inner layer prior to bag forming. Since the bag sealing process involves high temperatures, the topical treatment is easily evaporated, rendering the layer ineffective. Utilization of permanent antistatic materials for the inner layer produces the most effective bags. Permanent antistatic materials are formed with the antistatic property throughout the entire thickness rather than just on the surface, as is the case for those that are topically treated.

**15.7.5.1.4 Combination or converted bags.** Some bags are marketed that are combinations of existing bags and/or materials converted into

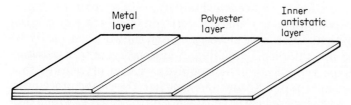

**Figure 15.2** Representative construction of a multilayer metallized bag. Common variations include a protective coating over the metallized layer, an added outer antistatic layer, or other means of adding some physical protection to this delicate but critical shielding layer.

new products. In general, these have consisted of carbon-impregnated bags combined with internal antistatic bubble-pack-type cushioning. Converted black conductive plastic bags with an internal antistatic bubble pack remain nontransparent. The sloughing problem is practically eliminated by virtue of the antistatic inner material. Shielding effectivity is improved somewhat by the added thickness provided by the bubble pack.

**15.7.5.1.5 Antistatic bags.** Antistatic bags are static-preventive rather than static-protective. That is, the antistatic property should be regarded as the ability to inhibit triboelectric generation. Such bags are ineffective as static shields. They are therefore ideal as an intimate wrap, that is, the bag that touches the sensitive item and avoids static generation during movement. These bags are generally used in combination with an outer static shielding container. Only in an environment free of static damaging potentials and fields would antistatic bags be regarded as sufficient. Antistatic bags are sometimes used for paperwork or even for insensitive items. This avoids ordinary polyethylene in the workplace.

Antistatic bags have been reported as having three known problems: loss of antistaticity (especially at low humidities), chemical contamination, and outgassing. The recently developed antistatic materials produced by electron beam radiation of plastics reportedly have eliminated all three of these problems. However, their usage history has been rather short and their qualities are regarded as unproven at this time.

## 15.8 Secondary Control Considerations

Secondary control measures are those supportive procedures or additional items necessary to ensure effective application of controls to meet the intent of the primary measures discussed in Sec. 15.5.

### 15.8.1 Procedures

Documented procedures are needed that define static control requirements in terminology that is clearly understood by all operators in the work area. Without such procedures primary controls can easily be misunderstood and improperly applied. Additionally, proper monitoring of disciplines and effectiveness requires written procedures that can be modified as required by new information. Operating procedures should describe proper application of static controls such as which protective packaging to use, proper wearing of wrist straps or conductive footwear, how to check the wrist strap and work surface, and the use of other controls.

It is extremely important to include descriptive and instructive information on any extraordinary control measures to be utilized. Maintaining a safe distance from necessary static generators is a classic example of an extraordinary rule. By definition this rule requires that operators know which items are hazardous and which items are susceptible. Additionally, the rule must be imposed in a manner that does not convey a message that unnecessary static hazardous materials at the workstation are condoned. Other extraordinary procedures might include operations requiring use of ionizing air nozzles, adjustments to pressure on spray operations, or limiting speed on processes prone to triboelectric generation.

### 15.8.2 Training

ESD training is considered by many to be about 90 percent of the battle in conquering static problems. It is necessary that all operators and supervisors be trained in this subject, which so often appears to defy intuition and logic, and has so many pitfalls. Operators should receive certification of training prior to assignment where susceptible items are to be processed. Even after individuals have received good ESD training, there seems to be a lapse into old myths and misconceptions that occurs with time. For this reason it is commonly believed that a good ESD control program has a half-life of about 6 months. It is therefore recommended that training be provided on a periodic basis of periods not longer than 1 year. Another consideration that might justify even more frequent training sessions is the need to train new employees and absentees from the initial sessions.

The content of an ESD training program must be tailored to meet the requirements of the facility or operation. At a minimum it should contain the following:

1. Instructions in the procedures for the operation
2. Instruction in proper usage of any ESD control item used
3. Some discussion on the principles of static electricity and static control
4. Illustrative examples of static problems including cost factors
5. Allotted time for comment and questions from attendees

Many videotapes are available to assist in ESD training. The use of actual demonstrations of static principles, use of controls, effectiveness of controls, and static damage are highly recommended topics.

### 15.8.3 Self-checks and audits

Monitoring of proper implementation of ESD control measures is required to assure effectivity. This is accomplished by periodic self-checks and audits performed by quality assurance or ESD specialists. Static awareness is greatly enhanced when operators become part of the monitoring system. Proper utilization of many control measures can best be accomplished by operator self-checks. The types of items to be checked and the frequency of checks are determined by the many variables discussed in Chap. 14. Typically included are wrist-strap continuity to ground, insulative item purge, reporting obviously improper packaging, resistance of footwear, and other designated items specific to an operation. Operators can greatly assist by being watch guards for infractions by others, appearances of static hazardous materials in the work area, improper packaging, and suspected static failures. The intent is not to have a selected cadre of tattletales but to correct situations in a friendly manner without involving management.

**15.8.3.1 Wrist-strap checks.** The primary self-check example is wrist-strap monitoring. Wrist-strap checks have been necessary because of past frequent incidence of open or intermittent wrist straps. Many have found it advisable to install electrical wrist-strap continuity and resistance checkers and have each operator verify operation of his or her wrist strap prior to the start of each shift. Data collected in this manner might justify less frequent checks as straps become more reliable.

**15.8.3.1.1 Types of wrist-strap checkers.** Wrist-strap checks are made in several ways with several types of instruments. Regardless of the instrument used, the measurement of resistance should include the connection to ground and the operator. Many problems of open circuit, intermittence, or high resistance are due to problems with the interface to the strap rather than the strap itself. Frequency of checks can be adjusted based on the history of data collected. At this writing most users have found at least a daily check to be necessary; however, product improvements might lead to less frequent checks in the future.

**15.8.3.1.2 Ohmmeter tests of wrist straps.** Many users have instituted periodic tests whereby an electronics technician checks the resistance across each wrist strap from the contact point at the band to the electrical plug at the end. Although this is better than not checking at all, there are severe shortcomings. A better method, recommended for any wrist-strap user, would be to measure the resistance to ground rather than just to the end of the strap. To include the operators wrist to band is impractical since ohmmeter checks to the human body are usually unstable.

**15.8.3.1.3 Checks upon entry or at workstation.** Typically, ESD controls are gradually improved as a user becomes increasingly aware and motivated about the seriousness of the problem. After institution of wrist straps, resistance checks might evolve from none to periodic (monthly or quarterly) ohmmeter checks to daily checks on entry to the workstation. Installation of a wrist-strap checker at entry is sometimes convenient and apparently practical. In this manner each operator on entry to the work area, whether at the start of a shift or at other times, would be required to test his or her wrist strap at an installed checker. Entry checks are recommended as an economical improvement over technician checks, but they have shortcomings. This method will check the operator's skin contact to band provided the tester selected has this provision. Most static control vendors now provide this essential feature.

Shortcomings are the lack of complete circuit verification at the station, the logistics involved with providing identification and storage of individual straps at entry and exit, as well as logging of checks. Where entry checks are not too inconvenient, they could be supplemented with periodic checks of station grounds. Clean-room entry might be an ideal application for this method.

**15.8.3.1.4 Workstation wrist-strap checkers.** Some type of wrist-strap checker at the workstation is becoming the rule rather than the exception. The tester should verify an appropriate resistance of the operator interface, the strap, and the connection to ground. This is usually accomplished by a tester with a metal plate that the operator touches to conduct a test. Most on the market today have three colored lights for resistivity ranges: red for too high, yellow for high, and green for within limits. Some testers have supplemental audible alarms for out-of-specification conditions. This type of tester also requires that a log be kept. Some users are now using constant monitors for wrist straps. The advantages are that the tester is independent of operator error and the need for logging is eliminated. Typical resistance limits are from 1 to 10 M$\Omega$.

**15.8.3.2 Audits.** Periodic audits of ESD control are recommended to ensure effectivity. Audits are typically done by quality assurance and/or ESD specialists. Audits by quality assurance specialists are usually limited to the scope of existing procedures. ESD specialists can add greatly by critiquing the procedures and possibly uncovering unknown problem areas.

**15.8.3.2.1 Audit agenda.** An ESD audit should begin with a review of existing in-house procedures. This may disclose a lack of knowledge in ESD control if improper methods or terminology is found. The facility should then be toured and audited considering both the criteria of in-house procedures and known proper ESD control measures. A representative ESD control audit agenda contains the following:

1. Procedures review and comment
2. Tour and audit
    a. Receiving
    b. Incoming inspection
    c. Incoming test
    d. Stores
    e. Kitting
    f. Lowest assembly level
    g. Lowest level inspection
    h. Repair and rework
    i. Higher assembly level
    j. Higher level inspection
    k. Final assembly
    l. Final inspection
    m. Shipping
3. Audit review and report

**15.8.3.2.2 Audit content.** The items assessed at all areas on the agenda would include the following:

1. Identification of sensitive items
2. Use and understanding of proper terminology
    a. Using "antistatic" materials for triboelectric inhibition rather than for shielding
    b. Not using "antistatic" work surfaces
    c. Using ionizers selectively rather than universally
    d. Clear definitions of types of protective packaging
3. Identification of susceptible items
4. Implementation of primary controls
    a. Personnel grounding
        (1) Proper wrist-strap usage
        (2) Wrist-strap checks
        (3) Proper conductive footwear usage with conductive flooring
        (4) Proper checks of operator resistance to ground
    b. Proper equipment grounding, especially work surfaces
    c. Purge of unnecessary static hazardous materials
    d. Use of protective package away from workstation
5. Implementation of secondary controls as needed
6. An apparent interest in good control on part of operators and supervisors

A data point on the importance placed on ESD audits by highly reputable firms is found in a paper[3] on the subject by Smith and Rief of National Cash Register. This paper indicates that ESD audits represent 15 percent of the total audits at their facility in Minnesota.

An ESD audit is a procedure of observing manufacturing and test operations for consistency with corporate ESD policy. Generally incidents of infractions will decrease with repeated audits. The auditor needs to have a technical background well founded in electrostatic principles, good communication skills, and a dedication to economically sound quality assurance.

### 15.8.4 ESD control certification

Control is sometimes more economically and effectively maintained by various types of certification. Most find certification of trained personnel to be necessary and worthwhile. Certification of ESD control materials on receipt and during self-checks or audits can also be worthwhile where ESD control compliance is critical. Authority to certify must be given only to knowledgeable individuals. An identification system such as stamping or labeling of materials is usually found to be necessary. Records must be maintained, so certification is recommended only as deemed necessary or economically advisable.

### 15.8.5 Vendor and subcontractor control

The same overall level of ESD control must be imposed on vendors and subcontractors to achieve a reliable product with a minimum of yield detractors. A major contractor should have enough knowledge to allow some variations in procedures that are consistent with the vendor/subcontractor's needs and yet are effective. The ESD audit is a good vehicle to monitor and evaluate such controls at off-site locations.

### 15.8.6 Additional control measures

The items discussed thus far have included essential primary controls and some secondary controls that could be considered as assurances for effective implementation of the primary controls. Certain other secondary controls are separate but supplemental to the primary controls and are generally used on a limited or selective basis.

**15.8.6.1 Ionizers.** Ionizers fit our definitions as a supplemental secondary control measure because of their usage to neutralize charged insulators. Use is advisable only when the insulative material cannot be purged from the process or controlled in a more cost-effective manner. Ionizers also fall under the definition of tertiary controls and are discussed in greater detail in Sec. 15.9.2.

**15.8.6.2 Topical antistat treatment.** Another method for controlling static generating insulative materials is to treat them with topical antistatic treatments. Topical antistats are liquid solutions that can be applied to materials to inhibit triboelectric generation. These treatments reduce friction by increasing lubricity and increase surface conductivity largely by absorbing a moisture-laden layer from the air. Antistats are often used on insulative materials that make up part of tooling or other items difficult to replace or discard. Sometimes the treatments are used as temporary stopgap measures until a more permanent fix can be installed. Properly selected, evaluated, applied, and monitored, they can be very effective on most materials. The antistatic agent and other factors about the carrier solution are generally held to be proprietary by the manufacturers.

**15.8.6.2.1 Topical antistat application.** Topical antistats can be purchased premixed or in concentrate and can be applied in several ways. Usually the primary consideration is to get the most complete coverage. Sometimes economics enters the application selection in the attempt to get the most coverage per dollar. Where possible, items dipped into a large container of the solution are likely to have excellent coverage. For routine workstation applications a combination of spraying and then wiping is common. Spraying can include various pressure and fine mist applications as determined best to fulfill the particular need.

**15.8.6.2.2 Problems with topical antistats.** Topical antistats are invisible. This requires special efforts to monitor, and labeling with date of treatment is recommended. Some antistats have been known to cause corrosion of device leads. This corrosion is believed to be caused by the presence of salts. Sometimes it can be due to use of tap water that has chlorine and other contaminants. Tap water should not be used by individuals mixing their own antistats from concentrate; however, some antistat vendors themselves may have used tap water in their bottled or canned product.

**15.8.6.3 Replacement materials.** Sometimes the most economical control measure for a generative material in the process is to replace it with a nongenerating substitute. The most practical approach is to test for triboelectric charging of any recommended substitute materials. The tests should include worst-case conditions using the normal interfacing items included in the process.

**15.8.6.4 Garments.** In the past, garment control has generally been instituted as a cleanliness requirement and ESD control features evolved later as a secondary issue. At present, and it is likely in the future, ESD control may be the primary reason for special garments. The electrostatic behavior of garments can be extremely important as they have been demonstrated to cause the destruction of parts as described in Chap. 13. ESD control consists of either replacement or antistatic treatment of existing garments, usually during the laundering cycle.

**15.8.6.4.1 Common garments available.** Garments available include those made from:

- 64 percent polyester, 34 percent cotton, and 1 percent stainless steel
- Ordinary insulative fabrics with special laundry-added antistats
- Disposable Tyvek
- Dacron with woven nylon conductive fiber
- Dacron with carbon-coated nylon fibers

**15.8.6.4.2 New conductive fibers.** A new method of attaining conductivity of a synthetic fiber was reported in 1987 by Minami[4] of Japan. The fiber is called Thunderon SS-N, and a dyeing technique reportedly forms a conductive layer by chemical bond. The chemical bond of copper sulfide with a cyanogen group is theoretically more durable than the usual method of physical absorption of a conductive liquid. This type of development requires further careful evaluation for effectiveness and potential problems.

**15.8.6.4.3 Problems with garment ESD control.** Some of the problems with ESD control of garments are related to the degree of cleanliness required. For example, disposable Tyvek is normally quite neutral or nongenerative as initially received, but it has too much loose particulate for most clean-room applications. However, the material is often highly generative following ordinary laundering. Laundering is discussed further in Sec. 15.8.6.4.4.

Most special ESD control garments contain some type of conductive fibers. There is a concern that a garment can shed such fibers, possibly resulting in an electrical short. Many garments are poorly constructed with conductive fiber discontinuities at seams or elsewhere. An ungrounded conductive garment could be a potential source of ESD when charged. Care must be taken to ensure that they are grounded at all times when susceptible items are handled. This is normally accomplished by attaching the wristbands tightly on both sleeves, with the operator grounded by wrist strap or footwear.

**15.8.6.4.4 Importance of laundering.** An important variable that must be closely monitored is the effect of laundering. For insulative garments depending on treatment during laundering, the effectiveness of the treatment must be monitored by static tests upon receipt and during usage, as discussed in Chap. 16. For special ESD control garments, laundering is equally important because excessively harsh handling can cause breakage of conductive fibers, reducing or destroying their effectiveness.

**15.8.6.5 Gloves and fingercots.** Special static control gloves and fingercots are commercially available. Most have been of an antistatic variety in order to eliminate the possible static generation that occurs with the usual insulative materials. As with other antistatic materials there are inherent problems with useful life and humidity dependence. More permanent conductive materials have had conductive particulate sloughing problems. Recently, conductive gloves and fingercots have been marketed that appear to have no conductive particulate sloughing characteristic.

**15.8.6.6 Floors or floor covers.** Conductive floors and floor covers such as mats can be used to supplement other personnel grounding measures. Since they are also sometimes used as the primary method of personnel grounding, detailed discussion is covered in Sec. 15.7.1.2. Generally conductive footwear should also be used with conductive flooring.

**15.8.6.7 Shunts.** Shunts are used to tie the leads of devices, printed wiring boards, or other assembly connector pins in common. The objective is to provide a low-resistance path for an electrostatic discharge rather than destroy a susceptible item. In order to be effective, a shunt must have significantly lower resistance than the item to be protected. A good rule of thumb is that the shunt should have no more than 10 percent of the resistance across the nodes to be protected. Ideally the shunt will be much lower.

The most obvious problem is excessively high resistance. Therefore the resistance of shunts and shunting foams must be specified and checked. A second problem has been the presence of chlorine contamination causing complete corrosion of device leads to the point of fracture. A somewhat less obvious problem is the fact that good shunts offer protection only between the shunted leads. Destructive paths often remain between other points and the shunted pins.

**15.8.6.8 Chairs and stools.** Static generating and dissipating properties of chairs and stools in a static controlled work area are frequently overlooked. Several factors relate significantly to the effectiveness of other primary controls in place. Unless altered or specially purchased, chairs and stools are apt to have insulative seats and/or backs. Sometimes modular connections, floor contact pads, lubricants, and even bearings are nonconductive. When conductive flooring is used, an operator can become ungrounded when the feet are raised. Thus the seat should be conductive, and there should be a continuous conductive path through the legs to the floor. An insulative seat is also a convenient resting place for susceptible items that presents an induction hazard. In computer applications soft failures are common from chairs discharging to metallic parts of the computer or nearby furniture that is electrically connected to the computer by visible or sneak paths.

Sometimes a conductive seat cover is all that is required to correct the operator for the feet-up condition. The remainder of a stool might be conductive to the floor. Continuity must be checked. Even in the event of an isolated conductive seat, mutual body to seat capacitance is usually sufficient to suppress the operator's potential to safe levels. Other possible modifications include conductive casters, drag chains or weights, and antistatic treatment on backrests. Extensive modification can become difficult to implement and monitor efficiently, and purchases of special conductive chairs may be more economical.

**15.8.6.9 Carts.** As mentioned in Chap. 13, carts can accumulate a static charge when being pushed. This presents a possible high static energy source if a susceptible item on the cart is touched by a person, whether grounded or ungrounded. ESD control of carts can become cumbersome and costly. The problem of grounding is particularly difficult. Hard-wired ground straps are out of the question. Reliable grounding to the floor requires, first of all, conductive flooring, which in itself can be expensive. Next, the problem of maintaining contact with the floor is full of challenges. Conductive casters can be purchased, but many lose effectiveness through lubricants or other insulative series elements. Some have found that a single ordinary drag strap or chain is also ineffective. Drag weights of greater than 5 pounds have been reported as successful. The recommended approach is to use shielding on items being transported with grounding provided at the loading and removal stations.

## 15.9  Tertiary ESD Considerations

### 15.9.1  Humidity control

Much data have been published showing the effects of relative humidity on static generation. Table 13.2 lists representative data of common activities with common materials. It is interesting to note that even at the high humidity extremes damaging potentials are attainable. So humidity control is considered a valuable passive control measure but is insufficient alone. There is no panacea.

The effectiveness of higher humidities in limiting charge accumulation is due to several factors. Most important is the fact that most materials tend to adsorb moisture in varying degrees, which renders their outer surfaces slightly conductive. This affects not only the charged item in question but also associated materials in series or parallel contact with the item. Thus charge tends to bleed off or decay more rapidly. A secondary benefit may be that increased conductivity and lubricity caused by the moisture layer might also inhibit triboelectric generation.

**15.9.1.1. Absolute versus relative humidity.** Most references in the literature are to the relationship of relative humidity to static accumulation. Blinde and Lavoie[5] have proposed a more complex relationship involving temperature and either relative humidity or absolute humidity. The decay time was determined to be an exponential of the form:

$$V(t) = V_{max} \, e^{-Gt} \qquad (15.1)$$

where $t$ is time and $G$ is the inverse of the time constant.

$G$ was found to be dependent on temperature and either relative humidity (RH) or absolute humidity (AH):

$$G = a_{RH} \times e_{RH}^{b} = a_{AH} \times e_{AH}^{b} \qquad (15.2)$$

where $a$ and $b$ are dependent on the temperature, material, and relative or absolute humidity as indicated by the subscript.

Of importance is the fact that temperature must be known as well as either RH or AH.

**15.9.1.2 Economical considerations in attaining higher humidity.** The cost of attaining relative humidity control increases exponentially with the percent targeted. This is especially true in dry atmospheres such as Arizona and the Arctic. For this reason the optimum ambient 40 to 50 percent RH is often compromised. McAleer et al.[6] report successfully attaining 45 to 50 percent relative humidity in the winter months at Abilene, Texas by injecting 15 lb/in$^2$ of steam into the heating duct system. Steam was piped to emitters in the central ducts. A humidity-monitoring feedback system was employed to maintain desired levels. Localized humidifiers for critical locations are commercially obtainable.

### 15.9.2 Ionization

Ionization is a very useful ESD control measure when properly utilized. Perhaps the greatest misunderstandings arise when ionization is viewed as a panacea that will passively eliminate not only all static problems but the need for additional controls as well. Ionizers will not do everything but are excellent choices for some applications. Their primary use is to remove or neutralize charges on insulative materials. In addition they have a limited effect in reducing static levels on other background materials both conductive and nonconductive. The neutralization of charged insulative materials is important not only from an ESD standpoint but in reduction of particle attraction as well. As device geometries are scaled down, the size of particulate capable of causing problems during processing is also reduced. Current near-micron technology means that particles as small as $\frac{1}{5}$ µm are problematic. Air ionization is a tool that has proved to be valuable in reducing charges on wafers and integrated-circuit chips, thus minimizing the adherence of particulate matter. Ionizers are of two primary types: high voltage electrical or nuclear.

#### 15.9.2.1 High voltage electrical ionizers.
High voltage electrical ionizers produce ionization of the air through high voltage (5 kV to 20 kV) applied to sharp metal needle-type emitters. The high voltage on a conductor with such a sharp geometry causes partial breakdown of the air around the points. Under normal conditions these ions would be within a fraction of a millimeter from the points. In order to produce ions of both polarities, pairs of emitters at equal + and − high voltages are used, or either alternating or pulsed dc voltages are applied to single emitters. Fans are generally used to project the ions.

High voltage type ionizers have a history of several problems, the most notorious of which is polarity imbalance resulting in charge accumulation rather than elimination. Attraction of particles to the vicinity of the needle emitters can result in dirt accumulation on ceilings, walls, or other nearby items. Emitters can deteriorate from corrosion or even sputtering of the needle material caused by electrochemical reaction with gaseous impurities. These problems mandate careful initial evaluation as well as continuous monitoring of ionizer effectiveness. Periodic maintenance in the form of calibration, needle cleaning, or replacement must become routine. In addition adjustments are often necessary after any significant changes in furniture or equipment placement in the area.

#### 15.9.2.2 Nuclear ionizers.
Alpha particle radiation results in collisions with air molecules that result in separation of an equal number of positive and negative ions. Commercial ionizers of this type typically use a polonium 210 source of alpha radiation. Positive and negative ions are produced from collisions of the alpha particles and air molecules. The alpha particle energy is rapidly dissipated from these collisions such that under ordinary condition the ions are concentrated within a few centimeters of the source. Therefore some type of blower is normally used for effective coverage of any appreciable area.

The radiation source is cause for concern as far as human health effects are concerned. The radiation level itself is generally considered to be below hazardous levels. Imperfect encapsulation seals, on the other hand, have resulted in release of the polonium 210 which is a carcinogen if ingested. The half-life of polonium 210 is about 138 days. Thus the expense and inconvenience of periodic (typically yearly) replacement of the element is required.

#### 15.9.2.3 Nuclear/pulsed voltage ionizers.
Air ions tend to recombine rapidly as unlike charges attract and the neutral or natural state of the molecules is the end result. In clean rooms laminar flow hoods are used to extract contaminants out of the area. Introduction of any other blowers to change air flow patterns is discouraged. Air velocity is limited to low rates in order to prevent dislodgement of deposited particles from walls and furniture. Thus the effectivity of ordinary nuclear ionizers is limited to a few inches.

A novel method of propelling the nuclear-generated ions was described by Robert Wilson.[7] A low-frequency pulsating field was produced on the metal case of a nuclear ionizer by a 1000- to 1500-V signal. As the case goes positive, virtually all negative ions in the proximity of the nuclear ionizer are attracted to the case and neutralized. All positive ions, however, are repelled away from the case by the force of the positive high voltage field. During the next (negative) high voltage half-cycle, the opposite occurs. Nearby positive ions are neutralized by attraction to the case and negative ions are propelled by the force of electrostatic repulsion. An appropriately located low-velocity laminar flow hood in the area will facilitate the movement of these ion densities of alternating polarity away from the ionizer. This alternating of spatial densities of opposite polarities reduces recombination and increases ion lifetime resulting in increased neutralization effectiveness.

Neutralization times from 1000 to 100 V were measured on a 6 × 6 in thin metal plate under varying conditions of air flow and high voltage ionization. The plate was parallel to and 4 in above the work surface. Decay results were

1. With no air flow and no high voltage it took 25 min.
2. With no air flow but with high voltage it took 160 s.
3. With air flow on and no high voltage it took 80 s.
4. With both air flow and high voltage it took 6 s.

Neutralization tests using both high voltage and air flow were also conducted with a 6 × 6 in plate at different locations on the work bench. Results listed for + and − 1000 V on the plate respectively were

1. At front center of bench it took 6 and 8 s.
2. At front corner of bench it took 7 and 8 s.

3. At rear center of bench it took 6 and 8 s.
4. At rear corner of bench it took 13 and 18 s.

Slower time at the rear corner is believed to be due to air flow stagnation at that location. The fact that negative ions have greater mobility than positive ions accounts for longer times to neutralize a positively charged plate.

The alternating high voltage can cause an induced voltage on ungrounded conductors in the area. The 6 × 6 in plate parallel to and at varying distances above the work surface was used to monitor this effect. Results for different high voltage pulse rates and plate distances below the ionizer were

1. At one cps and 4 in: 500 V swing
2. At one cps and 10 in: 232 V swing
3. At one cps and 24 in: 78 V swing
4. At two cps and 4 in: 180 V swing
5. At two cps and 10 in: 60 V swing
6. At two cps and 24 in: 20 V swing

**15.9.2.4 Bench-top ionizers.** The least controversial ionizers are the bench-top and other small-area ionizers such as blowers and nozzles. These are intended for specific locations where an insulative material prone to static accumulation could not be practically purged from the process. It is highly recommended that the use of these instruments be minimized, with removal or replacement of static hazardous materials as being the preferable control measure. At this writing many of the bench-top ionizers available have gained acceptance through product improvements and trouble-free experience on the part of users. Many now have feedback control circuitry that monitors and adjusts for ion imbalance.

**15.9.2.5 Room ionization systems.** Room ionization systems are much more controversial. If improperly selected, they can be a costly mistake. Ionization is usually accomplished in one of three ways: high voltage emitters at ceiling level are pulsing dc of both polarities or alternating with an ac signal, or paired dc emitters generate ions of each polarity. It is extremely important to measure offset voltages on detectors strategically placed in any room being ionized by any such ionization methods. The dynamic offset is most likely to be overlooked but is important in areas with static-sensitive items. Ionizers with voltage offsets of 0.25 × (the lowest sensitive item threshold) are not recommended. Therefore the good features of ionizers must be carefully evaluated against item sensitivity, cost, maintenance, convenience, and other factors. Airflow rate is an important variable. Degradation is often due to emitter point erosion, which in turn is dependent on the emitter material, potential, frequency, contamination, and other factors.

Room ionization usage has grown over the last few years, primarily for two different types of static control: reduction of particle contamination in clean room operations and reduction of damaging static potentials in ESD control work areas.

**15.9.2.6 Additional problems with use of ionization.** Some cautions are in order about other considerations on the general use of ionization. They are not effective in reducing the charge on human bodies, so the more conventional personnel grounding means will still be necessary. This is primarily due to the relatively slow time required for ionization to neutralize such a large physical capacitance. The possible rapid buildup of additional charge by a human must also be considered. Other problems with ionizers are the possibility of induced charging due to ion polarity imbalance with voltage type instruments, both ac and dc; and personnel hazard concerns of the radiation-type instruments. Perhaps the primary consideration to discourage ionization usage is high initial cost coupled with higher costs of monitoring, calibration, and maintenance. The most significant problem is polarity imbalance or the related dynamic imbalance sometimes referred to as space charging. The time required for ionization to neutralize charges must be integrated into plans for application.

**15.9.2.6.1 Ionizers can be used as generators.** A wariness of the possibility of problems caused by the use of ionizers may better be realized if one reflects on the fact that ionizers can be used to generate charge on initially uncharged items. Several well-known applications are precipitators, photocopy machines, and spray painting.

In conducting ESD awareness training seminars I have had an obvious need for a cheap, reliable, and easily portable generator capable of achieving several kilovolts. This need was critical to some of the many demonstrations utilized to illustrate basic principles and factory simulated conditions. Most music stores and record shops sell a small pistol-shaped ionizer used to remove charges from phonograph records. This inexpensive device is a piezoelectric transducer that floods the phonograph disc with ions as the trigger is pulled and released several times. In fact the ionizer releases positive ions from the needle point as the trigger is pulled and negative ions when the trigger is released. The needle point of a new unit typically reaches about ±8 kV. I have often used this instrument to charge a demonstration

1-ft-square aluminum plate to as high as 15 kV to either polarity. By pulling the trigger while aiming at the plate a positive charge is placed on the plate. The instrument is then aimed away as the trigger is released leaving the positive charge on the plate. Repeated applications will add to the positive charge until the desired potential is reached. The procedure can be reversed to accumulate a targeted negative potential.

**15.9.2.7 Alternatives to ionization.** What are the alternatives to ionization? Increased relative humidity and resultant moisture adsorption may render some materials sufficiently conductive to bleed off static potentials. Antistatic treatments can facilitate this option for materials not naturally prone to moisture adsorption. Other possibilities include replacement of the insulative materials or keeping sensitive items a safe distance away. The available options often have high cost or other problems associated with implementation. Therefore ionization can become the preferable choice, many times as a last resort.

### 15.9.3 Personnel motivation

Personnel motivation is an important factor in overall ESD awareness. The training programs can give the starting boost, but initial zeal begins to decay with a half-life of about 6 months without further shots in the arm. Posters in the work area are a big help in this regard if changed at least every 4 months or so. Newsletters on ESD items within the plant or division are also normally well received. Although I personally do not favor the idea, some employers have gone the route of disciplinary action for ESD infractions. Punishment varies from time off without pay to furlough or firing, typically after about three serious errors.

## 15.10 ESD Control in Automated Facilities

The electronics industry is currently at the knee of a curve in the transition to automation and robotics. This means that the ESD control parameters will change accordingly. To many, at initial consideration automation is viewed as being free from static electric problems. This is perhaps due to the common perception of metal robots and machines contacting the items processed, with the added misconception of metal as always being statically safe. In fact, for cost effectiveness and other reasons modern robotics involve extensive use of composite materials. Those having the good fortune to design their own machines and/or fixtures may influence the static propensity and interfaces of selected materials. Many are unable to alter the materials in purchased automated systems and must therefore assess the possibility of ESD and institute appropriate controls.

### 15.10.1 Possible ESD failure sources and control measures

Each facility must be analyzed for possible ESD problem sources and corrected accordingly. Some of the major considerations are discussed in this section.

Even with automation there is generally a requirement for human interface at some operations and during troubleshooting and maintenance. Since human interfaces are reduced drastically, it seems advisable to take extra care in ensuring grounded personnel. Since automation is normally found in modern facilities, grounded flooring may be a consideration in planning of some workstations. In this event conductive shoes and daily resistance checks of operator to ground are recommended. If wrist straps are used, checks before each shift or constant monitoring are recommended.

Isolated or ungrounded metal fixtures could conceivably become charged either by triboelectric or inductive means. Contact with sensitive items could result in direct ESD with energy levels determined by the capacitance of the metal. Therefore grounding of all such metal or other conductive materials is recommended.

The most likely problems to be encountered with automated movement of sensitive items are charged device model failures. In the literal sense *charged item model* is a more appropriate term, as the item might be an assembly rather than a device. The charged-device model could result from triboelectric generation on the part of its nonconductive carrier as it contacts various machine fixtures, whether conductive or nonconductive, grounded or floating. Rapid discharge and failure can then occur when a lead contacts a ground point, a low impedance test connection, or a floating conductor of sufficient capacitance. Recommended preventive measures include making material changes if possible, altering the position of parts or items during movement to change capacitance and/or contact areas, attempting to retain conductive portions of sensitive parts or items at ground, slowing movements, and using selective ionization.

McAleer et al.[6] reported that vacuum pickup units used extensively in automated systems have been known to leave devices with a residual triboelectric potential in the neighborhood of 500 V upon release. Effective corrective action included cycling vacuum pressure with machine cycles or using a puff of air to release the unit.

### 15.10.2 Field induction

Failures of voltage-sensitive parts or assemblies containing such parts can occur as a result of fields from other charged objects nearby whether part of the machinery or not. Unnecessary static generative materials should be purged from automated areas just as from any other static controlled work areas. Conductors prone to charging should be grounded. Cathode ray tubes should be strategically located to avoid proximity to sensitive items if possible. Shielding may be installed to block fields of sources that cannot be controlled otherwise. Ionization may be effective in eliminating voltages from some sources.

### 15.10.3 Conveyor belts

Conveyor belts impregnated with carbon to render them conductive can be obtained. The ground path must be verified. Typically it is through the metal conveyor chassis. Metal mesh designs are to be considered, but inductive effects could be a problem. Antistatic treatments are not likely to be effective with so much friction.

## 15.11 Clean Room Considerations

Special factors pertaining to clean rooms mean that the ESD control challenge differs significantly from control in other areas. Cleanliness requirements have dictated much of the material makeup of such rooms in the past. In the electronics industry clean rooms are representative of integrated-circuit fabrication or other sophisticated high-technology manufacturing facilities. Normally such facilities utilize special garments and insulative materials such as quartz. Contamination by static attraction is as large a consideration as ESD.

Special fabrics and/or treatments can eliminate any garment hazard presented. It is often more difficult to replace quartz containers and other insulative items and retain the required degree of cleanliness. Carbonized materials, and oftentimes even antistatic materials are unacceptable. Carefully selected and controlled ionization can often play a valuable role in clean-room facilities. Even whole-room ionization may be advisable if field fluctuations are less than the minimum susceptibility threshold in the room.

## 15.12 Engineering Concerns

Control of static electricity is not just a manufacturing or quality assurance duty. As was stated earlier, it is everybody's job. The engineering department is no exception. Engineering ESD duties revolve around design factors and providing a technical foundation for ESD control policies. ESD control is critical in engineering labs where design decisions are based on breadboard and prototype circuit tests. Otherwise parameters altered by ESD could have a profound effect on the end-product design.

The top ESD control document encompasses inputs from many disciplines, but engineering must provide technical input and review policies for technical merit. In most cases the ESD director will be required to provide the primary input, but engineering should critique and question items as appropriate. Interaction of this type will encourage engineers to apply their talents to this long-neglected area. In all the areas discussed, engineering input may be tempered by additional customer demands.

Whenever an ESD control program is implemented there must be some definition as to what is susceptible. The responsibility for part categorization rests with parts engineering. The same group must establish vendor controls and part qualification or screening criteria. Electrical designers must define which assemblies are susceptible. Generally, an assembly containing susceptible items is also regarded as susceptible unless proven to be otherwise. All categorization requires a definition of susceptibility threshold levels, usually broken down into classes.

In order to convey designations of susceptibility it is usual to require that drawings be marked. Where a given category of assemblies such as hybrids or printed wiring boards are recognized as susceptible, marking may not be necessary. Engineering must assess each situation and decide. Sometimes it is desirable to mark hardware at various levels depending on maintenance procedures. Engineering must be involved with what, how, and where to mark hardware.

Static electricity must be a part of design reviews. Consideration of this important aspect must be treated as a vital design parameter. After all, decisions at this stage can affect ESD immunity through part selection, placement, mechanical design, or other design alterations. The overall benefit can include improvements in manufacturability and reliability. Static damage can be subtle and even lead to latent failure. For this reason ESD failures should be of major concern to reliability engineering. Failure analysis statistics should be maintained, and corrective action should include identification and elimination of ESD control weaknesses. On high reliability programs failure analysis should include the identification of potentially secondarily damaged parts for possible replacement.

Sometimes the resources of specialized engineering laboratories will be required to perform qualification tests on ESD control items. Usually, the failure analysis laboratory is equipped to perform most of the evaluations required. The industry-wide problems in so many areas of ESD control will be solved only through the cumulative research of individual companies. Each bit of information learned in defining conditions that either cause or prevent static failures is helpful. Some suggested areas are latent failures, failure mechanisms, ESD models, ESD simulation circuitry, design for immunity techniques, and control measure evaluations. Involvement with professional activities or organizations involved with static is to be encouraged. Increased communications with a peer group dealing with this insidious problem can be extremely helpful to all concerned.

## 15.13 Safety Concerns

The extensive grounding for the sake of ESD control and increased conductivity of materials raises a question of personnel safety. Safety should always take precedence over ESD control, and the facility safety supervisor should be required to sign any ESD control documents.

## 15.14 Customer Image Enhancement

The benefit of improved customer image through good ESD controls cannot be overstated. For one thing, ESD control generally means better cleanliness, neatness, and consequent appearance. Additionally, a message is conveyed that the company cares about quality and reliability. If our supposition is correct, these parameters will show some improvement as well.

## References

1. Whitson J. Kirk, "Uniform ESD Protection in a Large Multi-Department Assembly Plant," *Electrical Overstress/Electrostatic Discharge Symposium Proceedings,* 1982.
2. Bossard et al., "ESD Damage from Triboelectrically Charged IC Pins," *Electrical Overstress/Electrostatic Discharge Symposium Proceedings,* 1980.
3. D. A. Smith and C. D. Rief, "Internal Quality Auditing and ESD Control," *Electrical Overstress/Electrostatic Discharge Symposium Proceedings,* 1986.
4. T. Minami, "Static Electricity Elimination Using Conductive Fiber by Dyeing," *Electrical Overstress/Electrostatic Discharge Symposium Proceedings,* 1987.
5. D. Blinde and L. Lavoie, "Quantitative Effects of Relative Humidity on ESD Generation/ Suppression," *Electrical Overstress/Electrostatic Discharge Symposium Proceedings,* 1981.
6. R. E. McAleer, G. H. Lucas and A. McDonald, "A Pragmatic Approach to ESD Problem Solving in the Manufacture Environment, A Case History," *Electrical Overstress/Electrostatic Discharge Symposium Proceedings,* 1986.
7. Robert Wilson, "A Novel Nuclear Ionization Device Employing a Pulsed Electric Field," *Electrical Overstress/Electrostatic Discharge Symposium Proceedings,* 1987.

# Chapter 16

# Control Product Considerations and Evaluations

During or following ESD control measure selection for a particular application, a need to evaluate a control product often arises. As previously mentioned, appropriate standards have not been developed for most control products. Control product vendors and governmental specifications have attempted to fill this void by utilizing existing standards designed for other materials and purposes. The correlation to usage effectiveness is often lost. The end result is that a defense contractor must first confirm that the product meets the governmental standards and then devise an evaluation test that is suitable to the application's life cycle. Commercial users get caught up in adherence to many of these same standards, not realizing that they do not always correspond to end item effectiveness. Until more comprehensive standards are developed the best approach seems to be to borrow techniques from existing non-ESD standards and try to understand and interpret the limitations appropriately. Supplemental testing more closely duplicating the interfacing conditions and materials in use is usually the most practical approach.

## 16.1 Factors Affecting ESD Control Product Tests

Whether testing for ESD susceptibility or evaluation of ESD control products, the sources of error in Table 16.1 are nearly overwhelming. Compounding the many variables involved are the associated lack of sufficient standards and poor understanding of important physical characteristics. A common inconsistency is that material standards are being applied to evaluate a product, not a material. ESD control consists largely of three factors: attempting to bleed away charge, pre-

TABLE 16.1 Sources of Error in ESD Tests and Measurements

Overlooked safety considerations
Contamination on sample under test, test apparatus, or associated equipment
Ambient relative humidity and/or temperature
Sample preconditioning
Apparatus preconditioning, if applicable
Degree of accuracy in simulation of usage conditions
Dielectric absorption
Waveform of simulated transient, if applicable
Stray capacitances, leakages, and inductances
Measurement changes due to stray fields or EMI
Repeatability of tests
Test perceptibility of degradation or weakness
Training, discipline, and experience of test personnel
Test sequence
Other

venting charge buildup, and protection from existing charges. Thus the parameters of resistance, triboelectric properties, and shielding effectiveness are involved with many control products. This chapter discusses some of the common evaluation measurements and some suggestions for innovation.

## 16.2 Resistance and Resistivity Measurements

Resistivity is the resistance per unit. For surface resistivity the unit is a square on the surface. For volume resistivity the unit is a cube of the volume. Oftentimes when an individual is measuring surface resistivity a certain component of the result is due to volume resistivity, so it is difficult to separate the two. Conceptually, surface resistivity under its best control of stray parameters is still a volume resistivity through a thin layer of material on the surface. These terms are addressed in more detail later. The intent of these resistivity measurements is to project the behavior of the control product in terms of charge bleedoff, decay, shielding, and to some extent triboelectric generation. Numerous factors affect the resultant measured values, and direct correlation to these other parameters is not always present.

## 16.2.1 Surface resistivity

Surface resistivity was defined in Chap. 4 as

$$R = \frac{kl}{W}$$

where $k$ = a proportionality constant
$l$ = length of the sample being measured
$W$ = width of the sample being measured

In the case of a square where $l = W$, the equation reduces to

$$R = k$$

This convention is that $k$ for a square is called surface resistivity with units of ohms per square. The fact that the size of the square is immaterial is illustrated in Fig. 4.5.

## 16.2.2 Volume resistivity

From Chap. 4, a solid material resistance can be expressed as

$$R = \frac{\rho_v l}{A}$$

where $\rho_v$ = volume resistivity
$l$ = length
$A$ = cross-sectional area

Since length $A$ and resistance can be measured, the expression is more commonly written in the form

$$\rho_v = \frac{RA}{l} \; \Omega \cdot cm$$

In the mks system of units, $\rho_v$ is numerically equal to the resistance between the faces of a 1-cm cube of the material. Of interest is the fact that the size of the electrodes used to measure volume resistivity is immaterial from a theoretical basis.

## 16.2.3 Electrodes

The most common electrodes consist of knife edges against the surface. They are separated by the same distance as their length so as to define a square for surface resistivity measurements. Other variations are painted electrodes. A handy electrode that I have used with reasonably accurate measurements is a conductive adhesive-backed copper tape electrode. Guarded electrode configurations are employed to eliminate undesired stray leakage currents.

Concentrically circular probes as shown in Fig. 16.1a are often

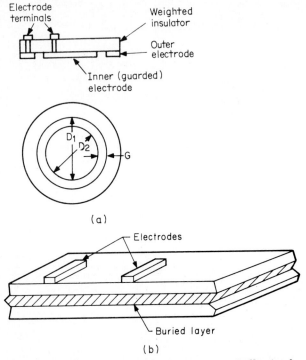

**Figure 16.1** Resistivity electrodes: (a) concentrically-circular probes; (b) conductive layer sandwiched between two high-resistance layers.

used to minimize fringing effects. Such electrodes can be used to measure either surface or volume resistivity. A cylindrical volume is measured for the latter. Volume resistivity is most accurately measured when the gap between the guard and the measuring electrode is kept highly insulative and arc-free but as small as possible.

$$\rho_s = \frac{\pi D_0}{G} \qquad (16.1)$$

where $\rho_s$ = surface resistivity
$D_0 = \frac{1}{2}(D_1 + D_2)$ = average diameter of the spacing or "annulus," cm
$G = \frac{1}{2}(D_1 - D_2)$ = width of the spacing, cm
$D_1$ = outer diameter
$D_2$ = inner diameter

Washer electrodes under screws are often used for resistance measurements of materials in the static dissipative to insulative range. Conductive patterns in the shape of combs are used to simulate printed wiring boards for the evaluation of conformal coatings. Bar electrodes are often used for measuring surface resistivity or very thin materials. The measurements taken in these manners are really combinations of surface and volume resistivity, and better means are necessary to separate the two. Tapered pins are sometimes inserted into reamed holes in insulating materials to be used as electrodes to measure volume resistance or to measure resistance between laminate layers.

### 16.2.4 Variables affecting resistivity measurements

A list of all variables affecting resistivity measurements would be exhaustive. The following list of the major variables is provided with a precautionary note that they can be affected by subsets of many additional factors:

1. Sample preparation
2. Nature and configuration of sample
3. Electrode configuration
4. Contact resistance
5. Environment
6. Measurement parameters

**16.2.4.1 Sample preparation.** Resistivity measurements will be adversely affected by uncontrolled variables of the sample. Contamination of the sample that either increases or decreases measurement results is a common detractor. Adsorbed or absorbed moisture as a result of incomplete or improper preconditioning is another. Lack of neutralization of preexistent electrostatic charges or polarization can also affect resistivity measurements. Often overlooked is contamination on the electrodes or other apparatus used. Control readings on known resistances in the range of concern are advisable.

**16.2.4.2 Nature and configuration of sample.** A material being measured for resistance or resistivity should ideally be homogenous. The desired parameters of resistivity depend on many factors such as the control apparatus being fabricated of the material, its configuration, characteristics of the item to be protected, and other factors present in the usage environment. For instance, a bipolar device may sometimes be best protected by limiting current in the event that a static discharge path involves the packaging material. An MOS device, on the other hand, would be protected by rapid dissipation of the voltage to ground avoiding voltages above gate breakdown potential at any gate.

Results will vary greatly between homogenous and inhomogenous materials, especially those with buried conductive layers. Since resistivity measurements are prone to such variability, larger samples provide opportunity for more measurements and thus better statistical significance. The relative area affects surface resistivity measurements. Thickness affects volume resistivity and can contribute to an erroneous volume resistivity component of surface resistivity measurements. Certain features of the sample impact contact resistance as discussed in Sec. 16.2.4.4. Sometimes configuration of materials prevents use of guard electrodes, which leads to inaccuracies in resistivity or resistance measurements.

**16.2.4.3 Electrode configuration.** The often followed ASTM D257 methodology contains 11 different electrode configurations. These include parallel bar electrodes, conductive paint bars, concentric circles, and variations of guard electrode placement, including no guard options. Many variables are inherent to any one electrode selection, and the possibilities for error are compounded by the multitude of choices.

Concentric circles are believed to be the most commonly used electrode configuration for quantitative evaluation of ESD control materials. Hopefully, a guard ring usage is the rule rather than the exception, although cited manufacturers' data typically does not specify qualification information in sufficient detail to know. Dimensions of the electrodes, including guard, as well as whether guarded or not have been shown to affect resistivity results.

**16.2.4.4 Contact resistance.** The interface between the electrodes used and the surface of sample will add a series resistance called *contact resistance*. Contact resistance can be significant. Where practical, efforts to minimize this variable within the allowable options of standards requirements are advisable. Proper preconditioning of the sample is a necessity in order to remove contaminants and moisture which affect contact as well as other resistances. The nature of the sample affects contact resistance. Sample characteristics relating to the interface contact area include hardness, porosity, smoothness, and compressibility. Contact interface of cellular foams, for instance, is minimal but greatly dependent on density. Compressibility generally tends to increase contacting interfaces. Similar characteristics of the electrode will affect interfaces and contact resistance accordingly. An important factor of the electrodes relates to applied pressure or weight, which generally will affect resistivity inversely.

**16.2.4.5 Environment.** Environmental factors can also affect resistivity measurements. The largest environmental influence is likely to result from variations of relative humidity. The reason is that sample materials have inherent tendencies for moisture adsorption that cover a wide range. Temperature also affects resistance of materials. A third major contributor especially in measuring high resistivities can come from affects of electromagnetic fields in the area. Thus shielding is advisable when measuring above $10^{10}$ Ω per square.

**16.2.4.6 Measurement parameters.** Experience has shown that certain measurement variables can significantly affect resistivity measurements. Sometimes a resistivity measurement observed for several minutes will exhibit instability and fluctuations. This is particularly likely at resistivities in the high end of the dissipative range and above. It is therefore recommended that resistivity measurements be observed for at least 3 min. In this manner, instabilities will be known and further investigation might reduce them. Minimally, an average rather than instantaneous resistivity might be taken as more meaningful.

The applied voltage is also important. Sometimes a certain voltage is required to break down inherent or external barriers prior to lower-level resistivity indications. The applied voltage is appropriately selected at the maximum level consistent with end-usage considerations. That level normally is the lowest threshold of susceptibility in the area in order to assure proper bleedoff of items that acquire voltages at or above damaging potentials. A reasonable level for many measurements is 100 V. This value is consistent with objectives of many ESD control programs as far as a voltage control limit is concerned. For safety reasons current limiting to less than 0.1 mA is recommended.

**16.2.4.7 Buried conductive layers.** The sheet resistance of buried layers can be measured by using an ac signal[1] that is capacitively coupled to the buried layer. Many static shielding bags are fabricated in a manner similar to that shown in Fig. 16.1$b$, where the conductive layer is sandwiched between two comparatively insulative layers. One or both of the outer layers may be truly insulative, or perhaps more frequently antistatic.

To illustrate the measurements principle, the example is analyzed for the case of two ideally insulative outer layers around the conductive layer. Each conductive electrode can be regarded as a capacitor plate with portions of the conductive buried layer forming the second capacitive plate. The generalized schematic circuit diagram is shown in Fig. 16.2$a$ and $b$ after reduction. For true insulators, $R_{vt}$, $R_{top}$, and $R_{bot}$ are infinite and can be discounted.

The impedance of this circuit is

$$Z = R_{buried} \frac{-i}{\omega C_{eq}} = Z \quad \text{at an angle } \theta \quad (16.2)$$

$$= \sqrt{R_{buried}^2 + \left(\frac{1}{\omega C_{eq}}\right)^7} \quad \text{at an angle } \theta \quad (16.3)$$

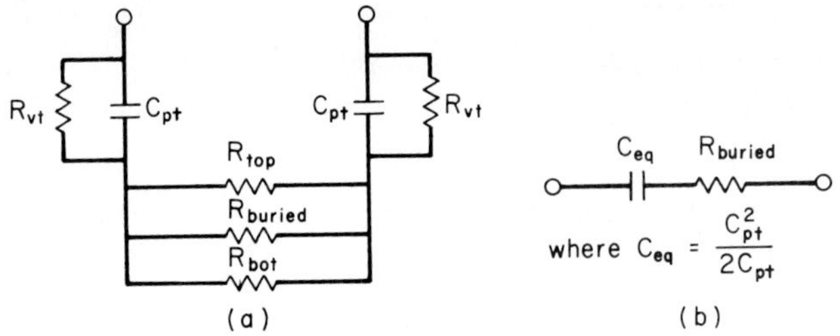

**Figure 16.2** (a) Circuit diagram of Fig. 16.1 configuration. (b) Reduced circuit.

$R_{vt}$ = Volume resistance of top layer
$C_{pt}$ = Capacitance from top to buried layer
$R_{top}$, $R_{buried}$, and $R_{bot}$ = resistance of respective layers
$C_{eq}$ = Equivalent capacitance

In the reduced circuit: $C_{eq} = \dfrac{C_{pt}^2}{2C_{pt}}$

where

$$\theta = -\tan^{-1} \frac{1}{\omega R_{buried} C_{eq}} \tag{16.4}$$

$$R_{buried} = |Z| \cos \theta \tag{16.5}$$

By measuring the current $I$ at given ac voltage $V$ and a given frequency $f$, $R_{buried}$ can be calculated. However instrumentation is available to readily read key parameters of complex impedances. For instance the HP 4275A LCR meter will give direct readings of impedance $Z$, phase angle $\theta$, effective series resistance $R$, capacitance, inductance, and other parameters.

**16.2.4.7.1 Measurement considerations.** In the case of a conductive layer that is sandwiched between two ideal insulators, accuracy of the measurement will improve as the frequency is raised. This is because the impedance of the series capacitance is inversely proportional to frequency and approaches a short circuit at higher values of $\omega$, which is equal to $2\pi f$. Significant error could be introduced if the instrument were referenced to ground because of the high probability of parasitic capacitive coupling of the buried layer to ground.

**16.2.4.7.2 Buried layers with parallel antistatic layers.** If the top and/or bottom layers were antistatic, or if an antistatic layer were added over the insulative layers, $R_{vt}$, $R_{top}$, and $R_{bot}$ are no longer infinite. $R_{top}$ and $R_{bot}$ are normally very high with respect to $R_{buried}$, and $R_{vt}$ is typically very high, so the effect on the measurement would be expected to be minimal. Inverting the bag to take the minimum resistance obtained from measurements of both sides may improve accuracy somewhat.

### 16.2.5 Insulation resistance of cables

Cable insulation resistance is measured from the center conductor to the shield or to a tank of water in which the cable is immersed. The technique is discussed in greater detail in Chap. 17 as it is used for wrist straps.

## 16.3 Triboelectric Generation Measurements

Characterization of materials used in a process is often desirable in order to identify possible contributors to static problems or to evaluate specially selected static inhibiting materials. Currently there are no effective standards to help fulfill this need. As a result evaluations are usually of a very quick and dirty type such as rubbing materials with the bare hand or with one readily available generative material like wool or nylon. Although this will often identify gross static generators, it is insufficient to predict material behavior in a factory setting. A better approach where practical is contact/separation or rubbing as applicable using all or at least a representative sample of materials involved in the process. An alternative approach is to choose at least two materials from the extreme positive and negative regions of the triboelectric series. A reasonable choice might be acetate or quartz from the positive end and Teflon from the negative. Even this approach is inferior to testing the actual process materials, however, as there are many other variables affecting charge generation. As mentioned in Chap. 4, contamination, surface finish, environmental history, and other factors can be even more significant than position in the triboelectric series.

Some inner antistatic linings on static shielding and other antistatic bags have been found to be generative and cause triboelectric charging on items enclosed. Antistaticity cannot be correlated to resistivity. This misconception may have arisen because of the likely occurrence of recombination when low resistivity measures are contacted and separated.

## 16.4 Static Decay Testing

Sometimes the primary interest is the speed at which a material is able to bleed away a charge. This is typically accomplished by measuring the time to discharge a known capacitance to a given percentage of the original potential. The tendency has been to use either a 1-ft-square or a 6-in-square aluminum plate as the charged capacitance. This test vehicle is often referred to as the *charged plate* electrode. Instrumentation and accounting for variables is critical and difficult. In particular stray mutual capacitances often affect measurements significantly.

Greater detail on decay testing is included in Chap. 17. Test methods are usually some derivation of Standard 101C, Method 4046.1. Decay time cannot be precisely derived from resistivity measurements. Full material characterization usually requires that resistivity or resistance measurements are also necessary in conjunction with decay tests. Considering all three desirable features, there is no real correlation between antistaticity, surface resistivity, or decay tests.

## 16.5 Dissipative Materials

Dissipative materials are most appropriately applied for protection of energy-sensitive items. A desirable feature is that the protective material absorb or use up a significant portion of the available energy in a transient. A fact often overlooked is that proper characterization of such materials will normally require volume as well as surface resistivity measurements.

## 16.6 Antistatic Materials

A serious problem with antistatic materials is the useful life limitation of the materials such that they become generative after time. Evaluations are necessary to ensure that they are not generative when received or during normal usage. Head[2] reported results of 1-year aging of 40 antistatic samples: 10 foams between ⅛ and ½ in thick, 10 plastic sheet materials between 25 and 250 mils thick, and 20 film materials of a few mils thickness. Upon receipt from the vendors, 9 of the 20 films exceeded the $10^{13}$-$\Omega$ per square upper limit of measurement capability. Initially the other 31 samples were measured: one each at $10^8$ and $10^9$; 11 at $10^{10}$; seven at $10^{11}$; eight at $10^{12}$; and three at $10^{13}$ $\Omega$ per square.

Parameters were remeasured after 1-year storage in a desk drawer with each sample paper clipped to a paper lab sheet. Results were that 33 had surface resistivities exceeding the upper measurement limit of $10^{13}$ $\Omega$ per square. In addition only two exhibited zero triboelectric generation after pulling from between two sheets of styrofoam with a 5-lb weight on top. Thirty-one of the samples generated in the range of 500 to 5000 V by this same test. Decay times to less than 25 V were measured while a grounded person held one edge of the 2-in sample under test. Only 11 samples decayed in 1 s or less. Fourteen samples decayed between 2 and 10 s. Three decayed between 13 and 30 s. Twelve retained levels in the range of 100 to 5000 V after 30 s.

### 16.6.1 Contaminated antistatic polyethylene

Problems with static control materials determined during failure investigations or material evaluations are of value to other users of the product. Review of problem histories should be incorporated into the evaluation process. The vehicle for this type of information is primarily the Government/Industry Data Exchange Program (GIDEP) discussed in Chap. 2. The following example is included to illustrate the value of this program, especially in light of the expense and skills required for sophisticated analysis of the type described.

GIDEP Alert Number E9-A-86-02 was written in January 1987 describing a contamination problem with antistatic polyethylene material used in antistatic packaging materials. The contaminant was an organic acid that reacted with and depleted the solder coating on part leads. Low magnification visual examination of affected parts revealed a white fibrous contaminant. Whiskers observed under the scanning electron microscope (SEM) are shown in Fig. 16.3. The result of this whisker growth and solder depletion is degraded solder ability of affected parts. X-ray diffraction identified the material to be lead di($n$-octanoate). The cause was then traced to $n$-octanoic acid contained in the antistatic agent used in the polyethylene. Packaging materials affected included blister packages, metallized bags with antistatic inner surface, and antistatic foam.

## 16.7 Voltage-Monitored Activity

When standards do not exist or appear inadequate, voltage monitoring of activity involving the item under question will generally deter-

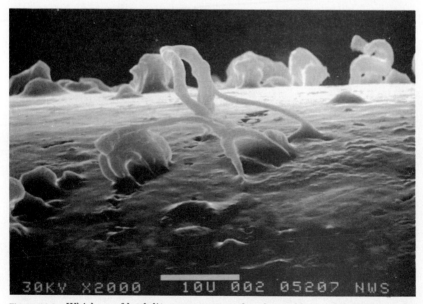

**Figure 16.3** Whiskers of lead di($n$-octanoate) produced on solder-coated resistor leads from contamination in antistatic package from a major supplier. (*Photo courtesy of Naval Weapons Support Center, Crane, Indiana.*)

mine effectivity. This method amounts to a basic look at the impact of a control measure under usage conditions. Generally, worst-case movement and environmental conditions are advisable to provide safety margins. Figure 16.4 shows typical instrumentation for voltage-monitored activity (VMA). The strip-chart recorder provides a permanent record of variations in voltage produced by certain movements. The charge plate is electrically connected to the item to be measured. Some example uses of this technique might include the following:

1. Human body voltage during shuffle test to evaluate footwear, flooring, or floor treatment
2. Tote box voltage during lifting and placing on work surface to evaluate tote box and/or work surface for voltage suppression versus static bleedoff
3. Voltage on cart while person simulates pushing by shuffle test
4. Voltage on chair seat as a result of rubbing other parts of chair
5. Voltage on an item during automated movement

The possible applications are virtually endless. Numerous variables must be well thought out in this type of measurement. Sometimes a quick comparison of control measures is all that is desired, and precision can be compromised in preliminary evaluations. The size of the charge plate should be minimal to reduce its capacitance. The strip chart and other apparatus must be carefully calibrated. An oscilloscope may be preferred for more accurate peak detection owing to better frequency response.

## 16.8 Evaluation of Flooring and Floor Finishes

### 16.8.1 Importance of footwear

In flooring or flooring finish evaluations which utilize personnel to do walking, shuffling, or similar motions, a good variety of footwear is

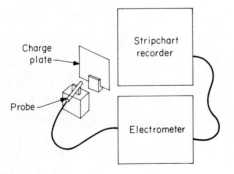

Figure 16.4 Voltage-monitored activity (VMA).

recommended. Footwear is often uncontrolled, and variations in the critical sole and heel materials can be great. In particular, synthetics and rubber varieties should be tested since their insulative properties are normally prone to charge generation. In the past evaluators were left on their own to devise a method to appraise so-called antistatic floor finishes. This situation has been alleviated somewhat by the new EOS/ESD standard and recent literature on the subject.

### 16.8.2 Shuffle test

Antistatic waxes or floor finishes are intended primarily to inhibit triboelectric generation on walking personnel. The application can be made to most floor materials, but vinyl asbestos tile is probably the most common. A good test is to compare the effects of workers shuffling their feet on treated versus untreated floor surfaces. An excellent technique is to use an electrometer or field meter coupled to a strip-chart recorder to preserve voltage waveforms during shuffling. This same evaluation can be extended to other activities and objects such as rolling carts or wheeled chairs.

Sample preparation is important in evaluation of any ESD control material in order to reduce effects from extraneous variables. Essentially this means that prior to evaluation tests all preexisting finishes must be thoroughly removed without leaving residue or contaminants that might affect results.

### 16.8.3 Comprehensive floor finish evaluation

A comprehensive methodology for evaluating antistatic floor finishes was described by Kolyer and Cullop.[3] The method included pertinent parameters beyond those that were strictly static-related.

**16.8.3.1 Sample preparation.** The samples used were $12 \times 12 \times 1/8$-in vinyl asbestos tiles. A nonionic detergent cleaner followed by tap water rinse was used to remove contaminants on the surface. Two coats were applied with 30-min and 24-h drying time for the first and second coats, respectively.

**16.8.3.2 Heel mark test.** A heel mark test was conducted with a rotating hexagonal chamber containing six 2-in plasticized rubber cubes and a 9-in tile sample as shown in Fig. 16.5. The chamber was rotated at 52 revolutions/min for 3 min in two opposite directions. The "heel" marks of a certain intensity are then counted as a measure of marking propensity.

**16.8.3.3 Friction, or slip, tests.** The reduction in friction afforded by floor finishes is pertinent to triboelectric generation. The Topaka slip

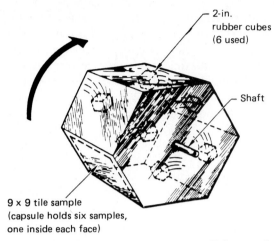

**Figure 16.5** Heel mark test. (*From Kolyer and Cullop,*[3] *with permission from the EOS/ESD Association.*)

test as depicted in Fig. 16.6 measures the dynamic coefficient of friction. The rubbing contact with the flooring is made by a special bond paper having 25 percent rag content. This paper is weighted with a soft bag filled with lead shot. The combination of bag, scale, and plastic scale base weighs 2500 g. The assemblage is dragged over the flooring at a speed of 3.14 in/s by an electrically driven windlass. The scale is read to give an indication of the force required for this movement.

**16.8.3.4 Gloss and scrub tests.** A gloss test consisted of measuring relative amounts of reflected light from an incident beam at 60° from the vertical. Scrub tests were conducted to determine a finish's ability to withstand detergent and/or water scrubbing. The 5.5-in-wide brush was mechanically reciprocated with a fixed pressure on the brush.

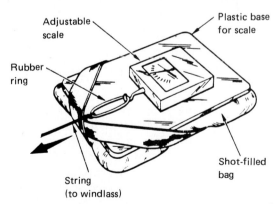

**Figure 16.6** Topaka slip test. (*From Kolyer and Cullop,*[3] *with permission from EOS/ESD Association.*)

**16.8.3.5 ESD tests.** ESD tests consisted of surface resistivity using a portable dc meter, rolling charge generation, drag charging, scuff, and walk tests. The roller test was as depicted in Fig. 16.7. The rollers were made of 10 different materials typically used in shoe soles. Accumulated rolling charge after initial neutralization was measured by a Faraday cup and nanocoulombmeter arrangement.

The drag test consisted of windlass dragging across the flooring sample of a 1.47-lb aluminum block that held a 1⅞-in × 2⅞-in shoe sole material sample. The speed was 3.14 in/s as in the Topaka slip test. Measurement was made of charge on the sole sample as it automatically dropped into a Faraday pail. Both scuff and walk test can be instrumented by using a charge plate and electrometer VMA arrangement as shown in Fig. 16.4.

**16.8.3.6 Hop test.** During actual walking there are several possible charge generation actions to consider:

1. Contact/separation of shoes and floor
2. Compression and decompression of shoes
3. Clothing friction
4. Friction from rubbing or scuffing of shoes against floor

After a series of walking measurements, Chase and Unger[4] determined that the major contributor to charge generation was contact and separation during normal walking. The frictional effects of purposely rubbing by shuffling or similar movements was ignored.

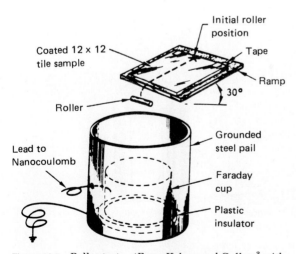

**Figure 16.7** Roller test. *(From Kolyer and Cullop,[3] with permission from the EOS/ESD Association.)*

A test method was then proposed that consisted of measurements of charge accumulated on a contact disk apparatus after a series of five hoplike contacts to a test tile. The procedure is depicted in Fig. 16.8. The contact disk was alternated between Teflon and quartz to represent triboelectric negative and positive extremes. The top of the 2-in disks was covered with copper foil to simulate the insulative footwear-conductive body model. A 5-in aluminum handle insulated by Teflon was used to move the disks. Three steel washers weighing about 100 g were placed on the copper by sliding over the handle in order to simulate actual footwear pressures on a floor. Representative tiles or treated tiles were placed on a ground plane for testing. An operator, grounded via wrist strap, would then move the disk for five hops across a 3- by 12-in sample. Accumulated charge was measured by dropping the pogo stick into a Faraday cup connected to an electrometer.

## 16.9 Protective Packaging Considerations

Imagine a sensitive item in a protective package as indicated in Fig. 16.9. The protective package can be considered in the general case, that is, a bag, tote box, DIP tube, or some other shipping or storage container. Likewise the sensitive item might be a part or an assembly containing a sensitive part or parts. What is required of the package to afford sufficient protection? By answering this question the parameters to evaluate can be determined. For a voltage-sensitive part, the package must provide the following:

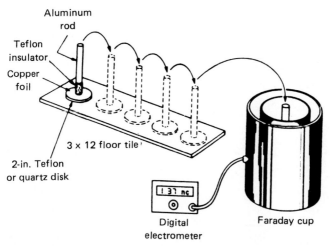

Figure 16.8 Hop test. (From Chase[4] with permission from the EOS/ESD Association.)

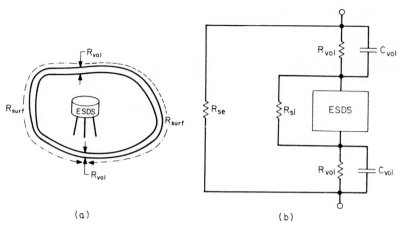

Figure 16.9 Susceptible item in protective package.

1. An external low-resistance path to ground to prevent breakdown potentials inside the package in the event of direct discharge to the outside of a grounded package
2. The same external low-resistance path to prevent breakdown potentials inside the package when a previously charged package is grounded
3. Sufficient shielding for protection from the influences of external fields exceeding the breakdown potentials inside the package

For an energy-sensitive item the package must provide the following:

1. Either a sufficiently low-resistance external path to prevent damaging energy levels from entering the package during direct discharge to the grounded package, or relatively high bulk resistance in order dissipate sufficient energy en route through the package to limit energies inside, or a combination of relatively low surface resistance compared with bulk volume resistance.
2. Conditions described in item 1 to prevent entry of damaging energy levels during discharge of a previously charged package.
3. Sufficient shielding to prevent entry of damaging potential energy levels from external fields. This need is based on a possible FID model discharge from a contained capacitance.

The type of protection needed differs depending on whether the item is voltage- or energy-sensitive. The bag can be regarded as the most critical configuration because of flexibility permitting intimate contact with the sensitive unit.

### 16.9.1 Shielding

As far as ESD is concerned, shielding can be regarded as two types: shielding (or more appropriately, protection) from direct discharge to the container, and shielding from electrostatic fields.

### 16.9.2 Bag evaluations

Many of the properties discussed elsewhere in this chapter apply to bag evaluations. Evaluation of shielding properties have not been addressed. For a number of years users were lulled into a false sense of security by the belief that even a low conductivity would provide a sufficient Faraday cage and keep all charge on the outside. The only parameters cited were likely to have been decay time to 10 percent of initial value in less than 2 s and surface resistivity $<10^{14}$ Ω per square, with or without nebulous reference to Mil B 81705B. Users who questioned the shielding properties of "protective" bags were left to their imaginations to devise testing techniques.

**16.9.2.1 Part in bag tests.** The many tests cited on bag materials do not give a user confidence that the bag will perform as intended. Proposed methods for bag evaluations are controversial. Therefore, as in evaluations of other control measures, the most practical approach may be to duplicate the end-use conditions. The most direct approach is to test a sensitive part in the bag. Generally a voltage-sensitive MOS device would be placed in the bag with the gate lead positioned on one inner side of the bag and the source and/or drain at the other, as shown in Fig. 16.10. For a margin of safety, a worst-case application is desirable. I have used a technique of inserting an unprotected discrete MOSFET such as a 3N128 into the capacitive field-sensing

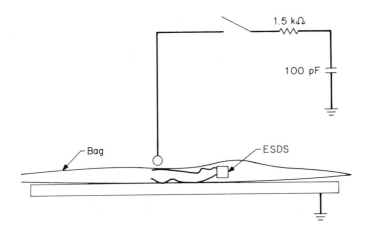

**Figure 16.10**  Part in bag test.

structure shown in Fig. 16.11. Energy-sensitive or other part types could be inserted in the same fixture as needed. This method is extremely worst-case and thus has an inherent margin of safety built in. With consistent part samples the method can provide good comparative results between bag types.

Kolyer and Anderson[5] conducted both radar and charged "peanut" exposure tests of an unprotected 2N4351 MOS device in a variety of bags. The radar test consisted of placing a bagged device directly in front of a 1.6-in × 2.1-in Ku band horn antenna radiating 6-kW peak power at 16 GHz for 10-s exposures. The charged peanut or vehicular bounce test consisted of packing a bagged MOS device in a larger plain polyethylene bag filled with polystyrene foam cushioning peanuts. The larger bag was then placed in a cardboard box and was shaken to simulate shipping movement. Although the peanuts charged to 5000 V, no failures occurred, even in two varieties of antistatic bags. The radar test also produced no failures.

### 16.9.3  Useful life of metallized bags

Users often disregard the fact that metallized static shielding bags have a limited life. The thin metal layer becomes discontinuous after repeated handling and crinkling. This can be readily demonstrated by performing a "snowball" test. To perform the snowball test repeatedly crunch a metallized bag as if making a snowball. Then fold the bag and place two coins at opposite side locations as shown in Fig. 16.12.

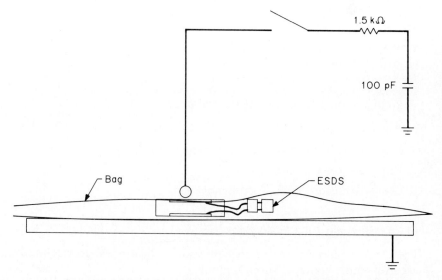

**Figure 16.11**  Capacitive field-sensing fixture (device leads are attached to 1-in copper sensor disks.)

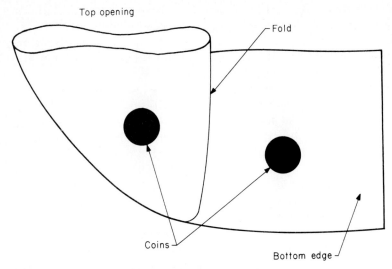

**Figure 16.12** Metallized bag continuity test.

An ohmeter check from coin to coin is likely to read "open circuit." Greater than 10 k$\Omega$ is generally considered unacceptable. If a carefully planned definition of usage cycle is not in place, open-circuited bags are likely to be found throughout a facility. The crinkled appearance need not be severe, so visual inspection may not be an adequate means of detection.

### 16.9.4 Other parameters

Evaluations of bags must consider important parameters other than electrostatic properties. These include the following:

1. Water vapor permeability per Federal Standard 101 Method 3030B, as required by Mil B 81705
2. Abrasion resistance per Federal Standard 141A Method 6192
3. Heat seal strength using a pull-type strain gauge
4. Puncture resistance
5. Cost, including reuse if applicable

### 16.9.5 Shipping Tube Evaluations

There are several types of special dual in-line package (DIP) tubes used for static control. These include:

1. Antistatic coated PVC
2. Antistatic impregnated PVC
3. Black carbon-loaded conductive
4. Metallized exterior with antistatic interior PVC
5. Aluminum
6. Carbon-loaded or aluminum with transparent plastic windows (Windows are generally antistatic, although some have been found to generate.)

The properties desired in shipping tubes could be considered to be antistaticity (meaning triboelectric charge inhibition), acceptable static dissipation or decay rate, and static shielding. Most of the techniques previously described can be applied. The first test to detect gross problems is simply to rub a tube by hand to see if triboelectric generation occurs. If any voltage is detected, other materials such as paper, nylon, and/or wool can be used to broaden the triboelectric range. More refined measurements can be attained by using the Faraday cup technique described in Chap. 8.

#### 16.9.5.1 Static shielding DIP tubes

Although static shielding is a desirable feature in DIP tubes, there are several problems associated with attainment of this feature:

1. Antistaticity of the interior is desirable and is to be retained to avoid CDM failures.
2. Many users believe that transparency is a highly desirable, if not absolutely needed, feature.
3. A highly conductive tube interior or exterior might contribute to incidence of CDM failures by providing sparkover opportunity.
4. The cost of shielding containers is high, especially with antistatic interiors.

Static shielding is needed only when susceptible items are away from static controlled areas. Based upon economic considerations, my preference is that, for large quantity shipping and in-plant storage and transit, static shielding be provided by outer bulk packaging means rather than being incorporated in DIP tubes. This opinion would be modified with the development of an economical shielding tube with an antistatic interior.

#### 16.9.5.2 Antistaticity

The ability to prevent or greatly inhibit triboelectric charging is the most important feature of DIP tubes. This need arises on two fronts:

1. To impede charging of sliding devices internal to the tube in order to prevent CDM failures
2. To limit charging of the tube itself from handling, parts sliding, and other movements in order to prevent field-induced failures and detrimental induction effects from tubes at high potentials

The considerations for antistaticity testing discussed in Sec. 16.2 apply to evaluation of DIP tubes as well. Certainly device sliding tests using the Faraday cup as described in Chap. 8 are appropriate. Additionally, rubbing tests with the bare hand and possibly other process-related materials would be in order.

#### 16.9.5.3 Conclusions for DIP tube evaluations

Based on the preceding discussion the recommended method for evaluating DIP tubes is to test for antistaticity and decay times. Static shielding is desirable but generally more economically provided by other outer pack means. Resistivity is not correlative to the other desired features and is not recommended.

### 16.10 Chairs and Carts

Evaluations of chairs and carts involve questions of charge buildup propensity and charge bleedoff rates. Charge buildup can be from triboelectric generation on the item itself or by transfer from other objects, such as the human body. In either case consistent bleedoff to ground at a reasonably fast rate is desired. Chairs or stools might have insulative seats or backs. Most carts are conductive, but occasionally insulative plastics or cushioning is used. Triboelectric tests can be conducted on these portions by rubbing with the bare hand, paper, wool, and nylon. If unacceptable voltage levels are achieved, the item can be replaced, modified, or treated with a topical antistat. Static bleedoff, even from conductive portions of the chair, stool, or cart depends on a conductive path to ground, normally via the floor. Evaluation requires that electrical continuity is verified to the base of the chair legs or to conductive casters on the cart. Construction sometimes contains hidden insulators that obstruct the path to ground. The third problem is contact resistance to the floor. A useful test for carts is to connect the cart to a charge plate via an insulated wire. The charge plate is then monitored by a field meter and recorded on a strip-chart recorder as a person wearing insulative shoes pushes the cart. If a good path to ground is not present, the charge plate will register large voltage transitions generated by the person.

## 16.11  Ionization Evaluations

Product ESD sensitivity must be determined in order to properly evaluate offset and induced voltage swing requirements as well as decay times. All static decay rates are not necessarily exponential as a result of ionizer neutralization. Standardization of the charged plate dimensions and placement is critical to comparisons of data from different sources.

### 16.11.1  Parallel plate ion counter

This method for measuring the output of ionizers consists of two parallel metal plates with a voltage applied between them. A fan is used to control the amount of air flow between the two plates. Ion concentration of the air between the plates results in measurable current that is indicative of the ion concentration. Properly constructed and instrumented this method will accurately determine the ion concentration or ion count of the air. A difficulty is relating the test conditions to factory conditions. In particular, consistency of air flow conditions is important.

### 16.11.2  Biased plate

A similar current can be measured by the effects of ions impinging upon a single biased plate. Inaccuracies can result from effects of extraneous fields. A biased plate surrounded by a grounded metal grid is an attempt to shield from the effects of these extraneous fields and still measure the current as an indication of ion concentration. An error built into this method is the unaccounted for effects of ions attracted to the grounded grid.

### 16.11.3  Static decay

This method measures the time to decay to a prescribed level from an established charge level. Normally a floating (ungrounded) metal plate is used for this purpose. This method is popular because of its simplicity and perceived similarity to actual usage. The problem of effects of extraneous fields remains a source of possible error.

## 16.12 Antistatic Treatments

Antistatic treatments must be measured periodically for triboelectric charge buildup after rubbing. Items selected for rubbing can be based on the application or an assortment such as used for chairs in the preceding paragraph. Where the treatment is intended to provide static bleedoff to ground, periodic resistance to ground measurements must also be made. Care must be made to make sure the treatment does not contain corrosive contaminants.

## 16.13 Miscellaneous

A good general guideline to follow in evaluating ESD control products is to keep the intent of usage in mind for any tests to be conducted. There are now standards specifically developed with ESD control in mind for such things as wrist straps, work surfaces, ionizers, garments, and flooring. Detailed discussions of these items are in Chap. 17.

## References

1. B. A. Unger et al., "Sheet Resistance Measurement of Buried Shielded Layers," *Electrical Overstress/Electrostatic Discharge Symposium Proceedings*, 1986.
2. Gary O. Head, "Drastic Losses of Conductivity in Antistatic Plastics," *Electrical Overstress/Electrostatic Discharge Symposium Proceedings*, 1982.
3. John H. Kolyer and Dale M. Cullop, "Methodology for Evaluation of Static-Limiting Floor Finishes," *Electrical Overstress/Electrostatic Discharge Symposium Proceedings*, 1986.
4. E. W. Chase and B. A. Unger, "Triboelectric Charging of Personnel from Walking on Tile Floors," *Electrical Overstress/Electrostatic Discharge Symposium Proceedings*, 1986.
5. John M. Kolyer and William E. Anderson, "Selection of Packaging Materials for Electrostatic Discharge Sensitive (ESDS) Items," *Electrical Overstress/Electrostatic Discharge Symposium Proceedings*, 1981.
6. Charles E. Jowett, *Electrostatics in the Electronics Environment*, Wiley, New York, 1976.

Chapter

# 17

# Standards

Numerous documents and standards are used in the field of ESD control. Unfortunately, standards have not yet been developed in many needed areas, and existing ones, intended for other purposes, are frequently improperly applied. Often the misapplication of otherwise good test methods is due to a combination of ignorance and the lack of more appropriate ESD-specific standards. The problem has been compounded by the use of raw-material standards to characterize finished ESD control products. This approach ignores variations in performance due to construction and/or compound material content. Some existing ESD-specific standards have improper or inadequate parameters specified. In time all these inconsistencies will be corrected through new and revised documents. In the meantime discretion is necessary to specify parameters to ensure effectiveness of control products. The following questions are listed to invoke thought and facilitate better ESD control product characterization regardless of the standards cited by customers or vendors:

1. If static neutrality or an "antistatic" property is desired, why does the vendor cite resistivity values?
2. If static neutrality or an antistatic property is desired, why does the vendor cite static decay test results?
3. Is ASTM D257, a standard intended for resistance and conductance characterization of insulators, appropriate to show "qualification" of antistatic, static dissipative, and even conductive materials?
4. If static shielding is desired, are resistivity or static decay appropriate parameters to qualify a product?
5. Even when the right parameter is measured, are the test conditions and sample configuration appropriate?

These types of questions will be answered through standards specifically developed for ESD control. Although progress is being made in this regard, some of the newer standards have been experiencing "growing pains" and are still somewhat controversial.

The subject of standards related to electrostatic discharge would require an entire book to give complete coverage. This chapter will highlight important aspects of some of the major standards involved with ESD control procedures and control product characterization. Readers are encouraged to refer to the actual standards for further information.

Additionally, the fact that standards development in the area of ESD control is a dynamic ongoing activity requires that the user must become familiar with the latest revisions and newly released documents.

## 17.1 Military Standards

The industry-wide impact of ESD standards developed for the military has been undoubtedly greater than from any other source. Criteria contained in these documents have been widely used, even in commercial applications. Information contained in them can provide excellent guidance in ESD control procedures, control measures selection, and control product acceptance parameters.

### 17.1.1 Military Standard 883, Method 3015, and Mil M 38510

Mil M38510 is the specification generally applied to the purchase of integrated circuits for military applications. This standard calls out tests for electrostatic classification according to Mil Std 883, Method 3015. The current version is notice 7. This standard classifies parts utilizing the standard 100 pF, 1.5 k$\Omega$ human body model. Parts sensitivity classes are as follows:

*Class 1:* > 0 to ≤999 V
*Class 2:* > 2000 to ≤3999 V
*Class 3:* > 4000 V and above

Owing largely to the results of ESD human body simulator comparisons discussed in Chap. 8, Method 3015 of Mil Std 883 now requires a current waveform test to qualify the HBM simulator circuit. The required waveform is shown in Fig. 17.1.

### 17.1.2 DOD Standard 1686, Military Standard 1686A, and DOD Hdbk 263

During the period between 1978 to 1980 the reliability engineering department of the Naval Sea Systems Command developed DOD Std 1686 and DOD Hdbk 263. These documents were formulated in order

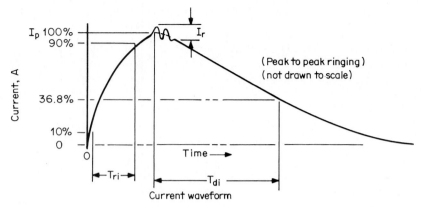

**Figure 17.1** The required HBM simulator current waveform required by method 3015 of Mil Std 883.

to influence the military services and their contractors to initiate comprehensive ESD control programs. The motivating force behind these documents was Toshio Oishi, then Manager of Reliability. After investigating shipboard reliability and equipment downtime, he became convinced that ESD was a major factor. Additionally he found little or no ESD control measures practiced on the part of navy personnel or contractors investigated.

The two documents are complementary. DOD Std 1686 contains the requirements for a comprehensive ESD control program while allowing a considerable amount of flexibility in selecting specific control measures. DOD Hdbk 263 contains tutorial and technical backup data to help educate the user as an aid in selecting and properly using ESD control options. The handbook provides definitions and other information necessary for implementation of DOD Std 1686. In August 1988 revised Military Standard 1686A was released. In summary, both DOD Std 1686 and Mil Std 1686A:

- Define the content of a comprehensive ESD control program
- Provide a method for ESD sensitivity classification
- State design requirements intended for greater ESD immunity
- Define ESD protected areas and protective packaging
- Require identification of ESD sensitive items, proper handling procedures, awareness training, quality assurance provisions, and audits

**17.1.2.1. Control program tailoring.** An important aspect of DOD Standard 1686 is the recognition of the need for control program tailoring. Paragraph 1.2 of the standard reads as follows:

> 1.2 Purpose. The purpose of this standard is to minimize the impact of ESD on equipment reliability and life cycle cost. This standard may be

tailored for various types of acquisitions and allows implementation options to users to effect a cost effective ESD control program.

Mil Std 1686A states "The selection of the applicable control program functions and elements of table 1 for each acquisition shall be made by the contractor and shall be approved by the acquiring activity."

The need for ESD control program tailoring can be emphasized by contemplation of analogous tailoring recognized as necessary in the specialty of reliability engineering. There are numerous texts available as well as countless military and other standards on virtually every aspect of the subject. These include Military Handbook 217 for predictions and a great many more. Most government contractors have an accumulation of the available documents to provide a valuable source of detailed information on any reliability related topic. Yet never, even for the most extreme high reliability requirements, are all these standards followed to the letter. A cookbook approach would be impractical, too costly, and ineffective. So the specialty of reliability engineering is needed to rise to the task and tailor a plan to best meet the requirements while contending with all the other variables associated with the contract.

Just as reliability programs are tailored for specific contracts, a certain amount of variability may be necessary in ESD control programs. As discussed in Chap. 14, the stringency of measures may be based on contract requirements, cost, susceptibilities, manufacturing environment, and specific operation needs. The fact that extensive ESD-related standards are not yet available precludes a cookbook approach even if desirable. Even where standards exist, tailoring is not only desirable but essential. Perhaps ESD control is not as complex as the many facets of reliability engineering. Nevertheless, the topic ESD is quite diversified when taken in its entirety as evidenced by past proceedings of the annual EOS/ESD symposia if not this text. The need for variability of controls with such a rapidly changing state of the art is another reason to establish a logical manner in which to tailor an ESD control program.

**17.1.2.2 Comprehensive ESD control program.** Both standards outline and require that a tailored but comprehensive ESD control program be in place. Table 17.1 is the outline of control program elements and functions of applicability called out in DOD Std 1686. The modified version from Mil Std 1686A is shown in Table 17.2.

**17.1.2.3 ESD sensitivity classification.** Both DOD Standard 1686 and Military Standard 1686A define sensitivity ranges or categorization classes. Both documents allow several methods to derive sensitivity

TABLE 17.1 Outline of Control Program Elements and Functions of Applicability Called Out in DOD Std 1686

| Functions applicable to an acquisition | Control program elements | | | | | | | | | | |
|---|---|---|---|---|---|---|---|---|---|---|---|
| | Identification & classification See 5.1 | Design protection See 5.2 | Protected areas See 5.3 | Handling procedures See 5.4 | Protective covering See 5.5 | Installation site See 5.6 | Training See 5.7 | Marking documentation See 5.8.1 | Marking hardware See 5.8.2 | QA provisions, audits, & reviews See 5.9, 5.10 | Packaging for delivery See 5.11 |
| Design | X | X | | | | | | X | | X | |
| Manufacturing | | | X | X | X | | X | | X | X | |
| Inspection (examination and test) | | | X | X | X | | X | X* | X | X | |
| Packaging | | | X | X | X† | | X | | | X | X‡ |
| Rework/repair | | | X | X | X | | X | | X* | X | |
| Failure analysis | | | X | X | | | X | | | X | |
| Training courses | | | | | | X§ | | | | | |
| Field installation | | | X | X | X | | X | | | X | |
| Field maintenance test | | | | | | | | | | | |

*If not previously performed.
†Internal to contractor's facility.
‡External to contractor's facility.
§For a shipbuilding contract, unless otherwise specified in such contract, the installation site element is the only element of this table that applies to ships and shipyards.

TABLE 17.2  Control Program Elements of Mil Std 1686A

| | Elements | | | | | | | | | | | |
|---|---|---|---|---|---|---|---|---|---|---|---|---|
| Functions | ESD control program plan (5.1) | Classification (5.2) | Design protection (excluding part design) (5.3) | Protected areas (5.4) | Handling procedures (5.5) | Protective covering (5.6) | Training (5.7) | Marking of hardware (5.8) | Documentation (5.9) | Packaging (5.10) | QA provisions, audits, & reviews (5.11–5.12) | Failure analysis (5.13) |
| Design | X | X | X | | | | | | X | X | X | X |
| Production | X | | | X | X | X | X | X | X | X | X | X |
| Inspection & test | X | | | X | X | X | X | X | X | X | X | X |
| Storage & shipment | X | | | X | X | X | X | X | X | X | X | |
| Installation | X | | | X | X | X | X | X | X | X | X | |
| Maintenance & repair | X | | | X | X | X | X | X | X | X | X | X |

classification. The voltage ranges differ significantly between the two documents, and there are slight variations on the allowed categorization methods. The high cost of actual part sensitivity testing has been acknowledged by the fact that neither standard requires actual testing even though detailed test procedures are provided.

**17.1.2.3.1 DOD Standard 1686 sensitivity classes.** DOD Standard 1686 specifies two classes for electrostatic-discharge-sensitive (ESDS) items:

Class 1 items are susceptible to damage from ESD voltage levels of 1000 V or less.

Class 2 items are susceptible to damage from ESD voltage levels greater than 1000 V but less than or equal to 4000 V.

Although the standard does not define any other classes, it does recommend that for high reliability or critical programs that less sensitive items be considered in the ESD control program. DOD Handbook 263 is cited for the definition of sensitivities between 4000 and 15,000 V. (This range is defined as class 3 by the handbook.)

**17.1.2.3.2 Military Standard 1686A classes.** Mil Std 1686A defines three classes in order to be consistent with Mil Std 883 method 3015:

Class 1 is susceptible to damage from ESD levels between 0 and 1999 V.

Class 2 is susceptible to damage from ESD levels between 2000 and 3999 V.

Class 3 is susceptible to damage from ESD levels between 4000 and 15,999 V.

Mil Std 1686A requires that class 3 items be included in the control program when equipment is designated as mission critical or essential.

**17.1.2.3.3 DOD Standard 1686 classification methods.** Appendix A (see Fig. 8.20) of the standard is provided with the intention of part sensitivity categorization without the need for testing. Other allowed means are to be the exception. "When there are reasons to believe a part is less sensitive than indicated by appendix A, the part may be reclassified based upon:

a. ESD voltage level when specified in the applicable military part specification such as the Mil-M-38510 V-Zap test levels or

b. ESD voltage levels from test data obtained in accordance with appendix B or a test method similar to that of appendix B and approved by the acquiring activity. Copies of the test data and test procedure shall be available for the acquiring activity.

Assemblies are classified according to the most sensitive parts contained therein. Reclassification of assemblies to lesser sensitivities can be done by circuit analysis.

**17.1.2.3.4 Military Standard 1686A classification methods.** The standard permits classification of ESD sensitivities of parts in four ways:

1. As specified in the applicable part specification
2. In accordance with appendix A test data contained in the reliability analysis center (RAC) ESD-sensitive items list (EDSIL)
3. In accordance with appendix B (see Tables 17.3 to 17.5)
4. When specified, or at the option of the contractor, by test according to appendix A of the standard

**17.1.2.4 ESD sensitivity identification.** Both DOD Std 1686 and Mil Std 1686A require that deliverable drawings and related specifications

TABLE 17.3 Mil STd 1686 Class 1 Parts

Appendix B
Table III. List of ESDS Parts by Part Type, MIL-STD-1686A, 8 August 1988

| Part type |
|---|
| Microwave devices (Schottky barrier diodes, point contact diodes, and other detector diodes > 1 GHz) |
| Discrete MOSFET devices |
| Surface acoustic wave (SAW) devices |
| Junction field-effect transistors (JFETs) |
| Charge-coupled devices (CCDs) |
| Precision voltage regulator diodes (line or load voltage regulation < 0.5 percent) |
| Operational amplifiers (OP AMPs) |
| Thin-film resistors |
| Integrated circuits |
| Hybrids utilizing class 1 parts |
| Very high speed integrated circuits (VHSIC) |
| Silicon controlled rectifiers (SCRs) with $I_0$ < 0.175 A at 100°C ambient |

**TABLE 17.4 Mil Std 1686A Class 2 Parts**

| Part type |
|---|
| Devices or microcircuits when identified by appendix A test data as class 2: |
|   Discrete MOSFET devices |
|   JFETs |
|   Operational amplifiers (OP AMPs) |
|   Integrated circuits (ICs) |
|   Very high speed integrated circuits (VHSIC) |
| Precision resistor networks (type RZ) |
| Hybrids utilizing class 2 parts |
| Low-power bipolar transistors, $P_T < 100$ mW with $I_c < 100$ mA |

**TABLE 17.5 Mil Std 1686A Class 3 Parts**

| Appendix B |
|---|
| Table III. List of ESDS Parts by Parts Type, MIL-STD-1686A, 8 August 1988 |

| Part type |
|---|
| Devices or microcircuits when identified by Appendix A (Fig. 8.20) test data as class 3: |
|   Discrete MOSFET devices |
|   JFETs |
|   Operational amplifiers (OP AMPs) |
|   Integrated circuits (ICs) |
|   Very high speed integrated circuits (VHSIC) |
| All other microcircuits not included in class 1 or class 2 |
| Small signal diodes with power $<1$ W or $I_0 < 1$ A |
| General-purpose silicon rectifiers |
| SCRs with $I_0 > 0.175$ A |
| Low-power bipolar transistors with 350 mW $> P_T >$ 100 mW and 400 mA $> I_c >$ 100 mA |
| Optoelectronic devices (LEDs, phototransistors, opto couplers) |
| Resistor chips |
| Hybrids utilizing class 3 parts |
| Piezoelectric crystals |

CAUTION NOTE
THIS EQUIPMENT CONTAINS PARTS AND
ASSEMBLIES SENSITIVE TO DAMAGE BY
ELECTROSTATIC DISCHARGE (ESD). USE
ESD PRECAUTIONARY PROCEDURES WHEN
TOUCHING, REMOVING, OR INSERTING.

**Figure 17.2** Mil Std 129 symbol and cautionary note.

identify items as ESDS without necessitating category level breakdowns. DOD Std 1686 requires that installation and interface drawings, technical manuals and training course materials are to identify items and external terminals as class 1 or class 2 as applicable. The latter types of documentation are to include or refer to documented protective handling procedures. Nondeliverable documentation used internally by a contractor is to identify items as class 1 or class 2 and include or refer to documented handling procedures. Mil Std 1686A does not require class breakdowns for internal documentation but leaves such breakdowns as the option of the contractor.

DOD Std 1686 hardware marking requirements consist of placing the Mil-STd-129 symbol on sensitive assemblies in a position that is readily visible when the assembly is incorporated in its next higher assembly level. Equipment enclosures are required to have the Mil-Std 129 symbol with a cautionary note as shown in Fig. 17.2.

(MIL-STD-1285)

(RS-471)

"CAUTION
CONTAINS PARTS AND
ASSEMBLIES SUSCEPTIBLE TO DAMAGE BY
ELECTROSTATIC DISCHARGE (ESD).

**Figure 17.3** Mil Std 1285 or the EIA RS-471 symbol and cautionary note.

Mil Std 1686A requires part marking per Mil Std 1285. Assemblies are to be marked with either the Mil Std.1285 or the EIA RS-471 symbol and cautionary note as shown in Fig. 17.3. Equipment enclosures are to be similarly marked adjacent to sensitive external terminals.

**17.1.2.5 ESD protective packaging.** Protective packaging is required by DOD Std 1686 for sensitive items during transportation and storage. Packaging for delivery must protect against triboelectric generation and corrosion. Sensitive items must be enclosed in an ESD protective material, "such as that conforming to Mil-B-81705 Type 1 to protect from electrostatic voltages of at least 4000 volts in the form of fields or direct discharge." The following cautionary note is required on the outer package:

CAUTION NOTE

OBSERVE PRECAUTIONS FOR

HANDLING ELECTROSTATIC

DISCHARGE SENSITIVE ITEMS

Mil Std 1686A requires protective packaging in accordance with Mil E 17555. In addition, protective caps are required on sensitive external connector terminals.

**17.1.2.6 ESD protected areas.** DOD STD 1686 requires that protected areas be limited to electrostatic voltages below the lowest voltage sensitivity level of items processed. The protected area "shall extend, as a minimum, 1 meter from the periphery of a Class 1 or Class 2 item work area." ESD protective handling procedures are required in ESD protected areas.

Mil Std 1686A also requires that voltages in protected areas be maintained below the lowest voltage sensitivity level of items processed. The extension of the area to 1 m from the periphery has been dropped. In addition the standard allows the use of "detailed alternative handling precautions and procedures" in unprotected areas where practicality precludes handling in protected areas.

**17.1.2.7 ESD design requirement.** DOD Std 1686 states that class 1 parts shall not be used where less sensitive parts are available. Where class 1 parts must be used, protective circuitry is to be incorporated at the lowest practical assembly level to limit sensitivity to class 2 as a minimum. External equipment cabinet terminals are to be protected to a sensitivity of 4000 V minimum. Where protective circuitry cannot be incorporated, a waiver must be obtained from the acquiring activity. Mil Std 1686A has essentially the same requirements worded more succinctly.

**17.1.2.8 Proper handling procedures.** Detailed documented handling procedures related to the item sensitivity levels are required. "The more sensitive the item, and the fewer controls afforded by the protected area, the more detailed the procedures shall be to provide the required protection from ESD."

**17.1.2.9 Control program discipline.** ESD control discipline is ensured by DOD Std requirements for

1. ESD awareness and handling training for personnel who specify, acquire, design, manufacture, assemble, process, inspect, test, package, repair, rework, install, or maintain class 1 or class 2 items
2. Quality assurance provisions including certification of protected areas with periodic monitoring
3. Scheduled design and program reviews to cover pertinent ESD related topics
4. Reserved rights of the acquiring activity to conduct audits, design reviews, and program reviews
5. Extension of items 1 to 4 to subcontractors and suppliers

Mil Std 1686A has similar requirements with the noted difference that ESD awareness training is specifically mentioned for supervisors of all functions listed in Table 17.2. Additionally a point is made that training shall be recurrent.

**17.1.2.10 Documentation.** DOD Std 1686 requires the following documentation:

1. Copies of susceptibility test procedures and data, as applicable
2. Circuit analysis to show lesser sensitivity levels of assemblies, as applicable
3. ESD handling procedures
4. ESD training records
5. Minutes of design and program reviews

Mil Std 1686A contains the same requirements with the following noted changes:

1. An approved ESD control program plan is required.
2. Sensitivity classification and rationale is to be included as part of the ESD control plan.
3. The test method or analytical technique to meet design requirements requires approval.
4. Internal quality assurance provision records are to be maintained.

**17.1.2.10.1 Deliverable data items.** DOD Std 1686 requires no additional deliverable data items other than inclusion or references to handling procedures as discussed in Sec. 17.1.2.10.

Mil Std 1686A requires also the following deliverable items:

1. Electrostatic discharge control plan
2. Electrostatic discharge sensitivity tests, as applicable
3. ESD handling procedures

**17.1.2.11 Comments on DOD Std 1686 and Mil Std 11686A.** The fact that, when imposed, these documents require tailoring to any specific program cannot be overemphasized. A natural response on the part of personnel who are not ESD specialists is to attempt to initiate each control program element to the fullest or most stringent extent. This approach does not meet the intent of either standard, which have decreased life-cycle costs as a primary objective. The needs of each program must be carefully analyzed with due regard to such parameters as reliability, maintenance, functional operating parameters, mission criticality, and costs as well as ESD sensitivity.

**17.1.2.12 DOD Handbook 263.** DOD Handbook 263 is a companion document to be used in conjunction with either DOD Std 1686 or Mil Std 1686A. This document is intended as a guide that contains definitions intended for implementation of an ESD control program to meet the requirements of the 1686 standards. This handbook contains a wealth of information helpful to anyone establishing a control program whether the standards are imposed or not.

It is anticipated that a revised version will soon be released to bring the information more in line with advances over the past decade. Of concern are the definitions of antistatic, static dissipative, and conductive materials. In the current nonrevised handbook the definitions are

*Antistatic material:* ESD protective material having a surface resistivity of greater than $10^9$ but not greater than $10^{14}$ Ω per square

*Static dissipative material:* ESD protective material having a surface resistivity of greater than $10^5$ but not greater than $10^9$ Ω per square

*Conductive material:* ESD protective material having a surface resistivity of $10^5$ Ω per square maximum.

The current thinking reflected in the August 1987 EOS/ESD Association Glossary of Terms and EIA 541 is that resistivity is not directly correlated to antistaticity, which is the resistance to triboelectric generation. The glossary defines

*Antistatic:* Those materials that resist triboelectric charging and produce minimal static charge when separated from themselves or other materials. A material's antistatic property is not necessarily correlated with its resistivity.

*Static Dissipative ESD Protective Materials:* Materials having a surface resistivity of at least $1 \times 10^6$ Ω per square, or $1 \times 10^4$ Ω · cm volume resistivity, but less than $1 \times 10^{13}$ Ω per square, or $1 \times 10^{11}$ Ω · cm volume resistivity.

The recently issued EIA 540 is more stringent as the surface resistivity upper limit for static dissipative is $10^{12}$ Ω per square.

### 17.1.3 Mll B 81705

Type I is a water vapor–proof, grease-proof, electrostatic, and electromagnetic protective laminate bag containing a metal-foil layer. Type II is a transparent, waterproof, electrostatic protective bag tinted any color. Bags meeting the standard must be clearly marked as to type, manufacturer, month and year of manufacture, and lot number in letters at least ⅛ in high. The words "antistatic" or "EMI/static shield" must be included in ½-in minimum letters. This standard contains a static decay test per FTMS 101, Method 4046, with a requirement of dissipation to 0 from 5 kV in less than 2 s. A 3- by 5-in sample of the material is tested with relative humidity at less than 15 percent and room ambient of 73°F.

### 17.1.4 Mil Std 2000 and Weapons Specification (WS) 6536E

These two documents are mentioned to indicate the widespread attention given to ESD controls within the military. These documents establish strict soldering standards in order to achieve quality work. The importance of ESD controls at stations where soldering is done is recognized specifically in both documents. DOD Std 1686 is referenced in each with pertinent requirements as noted in Secs. 17.1.4.1 and 17.1.4.2.

**17.1.4.1 Mil Std 2000.** Electrostatic discharge controls at stations where soldered connections are made for leads inserted into holes, surface mounted to lands, or attached to terminals shall be in accordance with Mil Std 1686. Specifically the standard also requires that bags, containers, tape, and reel materials used for static-sensitive solderable devices provide protection in accordance with DOD Std 1686. Related requirements addressing both electrical overstress as well as ESD apply to soldering equipment. The document specifies that "soldering irons, soldering machines and systems, and associated process equipment (including fluxers, preheaters, solder pots, cleaning system, and cleanliness test equipment) shall be of a type that does not compromise functional integrity by injecting electrical energy into the item(s) being cleaned." Soldering iron tips must have a resistance of 2 $\Omega$ maximum and a potential difference of 2 mV maximum, both measured with respect to ground while the tip is hot. More specifically, the standard requires three wire irons with grounded tips, designed in such a way as to provide zero voltage switching. Transformer-type soldering guns are not permitted.

**17.1.4.2 WS 6536E.** Electrostatic discharge protection called out in paragraph 3.2.12 follows:

> Electrostatic discharge protection shall be maintained in a manner that is routine, with controls exercised over parts during receipt and testing through the manufacture and inspection cycles, storage, and shipping. Electrostatic discharge protection criteria, as a minimum, shall be in accordance with DOD Std 1686 and DOD Hdbk 263. In addition to classes 1 and 2, class 3 static sensitive devices, as defined in DOD Hdbk 263 shall be protected from damage due to electrostatic discharge.

The last requirement, taken literally, could overshadow DOD Std 1686 or Mil Std 1686A. If the paragraph had clarified the requirement of protection of class 3 devices at the soldering stations, then the impact would be minimal. If the requirement is to be extended across the entire ESD control program, however, extensive costs could be involved with questionable benefit. The problems would not have to do with workstation protection. Those measures would be virtually unchanged because of the high likelihood that the class 1 and/or 2 items would already have mandated good ESD control measures. The added costs would be in the areas of class 3 sensitivity classification, extended hardware and document marking, protective packaging from vendors for class 3 items, and control discipline. Additionally there could be a negative psychological effect in fostering scorn at controls for parts not normally considered ESDS.

The extent of overall ESD controls, especially to include class 3 items, should appropriately be left to the contract statement of work (SOW) and DOD Std 1686 or Mil Std 1686, as applicable. These 1686 documents recognize the need to extend controls to class 3 as being the exception rather than the rule. Nevertheless, where WS6536E is invoked, the requirement stands unless a waiver is negotiated with the customer.

Soldering irons are not to inject damaging electric energy and must meet the requirements of DOD Std 1686 and DOD Hdbk 263.

The tips, when hot, shall have a resistance to ground less than 20 Ω and less than 2 mV rms. Transformer-type soldering guns, unplated copper tips, and plated tips with exposed basis metal are not allowed.

## 17.2 Existing Government and Industry Standards Applied to ESD Control Products

Numerous existing industry standards are currently being used to characterize ESD control products. Caution is advised in the use of these standards. The construction, materials and application of the item in question must be well understood to ensure appropriateness of a standard and to account for any inconsistencies. Highlights of some widely applied standards follow.

*AATCC 134:* This American Association of Textile Chemists and Colorists (AATCC) standard is a test method for static propensity of carpets. This standard's original purpose was to evaluate the potential shock discomfort of high static accumulated on personnel walking on carpets. The test is conducted at 20 percent RH and 21°C. Extraneous charged and/or grounded materials must be kept at least 61 cm distant from the test subject wearing Neolite soles and heels. The specification is very detailed but consists primarily of voltage-monitored scuff tests of a subject wearing Neolite, leather, and a third nonstandard sole/heel combination. Acceptance limits are not included.

*ASTM D257:* Called "DC Resistance or Conductance of Insulating Materials," this American Society for Testing and Materials (ASTM) standard was originally intended for characterization of insulators. This does not mean that some of the methods could not be properly applied to ESD control. The standard is lengthy and contains a wealth of information on the subject of resistance- and conductance-related variables. As an example of the complexity consider that electrode systems contained include binding posts, taper pins, metal bars, silver paint, sprayed metal, evaporated metal, metal foil, brushed-on colloidal graphite in water, mercury or other liquid metal, flat metal plates, or conductive rubber. For flat samples the electrode/sample configuration could be square, rectangular, or circular and various dimensions.

A problem exists in a control product vendor's claim that resistivity value was obtained per ASTM 257. Several questions should be raised:

1. Is resistivity a parameter of importance?
2. Was the resistivity value obtained on the material prior to subsequent processing steps in fabricating the product?
3. Is the end product a laminate material, and has this been accounted for in analysis of test results?

4. Which of the many electrode configurations and test apparatus of ASTM D257 were used?

5. What were the details of preconditioning, measurement environment, and other pertinent variables?

A copy of the test report should contain answers to the preceding questions. As ESD control standards are developed, the allowed variations should decrease, thus minimizing these concerns.

*ASTM D991:* Called "Rubber Property-Volume Resistivity of Electrically Conductive and Antistatic Products," this standard describes a four-electrode system for measuring volume resistivity of rubber used in electrically conductive and antistatic materials. The specification states that generally conductive rubber has a resistivity less than $10^6$ Ω at 120 V, and antistatic is generally in the range of $10^4$ to $10^8$ Ω. Contact resistance becomes more significant as resistivity decreases. The configuration shown in Fig. 17.4 is desirable to eliminate contact resistance errors and could be used to measure surface resistivity.

*EIA 541:* This Electronics Industries Association standard, formerly issued as Interim Standard 5A (IS 5A), is a standard on packaging materials for ESD-sensitive items. This standard is designed to evaluate bags and pouches, DIP tubes and slides, cushioning materials, cartons and tote boxes, rigid materials, and small irregularly shaped items. The standard incorporates: (1) triboelectric static generation evaluations utilizing the Faraday cup method, (2) static decay tests, (3) surface and volume resistivity, and (4) static shielding using a capacitive probe detector. This standard appropriately evaluates the finished ESD control product where practical. The most significant test is the static shielding test with a capacitive sensor inside a bag as depicted in Fig. 17.5. The voltage waveform detected across the sensor is monitored, and shielding effec-

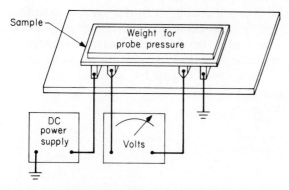

**Figure 17.4** Test configuration for volume resistivity (or surface resistivity).

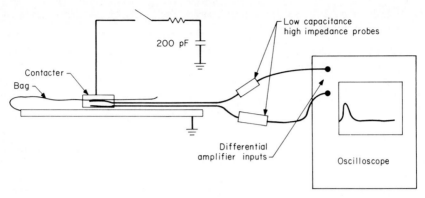

**Figure 17.5** Capacitive voltage sensor shielding test.

tiveness of different bags is compared. This test is not intended to be quantitative but only comparative between bags for the given conditions.

*FTMS 101, TM 4046:* Test Method (TM) 4046 of Federal Test Method Standard (FTMS) 101 describes a static decay test. The material sample under evaluation is grounded after having 5000 V applied. The time to discharge to 10 percent of the initial voltage (500 V) is measured. Generally, a limit of about 2 s is thought to be adequate for most ESD control applications.

*NFPA 56A:* This National Fire Protection Association (NFPA) standard primarily applied to measuring resistance of flooring. The original purpose was for hospital operating rooms where volatile anesthetic gases were used. The method utilizes a 5-lb, 2½-in cylindrical electrode attached to a megohmmeter. The standard demands a resistance of 25 k$\Omega$ to 1 M$\Omega$ when measured at 500 V between two electrodes 3-ft apart and resistance to ground from one electrode of greater than 25 k$\Omega$. The method can be used on static control flooring, work surfaces, and other flat materials with appropriate modifications. Users must account for the pliability and topography of the sample which affects contact area. These considerations and end-use conditions must be used to determine the appropriate voltage and acceptable resistance values to be used. Generally, a voltage maximum of 100 V may be desirable to avoid breakdowns and to be consistent with static control objectives.

## 17.3 Electrical Overstress/Electrostatic Discharge Standards

The Electrical Overstress/Electrostatic Discharge Association has an active standards committee that has been diligently addressing the industry need for ESD-specific standards. The nature of the task and required attention to detail mandate a seemingly slow rate of progress. The standards committee has released the following:

1. Glossary of terms
2. Wrist-strap standard
3. Garments standard
4. Ionization standard
5. Work surfaces standard
6. Preliminary HBM ESD sensitivity testing standard

In addition, standards on grounding and walking surfaces are near completion. These standards sometimes contain recommendations and precautions as well as detailed test methodology.

### 17.3.1  Wrist-strap standard

EOS/ESD Standard No. 1 "Standard for Protection of Electrostatic Discharge Sensitive Items: Personnel Grounding Wrist-Straps," sets the electrical, material, and mechanical requirements for an effective wrist strap. The standard resistance is 1 M$\Omega$ ± 20 percent. Nonstandard values require a prominent red identifying feature. The standard is comprehensive in addressing insulation resistance, solvent and salt resistance, and significant mechanical integrity requirements. Some of the mechanical parameters evaluated are cuff expansion, breakaway force, strain relief, connection integrity, extendability of retractable straps, and a bending life test.

### 17.3.2  Garments standard

EOS/ESD Standard No. 2, "Standard for Protection of Electrostatic Discharge Sensitive Items: Personnel Garments," establishes a test method and other criteria for ESD control garments. In particular the standard requires static decay measurement of a garment charged to 5000 V. The charge is applied with a 1- by 4-in probe clamped typically to the collar. This same probe is then grounded and the voltage decay is measured at three other locations typically the cuff, tail, and sleeve. A decay to 500 V is required in less than 2 s. Residual voltages at three separate garment locations must be less than 20 V. Additional requirements include solvent-resistant marking of manufacturer's logo, special laundering instructions, indications of permanency or limited usage, and a wrist-strap connection snap.

### 17.3.3  Ionization standard

EOS/ESD Standard No. 3, "Standard for Protection of Electrostatic Discharge Items: Ionization," defines test methods for evaluating ionization equipment. The standard contains safety criteria and a neutralization efficiency test by measuring the decay rate of a charged insulated metal plate. Safety requirements are high voltage corona point current limiting to 200 µA, the Occupational Safety and Hazard Administration (OSHA) limit of 0.1 ppm ozone over an 8-h period not to be exceeded, and manufacturers of radioactive ionizers required to obtain a license from either the Nuclear Regulatory Commission (NRC) or the NRC agreement from the state of manufacture. The requirements of 10CFR Part 20 and any other applicable government regulations must be met.

### 17.3.4. Work-surfaces standard

EOS/ESD Standard No. 4, "Standard for Protection of Electrostatic Discharge Susceptible Items: Worksurfaces," establishes methods for evaluating ESD control work surfaces by resistance measurement. The method essentially consists of measuring resistance on 10 in × 24 in test specimens and from three locations to the groundable point(s) of the work surface and then taking measurements between a point near the work-surface center and to points 2 in from each of the opposite sides as shown in Fig. 17.6. Five-pound cylinders with 63.5-mm (2.5-in) diameters are used for electrodes to contact the work surface. Similar measurements are defined for installed work stations. Guidelines for acceptance are $10^9$ Ω minimum resistance to the groundable point and $10^6$ Ω maximum from point to point on the surface.

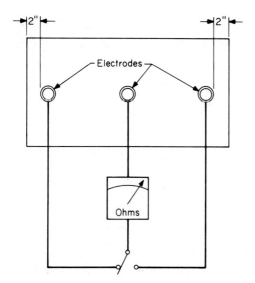

**Figure 17.6** Point-to-point work-surface-resistance measurement locations.

Owing to the preliminary nature, details of the HBM Testing Standard are not discussed herein.

## 17.4 Conclusion and Future Needs

The topics contained in this book were all intended to help the reader deal with ESD control. Yet "ESD control" was not directly addressed until Chap. 15, and it was not expanded on until the last two chapters. It is not wise to approach ESD control without proper preparation. ESD control cannot be accomplished with a "cookbook" approach or with learning the definitions of a few buzzwords. The many problems associated with ESD control procedures, products, evaluations, and standards require a good foundation in the many aspects of electrostatic discharge phenomena. The earlier chapters were intended to provide much of this needed background so that readers might deal with each new static control question from a better knowledge base.

The future holds many challenges in the field of ESD control. Some suggested areas of endeavor are continued standard development, on-board ESD protective methods, more reliable control products, improved overall ESD knowledge, research of electromagnetic interference (EMI) related upsets and/or permanent damage, more immune devices, and latent ESD studies:

*Continued standard development:* ESD control technology is changing at a rapid pace. The need for new and improved standards will continue.

*On-board ESD protection:* The problem of achieving ESD immunity at the assembly level when using electrostatic discharge-sensitive (ESDS) parts is impossible without compromising reliability, function, and speed, as well as size and weight. A breakthrough in ESD protection via some inexpensive easily applied material rather than a protective electronic device would be my first choice for a solution. In the meantime an inexpensive, highly reliable, low capacitance and low impedance, ESD immune-protective device will do.

*More reliable ESD control products:* The ESD control industry is to be acknowledged for the continuing advances made in new products. Yet it is disheartening to look at the state of extra expense and procedures required because of unreliability of the controls. The static problem has been around long enough that one might expect a wrist strap to last for a year without replacement. The confidence level should be so high as to dispense with daily, if not all, monitoring. Yet, many users are going to constant monitoring because of a real or perceived unreliability of wrist straps. A metallized bag can be used only once, after which it is likely to be degraded beyond effectiveness. Similarly, problems are encountered with shelf life of antistatic materials, ionizers' loss of balance, work-surface mat degradation, and improper construction of some products. Occasional lack of quality control results in products that were improperly processed with their effectiveness compromised. The task is not easy, but the reliability improvement has not kept pace with the functional enhancements achieved by control product manufacturers. The fact that users are likely to continue business with those control product vendors offering the best products and service will help. Application of some quality and reliability engineering principles by product developers might accelerate improvements.

*Improved overall ESD knowledge:* This is an area in which we all can help each other. Participation in interested professional organizations is encouraged to spread the knowledge base. Presentation of technical papers on new discoveries, problems, or developments will help advance the state of the art.

*Research on EMI-related upsets and/or permanent failures:* The problems due to EMI, in particular temporary upsets of digital circuitry, are costly and seem not to gain the proper level of attention. Studies in this area could lead to better devices as well as system immunity. This immunity might be realized through material and design changes not necessarily involving improved shielding, which seems to be the current preferred cure for this problem.

*More immune devices:* Great strides have been made in designing on-chip protection networks. Achievements have gone beyond the expectations of many. Perhaps this trend will continue as long as the semiconductor industry places a high degree of importance on the efforts.

*Latent ESD studies:* The lengthy discussion of latent ESD failures contained in Chap. 11 covers the bulk of known data on the subject to date. However, this database is small and is indicative a small probability of the occurrence of latent degradation due to ESD. The possibility remains that a higher likelihood of latency has not been indicated because the right device technologies or the proper depth of analysis has not been done. This is a frightful thought that I hope is not true. Further studies are needed to provide better insight into how to deal with ESD from a latent failure perspective. Additional part types should be tested, and different conditions might be tried to establish the initial undetected damage, in particular charged-device-type events. A most worthy project would be the study of field returns through detailed failure analysis in attempts to detect latent ESD failures. The could-not-duplicates, subtle, and intermittent failures should be included, as well as selected hard failures. In all such endeavors, understanding the physics of failure should be the predominant objective.

# Index

Absence of visible defects, 128, 130, 156, 211, 232, 236, 251
Adiabatic range of ESD, 143
Advanced Schottky TTL (transistor-transistor logic), 163
Air breakdown on insulator surfaces, 121
Amber, 30, 31
Analysis of data from NAVSEA and NASA studies, 284–296
Analysis of ESD failures, 6, 203–224
  analogous to a detective's sleuthing, 205
  approach to, 203
  characteristic traits, 221
  corrective action, 224
  fact gathering, 205
  failure-analysis elements, 204
  failure cause identification, 221
  importance of failure analysis to an ESD control program, 255
  key elements, 224
  notification, 205
  part analysis, 206
  (*See also* Failure analysis, case histories of)
Antistatic bags chosen by cost analysis, 10
"Antistatic" carpets, 346, 392
Assembly protection, 337–340
  electrical design protection, 339
  electrostatic shielding, 338
  EMI protection, 339
  mechanical design, 337
  selection of parts, 339
  strategic placement, 338
  transient suppressors, 340
Attractive force of static electricity, 30
Audit agenda, 403
Audit content, 404
Audits, 403
  customer, 374

Automated equipment, 344, 350
Automated facilities, 344, 350, 415
  conveyor belts, 417
  failure sources and control measures, 416
  field induction, 417

Back discharge, 109
Band structure of solids, 104
  forbidden, 104
  insulators, 104
  metals, 104
  semiconductors, 105
Batteries, first, 38
Bennet, Abraham, 42
Bipolar device, 7, 15
Bonds, 104
  covalent, 104
  electrovalent, 104
  ionic, 104
Boundary conditions of insulators, 99
Breakdown:
  of air and other gasses, 66, 121
  in fluids, 122

Capacitance, 58, 74
  concentric cylinders, 87
  concentric spheres, 79
  isolated sphere, 80
  mutual, 62, 97
  parallel plates, 93
  parasitic, 62
  self-, 59, 97
  stray, 62
Capacitive coupling, 62, 63, 180
Capacitive discharge, 3
Capacitive relationships between conductors, 96
Capacitor current, 98

Capacitors, 59
  circuit operation of, 62
  combinations of, 93
  first, 37
Capacity, 58
Categorization of materials, 53
Caution label, 10, 352, 380
CCD (charge-coupled device) failures, case history of, 251
  failure cause identification and corrective action, 251
  failure duplication attempts, 251
  400 degrees centigrade brought about recovery, 251
  no visible damage and no leakage path, 251
  part analysis, 251
Characteristic traits, 130
  aluminum-silicon alloy, 132
  confusion between ESD and EOS, 146
  cracked glass, 131, 146
  melting of intraconnects, 133
  EOS-caused junction failures, 132
  EOS failures of MOS devices, 128
  ESD-caused junction failures, 130
  ESD failures, 125
    of MOS devices, 126
  white spear, 132
Charge, 45, 47
Charge-coupled device [see CCD (charge-coupled device) failures]
Charge injection, 251
Charge recombination, 58
Charged device model (CDM), 169, 173–178
  Avery, 336, 337
  bulk of charge is mobile, 174
  correlation to Wunsch-Bell model, 176, 177
  damage characteristics, 178
  factors affecting failures, 176
  general case, 174
  internal inductive-capacitive coupling, 201
  LSI, 200
  parasitic components, 200
  protection against, 336
  schematic, 174
  waveform profile, 178
Cho-trap, 340
Clanton, John, 42
Clean room considerations, 417
Clothing-related problems, 345

CMOS (complimentary metal-oxide semiconductor) protection, 321
Cold climates, 345
Commitment, management, 20, 22, 25, 371
Comparative results of protective networks, 319, 320
Comparative uniform sheet resistivity, 53
Comparison of fields on different sized spheres, 76
Computer memory, 45
Computerized control system, 357
Concentric conductive spheres, 79
Conclusion and future needs, 465
  continued standard development, 465
  "cookbook" approach, 465
  future challenges, 465
  improved overall ESD knowledge, 466
  latent ESD studies, 465
  physics of failure, 466
  research, 466
  study of field returns, 466
Concrete sealers, 346
Conductive containers, 55
Conductive trays, 353
Conductive work surfaces, 347
Conductivity, 50
Conductors:
  in a field, 75
  and nonconductors, 34
Conformal coating, 349
Conservative nature of electric field, 71, 101
Contact electrification, 48, 108
  adsorbed gasses, 109
  contamination, 110
  effect of work functions, 110
  general variables, 109
  measurement, 109
  mutual capacitance, 109
  other variables, 110
  solid-to-solid, 109
  types of contact, 109, 110
Contact potential, 39, 107, 108
Contributors to static problems, 344
Control(s):
  additional personnel needed, 375
  cost of, as basis for ESD personnel, 372
  determination of needs, 365
  determining extent of, 370
  device manufacturer's sites, 351
  with director, 382

Control(s) (*Cont.*):
  without director, 378–382
  establishing a program, 382
  functional group for, 373
  funding, 364
  independent test laboratories, 351
  intangible benefits, 25
  knowledge and control decisions, 375–377
  line supervision, 372
  microelectronic distributors, 351
  minimal, 366
  moderate, 366
  primary, 386–388
  return on investment from assigned ESD personnel, 377
  rigorous, 367
  secondary (*see* Secondary controls)
  of static hazardous materials, 396
    insulative solvents, 397
    necessary static hazardous materials, 396
    purge of unnecessary hazardous materials, 396
    safe distance rule, 397
  tailoring, 364, 366, 384, 385, 401, 447, 448
  tertiary (*see* Tertiary controls)
  thorough, 367
  tradeoff decisions, 370
Control committee, 383, 384
Control conditions (*see* Frequently observed conditions)
Control consultants, 370
Control coordinators, 370
Control director, 370–373
Control management, 363–384
Control measures (*see* Supplemental control measures)
Control problem, 363
Control products:
  antistatic materials, 55, 399, 400, 430
  antistatic treatments, 392, 405, 433, 444
  chairs and carts, 409, 433, 442
  considerations and evaluations, 421
  contaminated antistatic polyethylene, 430
  dissipative materials, 430
  evaluation of flooring and floor finishes, 432–435
  factors affecting tests, 421–428
  ionization evaluations, 443

Control products (*Cont.*):
  miscellaneous, 444
  other factors affecting, 368
  protective material resistivity ranges, 457
  protective packaging considerations, 436–442
  (*See also* Protective packaging)
  resistance and resistivity measurements, 426–428
  shipping tube evaluations, 440–442
  static decay testing, 429
  triboelectric generation measurements, 429
  voltage monitored activity, 431
Conveyor belts, 44
Corona, 119
Coulomb, Charles Augustin de, 41, 67
Coulombs, 67
  defined, 48
Coulomb's law, 42, 68
Coupling capacitance, 180
Criticality of mission, 367
Cumulative damage, 164
Customer image enhancement, 27, 419
Customer requirements, 368

Damage, 125
  bipolar devices, 146
  confusion between ESD and EOS, 146
  differentiation of short EOS from ESD, 147
  gallium-arsenide devices, 162
  junction, 130
  masking of ESD failures, 147
  metallization, 139
  metallization burnout, 162
  MOS devices, 146
  passive device, 149
  reversible, 153
  Schottky diode, 161
  short duration EOS transient, 146
  wire, 147
Damage recovery, 153
  bake at 400 degrees centigrade, 251
  ultraviolet (UV) light, 154, 156
  X-ray, 154, 156
Decay time, 118
Degraded junction recovery, 157
Degraded junctions, 156
Department of Defense, U.S. (*see* DOD Handbook 263; DOD Standard 1686)
Depolarization, 113, 117

Design-for-immunity techniques:
  definition of, 309
  design hardening, 309
  negative tradeoffs, 309
  not well developed, 309
Design problem, 309
Devices monitored while damage occurs, 23
Dielectric, 59
Dielectric absorption, 113, 117
Dielectric constant, 68, 81, 117
Digital counter, 163
Diode protection, 311
  depletion layer width, 312
  high frequency characteristics, 312
  increase in capacitance, 312
  junction area, 312
  lower dynamic resistance, 312
Diode resistor combinations, 314
Dipoles, 104, 114
Discharging effects of points, 119
Disciplines participating in ESD control, 371
Discoveries from failure analysis, 6
Displacement field, 81
DOD Handbook 263 (U.S. Department of Defense), 11, 54, 446, 447, 457
  protective material resistivity ranges, 457
DOD standard 1686 (U.S. Department of Defense), 171, 191, 446, 447, 451, 454
  Appendix A classification, 192
  cautionary note, 454
  classification methods, 451
  and Military Standard 1686A:
    awareness training, 447
    comments, 457
    comprehensive ESD control program, 447, 448
    control program discipline, 456
    deliverable data items, 457
    design requirements, 447, 455
    documentation, 456
    ESD protective packaging, 455
    identification of ESD sensitive items, 447
    proper handling procedures, 447
    protected areas and protective packaging, 447, 455
    quality assurance provisions, audits and, 447
    sensitivity classification, 447, 448
    sensitivity identification, 452
    tailoring, 447, 448

DOD standard 1686 (U.S. Department of Defense) (*Cont.*):
  sensitivity classes, 451
  sensitivity classification testing, 191
Double implant field isolation device in well (DIFIDW), 329, 330
  high density, 1.2 micron protection circuitry, 330
  Hu, Yean-Shan, 330
  pseudocollector, 331
Doubt of ESD problem, 7, 15, 16
Du Fay, Charles, 36
  any material could be electrified, 36
  vitreous and resinous electrification, 36–37
Dust accumulation on disks, 45

ECL (emitter-coupled logic) output damage, case history of, 238
  analysis of failed parts with no visible damage, 238
  analysis of parts with visible damage, 239
  corrective action, 241
  failed parts with visible damage, 238
  failure cause identification, 239
  no visible damage, 238
Elastance, 95
  mutual, 95
  relationship of a system of conductors, 95
  self, 96
Electric field, 57, 68, 69
Electrical overstress [*see* EOS (electrical overstress)]
Electrical wind, 119
Electrically neutral, 47
Electricity, different types of, 36
Electrification, 36, 103, 111
  contact (*see* Contact electrification)
Electron affinity, 104
Electrophorous generator, 39, 63, 64
Electroscope, 42
  first, 34
Electrostatic discharge [*see* ESD (electrostatic discharge)]
Electrostatic force, 67
Electrostatic shielding, 55
Electrostatic voltmeter, first, 44
Elementary charge, 47, 48
Emitter coupled logic [(*see* ECL (emitter coupled logic) output damage]
Energy, 57, 74, 98
  and voltage relationships, 59

Energy sensitive, 60
Engineering concerns, 417
Environmental considerations, 17, 355, 356, 400
Environmental factors, 110, 426
　radiation, 110
　temperature, 110
EOS (electrical overstress), 132
　forward bias, 133, 134
　reverse bias, 132, 134
EOS/ESD Association, 11
　standards of, 457, 462–464
　　garments standard, 463
　　glossary of terms, 457, 463
　　grounding and walking surfaces, 463
　　HBM ESD sensitivity testing standard, 463
　　ionization standard, 463
　　protective material resistivity ranges, 457
　　work surfaces standard, 463, 464
　　wrist-strap standard, 463
　Symposium of, 8
Equipment grounding, 392
　charged device failure considerations, 393
　grounded work surfaces, 393
　miscellaneous equipment, 394
Equipotential surface contours, 76–79
ESD (electrostatic discharge), 1
　analysis of failures (*see* Analysis of ESD failures)
　control (*see* Control; Control products)
　definition of, 3
　early problems, 4
　event, 3
　extent of problem, 1, 2, 364
ESD awareness, 8, 15
　management's attention, 23
　threshold of feeling, 17
ESD awareness training, 8, 20–26, 344
　actual part failures, demonstration of, 26
　common myths to overcome, 22
　content and method of presentation, 22
　factory operator training, 21, 26
　line management training need for, 21
　need for, 20, 21
　ongoing ESD training and certification, 21, 26
　technical personnel training, 21, 26
　video programs, 26
ESD control management, 363–384

Evaluation of flooring and floor finishes, 432
　comprehensive, 433
　ESD tests, 435
　friction or "slip" tests, 433
　gloss and scrub tests, 434
　heel mark test, 433
　hop test, 435
　importance of footwear, 345, 432
　sample preparation, 433
　shuffle test, 433
Experience as the best teacher, 19
Explosions, 45

Factors affecting tests, 421
　buried-layer considerations, 427, 428
　contact resistance, 426
　electrodes, 423, 426
　environment, 426
　insulation resistance of cables, 428
　nature and configuration of sample, 425
　parameters, 427
　resistance and resistivity measurements, 422–428
　surface resistivity, 423
　variables, 425
　volume resistivity, 423
Failure analysis, case histories of, 8, 227
　avoidance of possible latent failures on IRAS spacecraft, 253
　charge-coupled device (CCD) failures in hybrid assemblies, 251
　CMOS devices contained in a metal chassis assembly, 228
　damage to op-amp inputs, 234
　emitter coupled logic (ECL) failures, 238
　hybrid breakdown/vaporized metal, 247
　shorted hybrid substrates, 241
　wafer level processing, 252
　(*See also* Analysis of ESD failures)
Failure cause identification, 223
Failure mechanisms, 204
　knowledge of, 204
　other than static electricity, 204
Faraday, Michael 42, 70
　ice pail experiment, 42, 44, 55, 56
　lines of force, 42
Faraday cage, 42, 55
Faraday cup, 155
Faraday shield, 55

Fermi-Dirac distribution, 105
Fermi levels, 105
Field-induced model (FIM), 169, 178–185
  failure cause by field strength alone, 179
  field induction from insulative portion of part, 180
  part-level field-induced damage, 178
Field induction discharge (FID) model, 169, 178
  ESD thresholds as function of series resistance, 182
  mathematical analysis of FID Board Tests, 183
  validation of assembly level model, 181
Field influence, 75
  on conductive sphere, 75
  on two separated spheres, 75
Field intensity, 69, 86
Field lines, 57
Field plate, 314
  walk-out of, 315
Field strength across thin oxides, 152
Field strength breakdowns, 4
Fields produced by selected charge configurations, 83
  any conductor, 90
  conductive cylinders, 85
  infinite charged conductive plate of finite thickness, 83
  infinite sheet of homogeneous charge density, 83
  line of charge, 87
  parallel conductive plates, 91
  spherical charge distribution, 89
  spherical shell of charge, 88
Filament, 159
Filamentary junction short, 156
Filamentary trait, 237
First batteries, 38
First capacitors, 37
First electroscope, 34
First electrostatic experiments to be recorded in history, 30
First electrostatic voltmeter, 44
First ESD control practice, 29
Floating (nonconnected electrically) lid, 23
Flooring, 344, 346
  evaluation of (see Evaluation of flooring and floor finishes)
Flow electrification, 113
Flux, 81

Flux density, 82
Footwear, 344, 345, 432
Force on a unit charge, 69
Franklin, Benjamin, 39
  first electric motor, 41
  kite experiment, 40
  Leyden jars connected in parallel, 40
  lightning, nature of, 40
  lightning rods, 40
  positive and negative charges, 39
  use of Leyden jar, 40
  verification of positive and negative polarities, 40
Freeze sprays, 344
Frequently observed conditions, 351–355
  assembly areas, 353
  carts, 354
  conformal coating, 353
  engineering labs, 355
  inspection, 354
    incoming, 351
  kitting, 352
  receiving, 351
  shipping, 355
  suppliers, 351
  test and troubleshooting, 354
  transient personnel, 354
  unusual observances, 355
  wave soldering, 353
Funds for control measures, 20
Furniture-related problems, 346
Future needs:
  for protective circuits, 153
  (See also Conclusion and future needs)

Galvani, Luigi, 38
Galvanic electricity, 39
Garments, 406–407, 463
  available fabrics, 407
  conductive fibers, 407
  EOS/ESD Association standard for, 463
  laundering, 407
  problems, 407
Gauss' law, 44, 82, 83, 101
Geometric considerations, 325
  channel length, 327
  contact designs, 325
  current crowding, 325
  metallization burnout, 327
  polysilicon resistors, 327
Gilbert, William, 31
  "electrics," 31
  versorium, 31

Gold-leaf electroscope, 42
Government-Industry Data Exchange
    Program (GIDEP), 12, 431
Government and industry standards,
    445, 457, 460–462
  AATCC 134, 460
  ASTM D257, 445, 460
  ASTM D991, 461
  EIA 541, 457, 461
  EIA protective material resistivity
    ranges, 457
  FTMS 101, Test Method (TM) 4046, 462
  NFPA 56A, 462
  (See also Military standards)
Gray, Stephen, 34
  electric "virtue," 35
  human body conductivity of, 36
  ivory ball, 35
  metals could be "electrified," 36
  transmission of attraction property of
    charged items, 35
Green's theorem of reciprocity, 93, 95
Grounding considerations, 395–396
  conductive work-surface grounding,
    395
  ground fault interrupters, 396
  separate ESD control grounds, 395
  (See also Personnel grounding)

Hauksbee, Francis, 33
  electric wind, 34
  glass sphere, 33
  inductive effect, 34
Hazardous items essential to a manufacturing process, 347
Hazardous materials (see Control, of
    static hazardous materials)
Hazardous situations, problem examples
    of, 356–362
  destruction of part by charged
    garment, 362
  masking tape induced failures, 358
  printed wiring assembly "cushioning,"
    356
  printed wiring layout templates, 358
  upsets of electric utility plant control
    system, 357
  verification of hazard from garments,
    360
  voltage suppression, 362
Helmholtz layers, 109, 111, 112
Hospital rooms, 44
  conductive floors, 44

Hospital rooms (Cont.):
  conductive shoes, 44
  grounded equipment, 44
  special garments, 44
Human body, static source, 344, 345
Human body discharge, 135
  body resistance, 170
  Bureau of Mines data, 170
  calculated human body capacitance,
    171
  development of the human body model
    (HBM), 169–173, 191
  Enoch-Shaw human body measurements, 172
  human body capacitance, 169–171
  human body resistances, 171
  standard human body model, 171
Human-body-related problems, 345
Human body simulators, 7, 193
  cumulative damage results, dependency on waveforms, 194
  current waveforms, 194
    from two simulators, 195
  effective stray capacitance, 198
  factors for consideration in HBM
    testing, 198
  failure criteria, 199
  high frequency circuit, 198
  inaccurate test results, 194
  Military Standard 883, method 3015.3,
    198
  parasitic components, 198
  repeatability, 198
  rise-time definition, 199
Humidity control, 55, 344, 409, 410
  absolute versus relative, 410
  economical considerations, 410
Hybrid, printed-circuit board, and field
    improvements, 10

I-V (current-voltage) characteristic curve,
    24, 235, 261
Ideal protection circuit, 310
Ignorance and myth, 309
Imbalanced ionizers, 344
Implanted resistors, 163
Importance of controls from all three
    categories, 388
Induced dipoles, 114
Inductance, 183
  stray, 183
Induction, 42, 58
  compound, 58

Induction (*Cont.*):
  effects of concentric spheres, 79
  induced potentials on isolated conductors, 184
  simple, 42, 58
Industry standards (*see* Government and industry standards)
Information sources, 11
Insulative clothing, 344
Insulative items essential to the process, 344
Insulative materials, 343, 344
Insulative work surfaces, 344
Intuition and electrostatics, 47
Inverse-square relationship, 41
Ionization, 65, 350, 410
  additional problems, 414
  alternatives, 415
  bench top ionizers, 413
  high voltage, 411
  nuclear ionizers, 411
  nuclear-pulsed voltage ionizers, 412
  room ionization systems, 413
  used as generators, 414
Ionizer evaluations, 443
  biased plate, 443
  parallel plate ion counter, 443
  static decay, 443
ITTRI/RAC, 11

Junction damage, 156
  initial damage, 156
  at sites of high-field concentration, 156

Kelvin, Lord, 43, 107

Latchup, 129, 322, 325
Latency (*see* Theories of latency)
Latency data, 284–296
  altered 54l04s that degraded, 288
  available data base, 284
  conclusions, 294
  control device behavior, 284, 290
  data base from NAVSEA and NASA studies, 289
  latent failure candidates, 291
  life test results of devices unaltered by ESD stress, 290
  Mac-1 or "classical" latent failures, 290, 291, 295
  Mac-1 and Mac-3 failures, 295
  Mac-2 failures, 294
  Mac-3 failures, 294, 295

Latency data (*Cont.*):
  marginally altered 3N128s that degraded, 281
  statistical analysis, 292–294
Latency studies:
  early, 258, 303
    commonality in earlier investigations, 262
    Branberg, 259
    Gallace and Pujol, 258
    Hasegawa, 258
    McCullough, Land, and Blore, 259
    Schreir, 259
    Syrjanen, 260
  other, 296
    British Telecom studies, 299
    Martin Marietta study, 296
    since 1980, 304
    RIT Research Corporation study, 298
    Sandia National Laboratories study, 298
    TriQuint semiconductor study, 301
    University of Southhampton study, 297
Latent, 165
  time-dependent latent failures, 257
  two common interpretations, 165, 257
Latent damage, 165
Latent effects, 165
Latent ESD failures, 25, 165, 257
  additional studies, need for, 306
  assessment of the controversy over, 303
  background of the latency question, 258
  definition of, 257
  early investigations (*see* Latency studies, early)
  general conclusions, 305
  other latency studies (*see* Latency studies, other)
  possible (*see* Possible latent failures)
  recommendations, 306
Leyden jar, 38, 40
Lightning, 29, 40, 123
Lightning rod, 40, 123
Lines of force, 42, 70

Machine model, 185–187
Machine-model discharge waveform, 186
Major static sources, 344
Management, 374
  commitment, 20, 22, 25, 371

Management (*Cont.*):
  control, 363–384
  control funding, 364
  convincing upper management, 371
  decision on ESD personnel assignments, 374
  training need for, 20
Maxwell, James Clerk, 43, 71
Maxwell's equations, 43, 101
Melamine, 346
Merged input protection circuits, 335
  Hu, Yean-Shan, 335
  shared pseudocollector and source, 336
MESFETs (metal semiconductor field-effect transistors), 163
Metal, 247
  in solution, 112
Metal chassis, case history of, 23, 228
  cholesteric liquid crystal isolation, 229
  corrective action, 233
  failure cause identification, 231
  failure duplication, 231
  microprobe damage site isolation, 228
  notification and fact gathering, 228
  part analysis, 228
Metal-oxide semiconductors [*see* MOS (metal-oxide semiconductors)]
Metallization, 139
  adiabatic, 140
  burnout, 162, 327
  energy absorbing element, 327, 328
  ESD-caused open, 147
  ESD opens at metallization steps, 143
  experimental proof of Adiabatic conditions, 139
  open(s), 139
    as a characteristic trait of EOS, 143
    due to ESD, 144
  theoretical calculation of $K_m$, 142
  vaporized (*see* Vaporized metallization)
Microminiaturization demands, 6
Military Standard 1686A, 11, 446, 447, 451–453, 455
  appendix B classification, 452, 453
  assemblies marked with Mil Std 1285 or EIA RS-471 symbol, 455
  cautionary note, 455
  classification methods, 452
  marking per Mil Std 1285, 455
  packaging per Mil E 17555, 455
  sensitivity classes, 451
  (*See also* DOD Standard 1686, and Military Standard 1686A)

Military standards, 446–459
  DOD-STD 1686 (*see* DOD Standard 1686)
  Mil-B-81705, 455, 458
  Mil M 38510, 446
  Mil-Std-129 symbol, 454
  Mil Std 883, Method 3015, 193, 446
    current waveform test, 446
    sensitivity classes, 446
  Mil Std 883C Method 3015, 193
  Mil Std 1686A (*see* Military Standard 1686A)
  Mil Std 2000 (soldering), 458
    ESD requirements in, 458
  WS 6536E (soldering), 459
    ESD requirements in, 459
Milliken, R. A., 107
MIM capacitors (metal-insulator-metal), 163
Models, 169, 178, 181, 187, 188
  body metallic model (BMM), 169, 187
  capacitive-coupled model (CCM), 169, 187
  charged-device model (CDM) (*see* Charged-device model)
  field-enhanced model (FEM), 169, 187
  field-induced model (FIM) (*see* Field-induced model)
  field induction discharge [*see* Field induction discharge (FID) model]
  floating device model, 189
  human body model (HBM), 169–173, 191
  machine (or Japanese) model (MM), 169, 185–187
  transient-induced model, 189
Molecular interfaces, 103
MOS (metal-oxide semiconductor), 6, 126
  buffered by other circuitry, 6
  charge trapping, 129
  complimentary (CMOS) protection, 321
  cross-sectional structure, 126
  damage, 6, 125
  gate-oxide shorts, 4, 127
  latchup, 129
  processing faults, 128
  threshold voltage shifts, 129
  unprotected, 60
Mutual capacitance, 62, 97
Mutual elastance, 95

NASA latent ESD study, 276–284

NASA latent ESD study (Cont.):
altered parameters:
degraded during life tests, 284
remaining stable, 284
showing improvement, 280
analysis of data, 284
ESD stress results, 277
life test results, 279
and NAVSEA study, 261–296
analysis of data from, 284–296
biased life tests, 266
differences in plan, 262
Mac-1 latent failures, 262
Mac-2 latent failures, 263
Mac-3 latent failures, 263
multiple pulse tests, 266
nonbiased bake, 267
single pulse tests, 266
parts used, 276
pin combinations and delta criteria, 277
unaltered parts, 279
NAVSEA (Naval Sea Systems Command) study, 267–276
altered parameters exhibiting improvement, 273
changed parameters exhibiting degradation, 274
ESD stress results, 268
life test results, 271
and NASA study [see NASA study, and NAVSEA study]
parts used, 267
unaltered parts, 273
Negative design tradeoffs, 337
added capacitive and resistive elements, 337
diminished functional complexity, 337
high-frequency limits, 337
"real estate," 337
Negative resistance, 133
NFPA (National Fire Protection Association) standard, 462
NMOS protection via thin or thick oxide transistor, 320

Operational amplifier, 7, 23, 131
Operational amplifier failures, case history of, 234
catastrophically damaged devices, 234
corrective action, 238
curve-tracer I-V characteristics, 235
lot acceptance test failures, 234

Operational amplifier, case history of (Cont.):
part analysis, 235
subtle beta degradation failures, 235
degraded parts analysis, 236
failure cause identification, 237
no visible damage, 236
partial chemical removal of the $SiO_2$ layer, 237
purposely degraded devices, 237
Ordinary work surfaces, 346
Oxide, 125
adiabatic nature of oxide ruptures, 126
field strength, 125

Packaging (see Protective packaging)
Parasitic elements, 320
diode, 235
lateral bipolar transistor, 129, 318
Part analysis, 206–221
chemical etching, 215
compared to a physician's analysis, 206
delidding, 209
reverification after, 209
electrical microprobing, 212
electron-beam induced current (EBIC), 220
external package analysis, 207
failure-site characterization methods, 213
failure-site isolation methods, 212, 213
internal visual, 209
liquid crystals, 217
microsectioning, 218
microsurgery, 214
physical dissection, 208
plasma etching, 215
preliminary analysis, 207
scanning electron microscopy, 219
SEM voltage contrast, 220
Paschen's law, 122
Pauli exclusion principle, 104
Permittivity of free space, 67
Personnel grounding, 389
antistatic carpet problems, 392
conductive flooring and footwear, 345, 391
applications for, 391
problems with, 391
conductive mat problems, 392
conductive tile problems, 391
conductive wrist-strap fibers, 390
"dead skin" layer, 390

Personnel grounding (*Cont.*):
  electrical safety hazard, 390
  floor finish problems, 392
  intermittent skin contact, 390
  metal contacter, 390
  metallic expandable watch-type bands, 390
  safety considerations, 389
  velcro bands, 390
  wrist-strap connection to ground, 391
  wrist straps, 389
    problems with, 389
    recommended, 391
Personnel safety, 30
Phantom emitter, 7, 331, 332
  effective schematic, 332
  Minear and Dodson, 331
  thermal runaway, 332
Photographic industry, 45
Piezoelectrification, 111
Pith balls, 47
  electroscope, 48
Plexiglass, 357
Polar and nonpolar molecules, 114
Polarities:
  different, on an insulator, 50
  of static electricity, 39
Polarization, 113
  atomic, 115
  in ceramic lid, 154
  dielectric constant variations, 117
  effects of, 116
    on resistivity, 117
  electronic, 115
  ionic, 115
  orientation, 115
  space-charge, 116
Polystyrene or styrofoam shipping trays, 11
Possible latent failures, case history of, 253
  corrective action, 255
  failure-cause identification, 255
  failure duplication, 255
  multiple ESD exposures, 254
  multiple-oxide breakdown sites, 254
  part analysis determined resistive shorts, 254
  single site accounted for each short, 254
Potential(s), 3, 74, 107, 112
  due to a cylinder, 86

Potential difference, 73
  as function of electric field, 73
  between concentric spheres, 79
  between plates, 93
Power, definition of, 57
Priestley, Joseph, 41
Primary controls, 387
  complications, 387
  considerations, 389
Principles of electrostatics, 47
Process variables, 326, 328
  aluminum-silicon diffusion, 329
  corrective measures, 328
  cracks, 328
  field oxide, 328
  overalloying, 328
  phosphosilicate glass, 329
  polysilicon strap, 329
  spikes, 328
  step coverage sites, 328
Processing highly insulative products, 45
Protected MOS (metal-oxide-semiconductor) circuitry, 6
Protection:
  assembly (*see* Assembly protection)
  circuit development, 310
  circuit tradeoffs, 337
  device applications, 319
  devices, 310–340
  diode (*see* Diode protection)
  resistive (*see* Resistive protection)
Protective packaging, 55, 397–442
  antistatic bags, 400
  antistaticity, 442
  bag evaluations, 438, 440
  combination or converted bags, 399
  conclusions for DIP tube evaluations, 442
  homogeneous material shielding bags, 398
  multilayer shielding bags, 398
  other parameters, 440
  shielding, 438
  shipping tube evaluations, 440–442
  static shielding, 398, 441
  static shielding boxes, 398
Punch-through devices, 316
Pyroelectrification, 111

RAC (Reliability Analysis Center) VZAP data base, 12
Reasons for disbelief, 16

Recoveries:
   of dielectric shorts, 160
   of ESD failures on subsequent exposures, 156
Recovery mechanism, 159
Recovery time, 110
Relating to ESD experiences of others, 364
Relative permittivity, 68
Relaxation time, 118
Reliability Analysis Center (RAC), 11, 12, 26
Reliability effects, 27
Requirements documentation, 383
Resistive protection, 312
   diffused resistor, 312
   distributed diode characteristic, 312
   distributed resistor, 312
   exponential attenuation, 314
   polysilicon resistors, 314, 327
Resistivity, 51–55
   comparative uniform sheet, 53
   surface, 52–55
      independent of size of square, 53
   volume, 52
      categorization of materials, 53
Resistors (see Thick-film resistors; Thin-film resistors)
Return on investment figures, 25
Richmann, Georg Wilhelm, 41
Robison, John, 41
Rubber gloves, 353

Safety concerns, 419
St. Elmo's fire, 30
Schottky diodes, 164
   and advanced Schottky device susceptibility levels, 163
Secondary controls, 388, 400
   audits, 401–405
   ESD control certification, 405
   procedures for, 400
   self-checks and audits, 401
   supplemental control measures, 405
   training, 401
   vendor and subcontractor control, 405
   wrist-strap checks, 402
Self-capacitance, 59, 97
Self elastance, 96
Sense of feeling, 16
Shorted and degraded junction recovery, 156

Shorted hybrids, case history of, 241
   CMOS chip experiments, 245
   corrective action, 247
   dissection of shorted substrates, 242
   fact gathering at the location of failure, 242
   failure duplication attempts, 244
   initial conclusions, 242
   part analysis and failure cause identification, 241
Shorted junctions, 158, 160
Silicon nitride, 126
SIT (static induction transistor), 333, 335
   under a contact pad, 334
   Turkman and Neelakkantaswamy, 334
   voltage current characteristics, 333
Slow reaction time of diodes, 328
Solder removing, 349
SOS (silicon on sapphire) protection, 322
   edgeless islands, 322
   edges of the silicon islands, 322
   Palumbo and Dugan, 322
   protective networks, 322
Sources of yield detraction data, 25
Spark gaps, 319
Sparkover of conductive points, 121
Sparks, 120
   from insulative materials, 121
Speakman model, 7, 135
   correlation with Wunsch-Bell model, 137, 138
   sample calculation, 137
Spray electrification, 111, 112, 344, 349
Standards, 1, 445
   improperly applied, 445
   lack ESD-specific standards, 445
   not developed in many needed areas, 445
   (See also Government and industry standards; Military standards)
Static electrification, 103
Static generating apparatus, 32
Static hazardous factory processes, 347
Static induction transistor (see SIT)
Static problems in other industries, 44
Static-sensitive items, 17
Static-sensitive parts, 9
   assemblies for, 18
   part types, 17
Static sources different from solid-to-solid contact, 344
Straw electroscope, 34
Stray capacitance to ground, 237

Index   479

Subtle damage paths, 19
Success stories, 9–11
  automotive industry, 9
  field repair, 10
  hybrid integrated circuits, 11
  incoming test, 10
  printed circuit boards, 10
  single control consisting of wrist
    straps, 11
Supplemental control measures, 405
  carts, 409
  chairs and stools, 408
  floors or floor covers, 408
  garments, 406
  gloves and fingercots, 408
  ionizers, 405
  replacement materials, 406
  shunts, 408
  topical antistat treatment, 405
Suppression, voltage, 61, 362
Surface inversion, 154, 155
Surface resistivity, 52–55
  conductive, 54
  definition of, 52
  independent of size of square, 53
Susceptibility, 18, 365
  determination of levels, 368
  known extremes, 365
  levels unknown, 365
  moderate, 365
  tailoring based on, 385
Susceptibility classification testing, 190
  charged-device model testing, 199
    problems with, 200
  DOD Standard 1686, 191
  human body model testing, 191
  Mil-M-38510, 191
  Mil Std 883C, Test Method 3015.1, 191
  Mil Std 1686A, 191
Susceptible parts and/or assemblies, 344
Synthetic materials, 3
System of charges, 69

Tailoring, 366, 385
  based on susceptibility, 385
  life cycle, 386
  progressive controls, 385, 386
Tape application and removal, 349
Temperature coefficient changes, 151
Tertiary controls, 388, 409
  humidity control, 409
  ionization, 410
  personnel motivation, 415

Textile industry problems, 45
Thales, 30
Theories of latency, 260–261
  charge trapping, 261
  gate-oxide degradation, 260
  gate-oxide healing or shorting, 260
  intermittent failures, 261
  metallization failures, 261
  PN junction degradation, 261
  protective circuitry damage, 260
Thick-film resistors, 5, 150
  failure mechanisms, 151
Thick oxide enhanced punch-through
    device, 318
Thick oxide transistor, 317
  parasitic lateral NPN transistor, 317
  problems with, 318
    metal-silicon alloy shorts, 319
    silicide cladding, 318, 319
Thin-film resistors, 5, 149, 150, 162
  failure mechanisms, 150
  resistance changes, 5
Thin oxide transistor, 316, 317
Threshold shifts in thin oxide MOSFETs,
    153
Topical antistats, 405
  application of, 406
  problems of, 406
Transistors (see Thick oxide transistor;
    Thin oxide transistor)
Triboelectric generation, 48, 49, 108, 347
  charge transfer rather than genera-
    tion, 49
  overlapping circular polar relationship,
    50
  variables affecting, 49
Triboelectric series, 49, 63
TTL (transistor-transistor logic), 132, 133
  advanced Schottky, 163
  hex-inverter, 7

Unsealed concrete, 346
Upset occurrences, 357

Van de Graff, Robert J., 44
Van de Graff generator, 64, 65
Van der Waal's forces, 103
Vaporized metallization, case history of,
    247–251
  conclusions, 251
  corrective action, 251
  failure cause identification, 249
  part analysis, 247

Variety affected without trend indicated, 17
Varley, C. F., 44
Varnishes, 346
Very thin oxides, 152
VHSIC (very high-speed integrated circuit), 152
Video display tubes, 344, 349
Video programs, 27
VLSI (very large-scale integration), 130, 152, 324
Volt, 73
Volta, Alessandro, 39, 63, 107
Volta electrification, 108
Volta potential, 107, 108
Voltage, 57
Voltage-sensitive devices, 60
Voltage suppression, 61, 362
Voltaic cell, 39
Voltaic pile, 39
Volume resistivity, 52, 53
   aluminum as function of temperature, 140
Von Guericke, Otto, 32
   sulfur sphere, 32
Von Kleist, Ewald, 37
Von Musschenbroek, Pieter, 37
VZAP data base, 12

Wafer-level controls, 10

Wafer level processing, case history of, 252
   "Bulls-eye syndrome," 252
   corrective action, 253
   failure duplication, 252
   water-spray clean operation, 253
Wafer spin drying, 10
"Walking wounded" devices, 166
Wave solder machines, 344
Waveforms, 175
Waxes, 346
Wimhurst, James, 44
Wire damage, 147
Work, definition of, 57
Work function, 106, 107
Work to move charges, 71
Workplace considerations and problem examples, 343
Wrist-strap checks, 402
   upon entry or at workstation, 402
   ohmmeter tests of wrist straps, 402
   types of wrist-strap checkers, 402
   at workstation, 403
Wunsch-Bell model, 133, 176, 177
   adiabatic range, 134
   quasiadiabatic region, 134
   steady-state conditions, 135

Zeta potential, 112